Automatic Control Systems

Automatic Control Systems

George J. Thaler

Distinguished Professor of Engineering
Naval Postgraduate School

WEST PUBLISHING COMPANY
St. Paul/New York/Los Angeles/San Francisco

Copyediting: Lorretta Palagi
Design: Roslyn Stendahl, Dapper Design
Art: George Morris, Scientific Illustrators
Composition: Polyglot Pte. Ltd.
Cover: David Farr, Imagesmythe Inc.

Printed in the United States of America
96 95 94 93 92 91 90 89 8 7 6 5 4 3 2 1 0

Library of Congress Cataloging-in-Publication Data

Thaler, George J. (George Julius), 1918–
    Automatic control systems/George J. Thaler.
        p. cm.

    Includes bibliographies and index.
    ISBN 0-314-43042-3
    1. Automatic control.   I. Title.
TJ213.T475 1989   88-19033
629.8--dc19   CIP

Contents

1 BACKGROUND   1

1.1 Introduction   1
1.2 When Do We Need Feedback?   3
1.3 Basic Procedures   4
1.4 Tools for Linear System Analysis and Design   4
Bibliography   6

2 MATH MODELS, BLOCK DIAGRAMS, AND STEADY- STATE ACCURACY   7

2.1 Modeling   7
2.2 The Laplace Transformation   7
2.3 Transfer Functions   9
2.4 Block Diagrams   16
2.5 Signal Flow Graphs – Mason's Gain Rule   19
2.6 Accuracy in Feedback Control Systems   26
2.7 Accuracy in a Type-Zero System   30
2.8 Accuracy in a Type-One System   31
2.9 Accuracy in a Type-Two System   32
2.10 Accuracy Effects of Load Disturbances and Nonlinearities   34
2.11 Dynamic Error Coefficients   39

2.12 Summary 41
Bibliography 42
Problems 42

3 STABILITY AND DYNAMIC RESPONSE 63

3.1 Definitions and Fundamentals 63
3.2 Determining the Characteristic Equation 65
3.3 The Routh Criterion 67
3.4 Some Results from Theory of Equations 72
3.5 Cauchy's Principle of Argument and the Nyquist Criterion 74
3.6 Nyquist Interpretation of Nonminimum Phase Systems 85
3.7 Relative Stability: Gain Margin and Phase Margin 87
3.8 Some Constraints and Cautions 92
3.9 Closed-Loop Frequency Response 93
3.10 Root Motion on the s-Plane and Locus of Roots 97
3.11 Summary 97
References 97
Bibliography 98
Problems 98

4 PERFORMANCE CHARACTERISTICS OF A SECOND-ORDER SYSTEM 105

4.1 Introduction 105
4.2 Analysis Performance from Root Patterns: Dominance 106
4.3 The Second-Order System: Time Response 107
4.4 The Second-Order System: Frequency Response 110
4.5 Uses and Cautions 113
4.6 Examples 114
4.7 Summary 126
Bibliography 126
Problems 126

5 FREQUENCY RESPONSE ANALYSIS   129

5.1 Introduction   129
5.2 The Bode Diagram: Magnitude versus ω 130
5.3 The Bode Diagram: Phase Angle versus ω 137
5.4 System Stability Analysis from the Bode Diagram   140
5.5 System Transient Analysis from the Bode Diagram   143
5.6 Evaluating the Closed-Loop Frequency Response: The Nichols Chart   151
5.7 Summary   156
      References   156
      Bibliography   156
      Problems   157

6 ROOT LOCUS METHODS FOR ANALYSIS   169

6.1 Introduction   169
6.2 Mathematical Basis for Construction of Root Loci   170
6.3 Development of Rules for Constructing the Root Loci   171
6.4 Drawing the Loci   183
6.5 Locating the Roots on the Loci   188
6.6 Root Loci for a Parameter Other Than K: Partitioning   193
6.7 Two Parameter Studies: Families of Root Loci   196
6.8 Summary   198
      References   198
      Bibliography   198
      Problems   199

7 STATE VARIABLE ANALYSIS   207

7.1 Introduction   207
7.2 Defining State Variables   208
7.3 Transfer Functions from State Variables   221
7.4 Analysis with State Variables: Stability, Controllability, and Observability
7.5 Observers   226
7.6 Summary   230
      Bibliography   230
      Problems   231

8 DESIGN   235

8.1   Introduction   235
8.2   Analysis of the Basic Approaches to Compensation   236
8.3   Cascade Compensation: General Concepts   237
8.4   Cascade Lead Compensation   240
8.5   Cascade Lag Compensator Design   248
8.6   Cascade Compensation with Several Sections of Filter   254
Including PID Controllers
8.7   An Analytic Approach to Cascade Compensation   261
8.8   Feedback Compensation   264
8.9   Root Locus Studies: Velocity and Acceleration Feedback   266
8.10   Bode Diagram Methods for Feedback Compensation   278
8.11   Effect of Feedback Compensation on Nonlinearities and
Parameter Variations   294
8.12   External Disturbances   295
8.13   Compensation by State Variable Feedback   296
8.14   Summary   306
References   306
Bibliography   306
Problems   306

9 DIGITAL AND SAMPLED DATA CONTROL SYSTEMS   321

9.1   Some Background   321
9.2   The Sampling Process: Methods and Rates   322
9.3   The Sampling Process: Mathematical Analysis   324
9.4   Feedback System Analysis in the Z Domain   336
9.5   Design of Compensation: Some Simple Cases   339
9.6   Summary   346
References   346
Bibliography   346
Problems   347

10 NONLINEAR SYSTEMS: DESCRIBING FUNCTIONS AND LIMIT CYCLES   351

10.1   Nature and Effects of Nonlinearities   351
10.2   Tools Available for Engineers   353

10.3   The Describing Function   354
10.4   Stability and Limit Cycles   359
10.5   Design of Limit Cycle Operation   364
10.6   Nyquist Analysis for Stability and Limit Cycles   371
10.7   Describing Functions on the Nichols Plot   374
10.8   False Indications   381
10.9   More on Describing Functions   387
10.10 Summary   394
         References   395
         Bibliography   395
         Problems   396

11 NONLINEAR SYSTEMS: THE PHASE PLANE AND DISCONTINUOUS SYSTEMS   399

11.1   Introduction   399
11.2   Slopes, Isoclines, Singular Points, and Limit Cycles   400
11.3   Isoclines and Eigenvectors   407
11.4   Switching Lines and Dividing Lines   412
11.5   Introduction to Discontinuous Systems   429
11.6   Bang-Bang (Minimum Time) Control   432
11.7   Switched Damping   437
11.8   A Second-Order Linear System with Discontinuous Velocity Feedback   439
11.9   Some Extension of Bang-Bang Control   442
11.10 Summary   443
         Bibliography   443
         Problems   444

Appendix A MATRIX THEORY   449

Index   453

Preface

Experience is a good teacher. Unfortunately, we cannot teach experience, but we can use it to guide our teaching. Forty years of teaching control theory and 25 years of consulting have guided the development of this text. While most of the theory is necessary for understanding the nature of feedback control, some features are much more useful in practice than others and a conscious effort has been made to emphasize what is useful. In addition, numerous illustrative examples demonstrate how to use it.

The advent of the digital computer, and in particular the personal computer, has had a significant impact on the practice of control engineering. Both the practicing engineer and the student find it an invaluable tool. This is clearly the trend of the future; soon all engineers will find the personal computer indispensable. In keeping with this trend, a computer-aided package for analysis and design accompanies this text. It is menu-guided, very easy to use, and intended for the convenience of the student so that many homework problems can be solved with minimal effort. This software accepts transfer function input, and will

1. Carry out block diagram manipulation.
2. Calculate and display system roots.
3. Calculate and plot:
   a. frequency response
   b. root locus
   c. time response.

It does not include software for matrix methods and state variables.

This book is concerned with the basic requirements for control systems, which are accuracy, stability, and dynamic (transient) response. The first seven chapters develop the tools for analysis, which are also the basic tools for design. The classical methods (frequency response, root locus) require curves, and the computer will draw them. The modern methods (state variables, matrices) *require* computer solution if the system order is four or higher, but in this text only low-order examples are used so that long-hand solutions are feasible.

As is usual, this text develops graphical procedures for constructing the frequency response curves and the root loci. Since the computer calculates and draws these curves, it is fair to ask why the graphical procedures are used at all. There are two kinds of answers to this question: 1) for purposes of system analysis the graphical procedures are clearly obsolete and are not needed. The computer drawn curves are adequate and perhaps even better, 2) For purposes of design the graphical methods are very helpful. They provide means of choosing values for any poles and zeros needed to stabilize the system or improve its transient response. The use of these graphical procedures in design is developed in Chapter 8.

The first eight chapters constitute the basic material for a first course in linear feedback control theory. They can be covered in a one-quarter course (three lectures per week for 11 weeks) though a tight schedule is needed and it may be difficult to include all of Chapter 8. A one-semester course is much more comfortable, and the instructor may wish to augment the coverage with material from Chapters 9, 10, and 11. Actually, Chapters 10 and 11 provide adequate material for a separate short course on nonlinear controls, if such is wanted.

Many people have contributed to the development of this text. First of all, students at the Naval Postgraduate School who developed some of the material in their theses, others who pointed out errors, made constructive suggestions, or just complained when my presentation did not seem clear. In like manner, I have had beneficial feedback from engineers in my "early bird" courses at the University of Santa Clara and in on-site courses taught at IBM, San Jose. Still others contributed as I worked with them (as a consultant) on specific control problems. I wish to thank all of them.

I also wish to thank the following reviewers for their suggestions:

Samri B. Atta
California Polytechnic, San Luis Obispo
Harold P. Boettcher
University of Wisconsin, Milwaukee
Henry D'Angelo
Boston University
M. Samy El-Sawah
California Polytechnic, Pomona
Lincoln Jones
San José State University
Gordon K. F. Lee
Colorado State University
Dan Lingelbach
Oklahoma State University

Richard Longman
San Jose State University
Gene Moriarty
San Jose State University
Charles Neuman
Carnegie-Mellon University
Peter Reischl
San Jose State University
W. P. Schneider
University of Houston
James Simes
California State University, Sacramento
M. K. Sundareshan
University of Arizona

The CAD software package was developed by Lt. Roy Wood, initially as a master's thesis, and with subsequent revision and updating. Thank you, Roy.

I also wish to thank Mrs. Robert Limes who did a magnificent job of typing the manuscript.

George J. Thaler
Naval Postgraduate School

Automatic Control Systems

1 Background

INTRODUCTION

What do we mean by the word *control*? One of the various dictionary definitions is a "device for regulating a machine." This is a reasonably good definition, although it implies that the "device" is a separate piece of hardware, and the word "regulating" should not be interpreted in its strict technical sense. Perhaps it would be better to say that a control is a scheme (or an algorithm) that makes a machine behave according to our wishes.

There are many quantities that we can control, in the sense that the machines used to manipulate these quantities are operated in such a way that we get what we want. Such quantities, often called the *controlled variables*, include:

> temperature
> voltage
> frequency
> flow rate
> current
> position
> horsepower
> speed
> illumination
> altitude (or depth)

and many others. The methods used to accomplish the control objectives can be classified, very generally, as open-loop controls and closed-loop or feedback controls.

Open-loop controls are, in general, calibrated systems. They are adjusted to provide the desired result and are expected to duplicate that result as needed because of

1

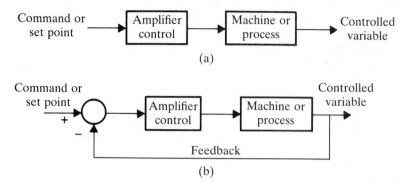

(a)

(b)

FIGURE 1.1 Block Diagrams for Basic Control Systems: (a) Open-Loop Control and (b) Closed-Loop Control.

the adjustment. Figure 1.1(a) shows a basic block diagram for open-loop control. Examples are:

1. *Stepper motor positioning systems*: The actual position of the output is not usually monitored. The motor controller commands a certain number of steps by the motor to drive the output to a previously determined location. It is assumed that the motor actually executes the steps and that the result is as predicted.

2. *Assembly robot*: Such machines do a routine, repetitive task such as picking up a part from a conveyer belt and placing it in a bin or inserting it in another part. The robot is "taught" the proper move, which it can repeat 24 hours a day. If the robot's ability to perform its task becomes impaired, there is no self-correction means. Note that it is possible to provide sensors for the robot (i.e., optical, thermal, electric, or magnetic devices, even television monitors) that monitor the controlled variable, such as gripper position. The signals from these sensors can be used to provide self-correction, but the control then becomes feedback control.

3. *Automatic washers and dryers*: These devices change operating modes and eventually shut themselves off according to a preprogrammed schedule, which is usually based on elapsed time.

Closed-loop or feedback controls operate according to a very simple principle:

1. Measure the variable to be controlled.

2. Compare this measured value with the desired (commanded) value and determine the difference (error).

3. Use this difference (amplified if necessary) to adjust the controlled variable so as to reduce the difference (error).

Figure 1.1(b) illustrates this principle in block diagram form. Examples of feedback control are numerous: Thermostatic control of room temperature, of refrigerator or freezer temperature, oven temperature, etc., all of which measure the temperature, compare it with the desired value as set on the control dial, and turn on or off the heat source (or sink) as needed. Other examples are electric supply voltage and frequency, which are measured at the generating station and compared with desired values. If the voltage is in error, the field current of the generator is adjusted; if the frequency is not correct, the turbine speed is adjusted.

Autopilots on aircraft, submarines, etc., are feedback controllers and satellite tracking antennas are automatic positioning systems. In computer systems, the magnetic disk memory requires a very accurate feedback controller to position the read/write head over the proper memory track.

1.2 WHEN DO WE NEED FEEDBACK?

From the preceding examples it is clear that both open-loop controls and feedback controls are used to control the same kinds of variables. What, then, are the advantages and disadvantages of each? Why do we need feedback control and when should we use it?

In general, the open-loop control is much simpler and less expensive. No sensors are needed to measure the variables and provide the feedback. Also it usually does not require much amplification. However, it is not very accurate (compared with feedback control), and its accuracy can vary without this variation being detected by the system.

For example, consider a sprinkler used to water the lawn. The system is adjusted to water a given area by opening the water valve and observing the resulting pattern. When the pattern is considered satisfactory, the system is "calibrated" and no further valve adjustment is made. The pattern will be maintained reasonably well if there is no change in water pressure. When a tap is opened inside the house, reducing pressure, the pattern changes, i.e., the open-loop control has no means of maintaining the desired steady-state condition.

Inspection of the previously listed examples indicates, correctly, that most automatic controls are used primarily to maintain the controlled variable at a fixed steady-state value. However, most systems are subject to disturbances that perturb the controlled variable, and a major reason for the use of feedback control is to reduce the effect of these disturbances. For example, when there is an environmental change such as an increase in ambient temperature from 40° at 6 a.m. to 75° at noon, an open-loop house heating system is not satisfactory, but a thermostatically controlled heater can maintain the desired room temperature. In most systems, the desired values of the variable may be changed (we often change the temperature setting on our room thermostat). When the controlled variable is disturbed, the recovery time (settling time) is often slow, and in most cases not much can be done to speed it up.

The closed-loop system is inherently more accurate than the open-loop system because of its principle of operation. In addition, it can be designed to provide extreme accuracy in the steady state (for almost all steady-state operating conditions). For example, in modern magnetic disk memories the width of a memory track may be less than 0.002 in. and the positioning controller must align the read/write head over the track center with an error of less than 10 micro inches. Such accuracy is not possible with an open-loop controller.

The closed-loop system has an additional advantage in that its response time can be adjusted by appropriate design. Normally the response time can be reduced substantially below that available from an equivalent open-loop system.

Unfortunately, the feedback link introduces a serious problem—stability! Feedback systems readily become oscillators; that is, the controlled variable fluctuates continuously at some periodic rate (frequency) and never reaches the desired steady-state condition. For any specific system, stable operation can be achieved by proper design, but the design problem is not always an easy one to solve.

1.3   BASIC PROCEDURES

When we wish to develop a feedback control system for a specific purpose, the general procedure may be summarized as follows:

1. Choose a way to adjust the variable to be controlled; e.g., the mechanical load will be positioned with an electric motor or the temperature will be controlled by an electrical resistance heater.
2. Select suitable sensors, power supplies, amplifiers, etc., to complete the loop.
3. Determine what is required for the system to operate with the specified accuracy in steady state and for the desired response time.
4. Analyze the resulting system to determine its stability.
5. Modify the system to provide stability and other desired operating conditions by redesigning the amplifier/controller, or by introducing additional control loops.

This text develops the tools and procedures which are useful in carrying out the analysis and design steps. These tools are derived from the mathematics and provide analytic, graphical, and computer-aided techniques.

1.4   TOOLS FOR LINEAR SYSTEM ANALYSIS AND DESIGN

To study any system using mathematical methods, one must first obtain a mathematical description of the system. This process is generally called *modeling*, and for control systems (which are dynamic systems) we must obtain the differential equations of the system. For typical physical systems, we need only the basic laws of nature. For example, Newton's laws, Kirchoff's laws, et al.

Having obtained the differential equations and, where possible, the numerical values of the parameters, the next step is analysis: Will the system operate as we wish? If not, what is wrong? One approach to analysis is simulation in the computer. This technique can be used whether the equations are linear or nonlinear, and the answers tell us whether the system will work or not; however, they do not tell us what is wrong—not in the sense of pointing out what needs to be changed.

Instead of solving the differential equations to obtain the time response of the system, an alternative approach is as follows: Use the Laplace transformation (with all initial conditions assumed zero) to convert the differential equations to algebraic equations. This also converts the problem from the time domain to the frequency domain. We then rearrange the algebra into transfer function form and use frequency response methods for analysis. Commonly used techniques are based on the Nyquist stability criterion or on the root locus method. Of course, what we are really interested in is the time response of the system, but we can predict this from the frequency response data. The Fourier transform pair

$$f(t) = \frac{1}{2\pi} \int_{-\infty}^{\infty} F(j\omega)e^{j\omega t} \, d\omega \tag{1.1}$$

and

$$F(j\omega) = \int_{-\infty}^{\infty} f(t)e^{-j\omega t} \, dt \tag{1.2}$$

guarantee that if we know the frequency response we can calculate the time response and vice versa; we seldom do the actual calculations, but Eqs. (1.1) and (1.2) provide justification for the semi-empirical correlations that are used.

The advantages of the graphical presentations (frequency responses and/or root loci) are twofold: They provide a sound and sufficiently accurate analysis while pointing out the sources of any problems, and they also provide guidance for design changes needed to "make it work."

Computer evaluation of frequency response, transient response, root loci, and roots of polynomials has long been available from mainframe computers, though the programs were not necessarily convenient to use. Such programs have now been made interactive and user friendly. They are readily available for personal computers as well as the mainframes. Their use makes feedback system analysis and design simple and less laborious. The theory is much easier to apply, trial and error design procedures converge more rapidly, and the analyst/designer is able to consider details that might otherwise be ignored.

To use the computer as an aid to analysis and design, the analyst requires understanding of the theory—in depth. Most of the graphical procedures associated with classical control theory are not needed—the computer draws the curves. Knowledge of the graphical method is helpful in problem preparation, in checking computer results (one still worries about "garbage in/garbage out"), and in choosing design modifications.

This text uses computer aids wherever possible, but it also develops the graphical methods of classical theory and combines the useful features with the computer results. The modern control approach is also developed with emphasis on the use of state feedback for pole placement.

Chapters 2 through 7 develop the mathematical and graphical tools commonly used for analysis. These analytical tools are applied to the design problem in Chapter 8, where a number of different design techniques are developed and demonstrated.

Since the advent of the microprocessor chip, the use of digital controls has become economically feasible for many applications. It is clear that digital control is very important and will soon dominate the control field. A detailed exposition of digital control is beyond the scope of this book, but Chapter 9 provides a basic (though minimal) introduction.

All physical systems incorporate some kinds of *nonlinear* effects. A few of these that impact accuracy are mentioned in Chapter 2, but nonlinearities also impact dynamic performance. Chapters 10 and 11 consider some basic nonlinear conditions, providing some guidelines for analysis and design.

BIBLIOGRAPHY

Eckman, D. P. *Principles of Industrial Process Control.* New York: John Wiley and Sons (1945).

Maxwell, J. C. "On Governors." *Proc. Royal Soc. London* 16: 270–83 (1968).

Mayr, O. *The Origins of Feedback Control.* Cambridge, Mass.: The MIT Press (1970).

2 Math Models, Block Diagrams, and Steady-State Accuracy

MODELING

To analyze and design control systems, we must formulate a quantitative mathematical description of the system. The process of obtaining the desired description is called *modeling*. The basic models of dynamic physical systems are differential equations obtained by application of the appropriate laws of nature. These equations may be linear or nonlinear depending on the phenomena being modeled: In this text, only linear models are considered, except in Chapters 10 and 11.

For most analysis and design manipulation, the differential equations are inconvenient, so the normal practice is to apply the Laplace transformation, which converts the differential equations into algebraic equations. Once the model is in algebraic form, further manipulations may be desirable. The algebraic equations may be put in transfer function form, and the system modeled graphically as a transfer function block diagram. Alternatively, a signal flow graph may be used. Either graphical form permits definition of state variables and a set of state equations.

This chapter is concerned with differential equations, transfer functions, block diagrams, signal flow graphs, and the interpretation of the steady-state accuracy of the control system. Other procedures are developed in later chapters as needed.

2.2 THE LAPLACE TRANSFORMATION

The Laplace transform of any time function, $f(t)$, is defined by the following integral equation:

$$F(s) = \mathscr{L}[f(t)] \triangleq \int_0^\infty f(t)e^{-st}\,dt,$$

where s is the complex variable (complex frequency):

$$s = \sigma + j\omega.$$

In order for the Laplace transform to exist, it is necessary and sufficient that:

$$\lim_{t \to \infty} e^{-\sigma t} f(t) = 0$$

for some finite value of σ.

TABLE 2.1   Table of Laplace Transforms

Type	Time Function	Transform
Unit impulse	1. or $\delta(t)$	1.
Constant	A	$\dfrac{A}{s}$
Step	$Au(t)$	$\dfrac{A}{s}$
Delayed step	$Au(t - T)$	$\dfrac{A}{s} e^{-sT}$
Ramp	Bt	$\dfrac{B}{s^2}$
Delayed ramp	$Btu(t - T)$	$\dfrac{B}{s^2} e^{-sT}$
Parabola	Pt^2	$\dfrac{2P}{s^3}$
Exponential decay	$e^{-t/T}$	$\dfrac{1}{s + \dfrac{1}{T}}$
Sine wave	$\sin \omega t$	$\dfrac{\omega}{s^2 + \omega^2}$
Cosine wave	$\cos \omega t$	$\dfrac{s}{s^2 + \omega^2}$
Damped sine	$e^{-t/T} \sin \omega t$	$\dfrac{\omega}{\left(s + \dfrac{1}{T}\right)^2 + \omega^2}$
Damped cosine	$e^{-t/T} \cos \omega t$	$\dfrac{s + \dfrac{1}{T}}{\left(s + \dfrac{1}{T}\right)^2 + \omega^2}$

TABLE 2.2   Laplace Transform Theorems

Description	Time Function	Transformed Function
Superposition	$f_1(t) \mp f_2(t)$	$F_1(s) \mp F_2(s)$
Real differentiation	$\dot{f}(t)$	$sF(s) - f(0+)$
	$\ddot{f}(t)$	$s^2 F(s) - sf(0+) - \dot{f}(0+)$
	$f^N(t)$	$s^N F(s) - s^{N-1} f(0+) \cdots$
		$s^{N-2} \dot{f}(0+) - \cdots f^{N-1}(0+)$
Real integration	$\int f(t)\,dt$	$\dfrac{1}{s}\left[F(s) + \int f(t)\,dt \Big|_{0+} \right]$
Initial value	$\lim_{t \to 0} f(t)$	$\lim_{s \to \infty} sF(s)$
Final value	$\lim_{t \to \infty} f(t)$	$\lim_{s \to 0} sF(s)$
Real translation	$f(t - T)$	$e^{-sT} F(s)$

Evaluation of the function $F(s)$ by carrying out the integration has been accomplished for most functions of interest and the results are available in tables. Table 2.1 is a short list of Laplace transforms; Table 2.2 gives a list of important theorems related to the Laplace transform. Of particular note in Table 2.2 is the final value theorem, which is very helpful in establishing relationships for the steady-state accuracy of control systems.

2.3   TRANSFER FUNCTIONS

Whenever a system is perturbed by some input signal, all of the dependent variables in the system vary as a result. If we call one of these variables the *output*, the transfer of a signal through the system may be described by the ratio of the Laplace transform of the output signal to that of the input signal. This ratio is called the system *transfer function*. This function describes the cause and effect relationship between input and output and therefore is defined for *all initial conditions zero.*

 EXAMPLE 2.1

Consider the simple *R-L-C* filter of Figure 2.1. Assume that the source impedance is negligible and the load impedance is very large. Applying Kirchhoff's law, the applied voltage (input) is equal to the sum of the voltage drops around the circuit:

$$E_i = iR + L\frac{di}{dt} + \frac{1}{C}\int i\,dt \qquad (2.1)$$

and

$$E_o = \frac{1}{C}\int i\,dt. \qquad (2.2)$$

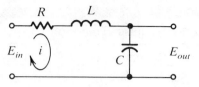

FIGURE 2.1 An *R-L-C* Filter.

Equations (2.1) and (2.2) are the basic mathematical model of the filter. Applying the Laplace transform, with all initial conditions zero:

$$E_i(s) = I(s)\left(R + sL + \frac{1}{sC}\right) \tag{2.1a}$$

and

$$E_o(s) = I(s)\left(\frac{1}{sC}\right). \tag{2.2a}$$

The transfer function is obtained by dividing Eq. (2.2a) by (2.1a):

$$\frac{E_o(s)}{E_i(s)} = \frac{I(s)\left(\dfrac{1}{sC}\right)}{I(s)\left(R + sL + \dfrac{1}{sC}\right)} \tag{2.3}$$

and manipulation of the algebra provides

$$\frac{E_o(s)}{E_i(s)} = \frac{1}{s^2LC + sCR + 1}. \tag{2.3a}$$

EXAMPLE 2.2

The armature-controlled dc motor of Figure 2.2 may be modeled by application of both Kirchhoff's and Newton's laws. A mechanical equilibrium is required by Newton's law, i.e., the drive torque and the reaction torque must be equal:

$$T_{\text{drive}} = k_T I$$
$$T_{\text{reaction}} = J\ddot{C} + f\dot{C}; \tag{2.4}$$

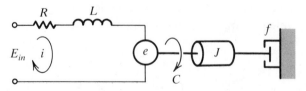

FIGURE 2.2 An Armature-Controlled dc Motor.

thus,

$$k_T I = J\ddot{C} + f\dot{C},$$

where

k_T = motor torque constant
J = inertia of motor and load
f = viscous friction coefficient
$C, \dot{C}, \ddot{C}$ = angle, velocity, and acceleration of the output shaft, respectively
I = motor armature current.

To model the armature circuit, Kirchhoff's law is used:

$$V = IR + L\dot{I} + e$$
$$= IR + L\dot{I} + k_B\dot{C}, \tag{2.5}$$

where

V = voltage applied to the armature
R, L = armature resistance and inductance, respectively
e = motor back electromotive force (emf)
k_B = back emf constant.

Transforming for all initial conditions zero:

$$k_T I(s) = (Js^2 + fs)C(s) \tag{2.4a}$$

and

$$V(s) = (sL + R)I(s) + k_B s C(s). \tag{2.5a}$$

To obtain the transfer function from voltage input to shaft angle, C, as output, $I(s)$ is eliminated by substituting Eq. (2.4a) in (2.5a):

$$V(s) = \frac{(sL + R)(Js^2 + fs)}{k_T} C(s) + k_B s C(s) \tag{2.6}$$

from which

$$\frac{C(s)}{V(s)} = \frac{k_T}{(sL + R)(Js^2 + fs) + k_T k_B s}. \tag{2.6a}$$

In the preceding examples, the input signal was not defined as a specific function, but was treated instead as an arbitrary function. This is convenient but not necessary. A commonly used input signal is the unit impulse, for which the transform is simply 1.0.

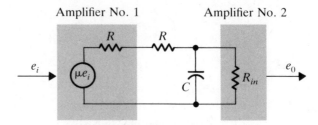

FIGURE 2.3   Filter Between Amplifier Stages.

This leads to an alternative definition for a transfer function:

> The transfer function of a linear system is the Laplace transform of its response to a unit impulse input.

When a device is part of a larger system, there are two basic constraints on the derivation of its transfer function:

1. The device must not load the source that provides its input signal; ideally the source should have zero (or low) internal impedance.

2. The output of the device must not be loaded by the component that receives its output signal.

If either of these conditions is violated, an incorrect transfer function results. For example, consider a simple filter located between two amplifier stages as shown in Figure 2.3. If we consider only the filter, a simple transfer function is obtained. If $R \gg R_1$, the internal drop in amplifier 1 is negligible and $R + R_1 \approx R$ so the filter time constant is unchanged. However, if R_1 cannot be neglected, then the filter *loads* amplifier 1 and the transfer function of the isolated filter is incorrect. In like manner, if the input impedance of amplifier 2 is very large ($R_{in} \cong \infty$), it does not load the filter and its effect may be neglected; but if R_{in} is not large, it provides a path in parallel with the C of the filter. Obviously, this would alter the filter transfer function. Such loading effects occur with electrical, mechanical, electromechanical, and fluid devices and must be taken into account in deriving transfer functions.

It is frequently desirable to *reduce* a block diagram; i.e., to reduce the number of blocks and the number of closed loops. For many of the analysis and design procedures to be developed in this text, it is necessary for the block diagram of the control systems to be reduced to a single feedback loop with unity feedback and manipulation procedures as developed in Figures 2.4 and 2.5. Consider the system of Figure 2.4(a) with the quantitative diagram as in Figure 2.6(a). The tachometer feedback forms an interior or *minor* loop, which would be reduced to one equivalent block in the forward path as shown in Figure 2.6(b). The transfer function blocks in the forward path are then combined to provide the desired single loop with one equivalent block in the forward path as shown in Figure 2.6(c).

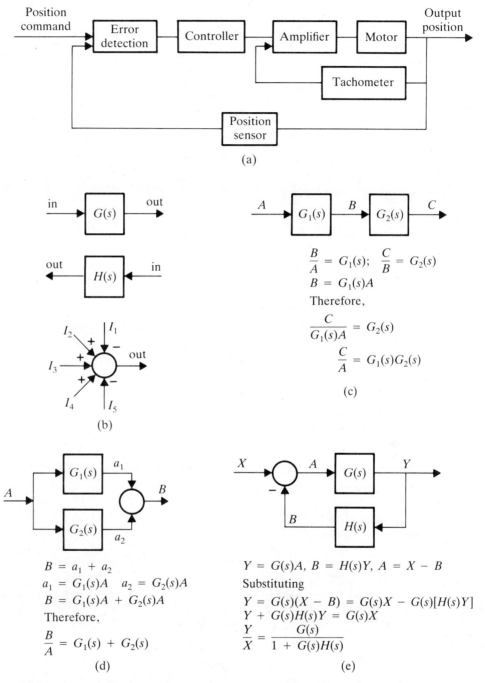

FIGURE 2.4   (a) Block Diagram of a Positioning System and (b), (c), (d), and (e) Basic Definitions for Block Diagrams.

14

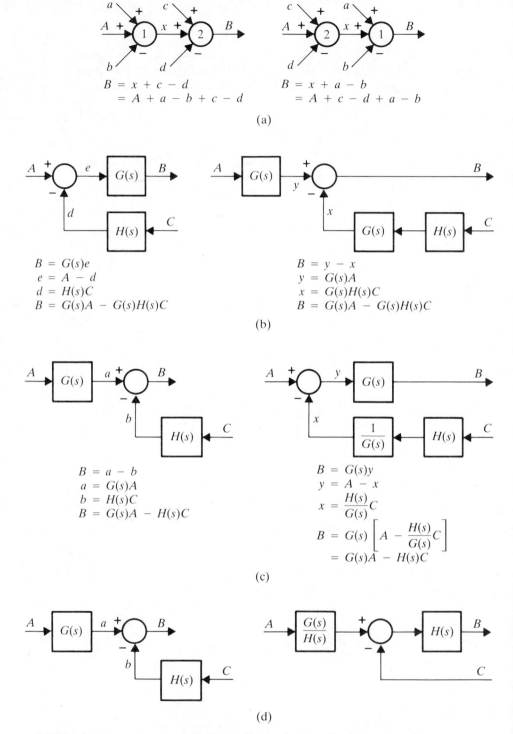

$$B = x + c - d$$
$$= A + a - b + c - d$$

$$B = x + a - b$$
$$= A + c - d + a - b$$

(a)

$$B = G(s)e$$
$$e = A - d$$
$$d = H(s)C$$
$$B = G(s)A - G(s)H(s)C$$

$$B = y - x$$
$$y = G(s)A$$
$$x = G(s)H(s)C$$
$$B = G(s)A - G(s)H(s)C$$

(b)

$$B = a - b$$
$$a = G(s)A$$
$$b = H(s)C$$
$$B = G(s)A - H(s)C$$

$$B = G(s)y$$
$$y = A - x$$
$$x = \frac{H(s)}{G(s)}C$$
$$B = G(s)\left[A - \frac{H(s)}{G(s)}C\right]$$
$$= G(s)A - H(s)C$$

(c)

(d)

FIGURE 2.5 Some Block Diagram Manipulations: (a) Interchanging Summers,
(b) Moving a Summer Past a Block, (c) Moving a Summer Past a Block (Against Signal Flow),
(d) Moving a Feedback Block Past a Summer, and (e) Moving a Feedback Block Past
a Summer.

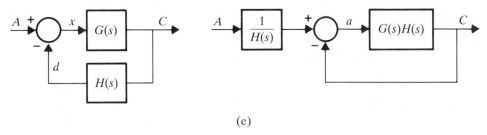

(e)

FIGURE 2.5   (*Continued*)

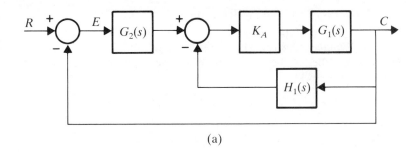

(a)

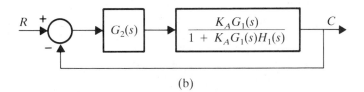

(b)

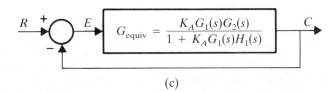

(c)

FIGURE 2.6   Manipulation of a System Block Diagram: (a) Positioning System of Figure 2.4 (a), (b) Reduction of Minor Feedback Loop to One Equivalent Cascade Block, and (c) Cascaded Blocks Combined to Give One Equivalent Block with Transfer Function G_{equiv}.

2.4	BLOCK DIAGRAMS

The interconnection of components to form a system is conveniently shown by blocks arranged in some sort of diagram such as in Figures 2.4(b) through (d).

When transfer functions can be derived for the components, they may be inserted in the appropriate blocks. The diagram then becomes a quantitative description of the system and is called a *transfer function block diagram*. Mathematical manipulations may be carried out as follows.

As shown in Figure 2.4(b), a transfer function block represents signal transfer in *one direction only*. A block may have only *one input* and *one output*. The direction of signal flow is indicated by arrowheads and $G(s)$ is the transfer function of a block in a *forward* path while $H(s)$ is the transfer function of a block in a feedback path.

A circle is used to represent the summation of signals and any number of input signals are allowed but only *one* output signal. *Signs* to be used in summing are specified at the circle (they are not defined by the transfer function).

2.4A   Basic Block Diagram Combinations

From Figure 2.4(c) for two blocks in cascade:

$$\frac{C}{A} \triangleq G_{\text{equiv}}(s) = G_1(s)G_2(s).$$

Thus, the equivalent transfer function of N blocks in cascade is simply the product of the N transfer functions. From Figure 2.4(d), for two blocks in parallel, summing:

$$\frac{B}{A} = G_1(s) + G_2(s).$$

From Figure 2.4(e), for a negative feedback loop:

$$\frac{Y}{X} = \frac{G(s)}{1 + G(s)H(s)}.$$

2.4B   Some Rules for Manipulating Block Diagrams

The most commonly used manipulations of block diagrams have been shown by the preceding *combinations*. The obvious intent is to reduce the complexity of the diagram by combining blocks. Further benefit is obtained because the block manipulations automatically combine the algebraic equations. Block diagrams of complex systems, however, may contain combinations of blocks that cannot be combined by these simple relationships.

With the aid of a few simple rules, portions of the diagram may be moved to obtain combinations that can be reduced by the basic combination rules. To justify additional rules, we observe that we are treating linear systems, so any manipulation that does not change the output signal is valid.

1. Two adjacent summing junctions may be interchanged. A summer may be moved past another summer, as shown in Figure 2.5(a). The output signal B is unchanged by the manipulation.

2. A summer may be moved past a block in the direction of signal transmission as in Figure 2.5(b). The basic principle is that the output signal B must not be changed by manipulations. If the summer is moved past the block $G(s)$, the proper signal is maintained at B if the block $G(s)$ is placed in *both* inputs to the summer.

    In general, if a summer is moved past a block *in the direction of signal transmission*, the block must be duplicated in all paths that supply inputs to the summer.

3. A summer may be moved past a block, against signal flow [see Figure 2.5(c)]. When the summer is moved past a block against signal flow (or one may say the block is moved past the summer with the signal flow), then the *reciprocal* of the block must be inserted in all paths which are inputs to the summer.

    This rule also applies to movement of a feedback block past a summer as in Figures 2.5(d) and (e).

These rules are sometimes convenient in reducing a block diagram to obtain the desired equations.

It is usually sufficient to model each component of a system with a single transfer function block such that the natural physical signals are represented as input and output. When this is done, the internal behavior of the system is not considered, nor is it readily available from the mathematics. In like manner, when a system block diagram is reduced to simpler form, only the input/output information is explicitly available.

Occasionally, some component in a system is subject to several inputs or to some type of load disturbance that must be included in its mathematical model. In such cases, the usual transfer function, though correct, is not complete and the usual block diagram representation is not complete.

The differential equations of the system are, of course, the proper starting point for analysis and, for linear systems, the Laplace transformation and the principle of superposition are applicable, allowing us to derive correct transfer functions from each input to the output, or in some cases from any input to each of several signals designated as outputs. Sometimes it is possible to obtain a more detailed block diagram directly from the differential equations. Such models portray internal signals explicitly, and these may then be used as needed in the analysis. The following example treats Eqs. (2.4) and (2.5) of a dc motor in this fashion.

 EXAMPLE 2.3

A dc motor is used to drive a satellite tracking antenna. When the wind blows, a load torque develops, causing an error in position. We must develop a model that includes the loading torque caused by the wind.

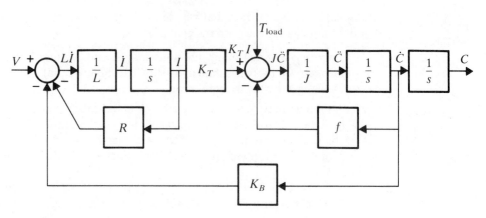

FIGURE 2.7   Expanded Block Diagram of a dc Motor.

Considering Eq. (2.4), rewrite it in the form:

$$\ddot{C} = \frac{k_T I - f\dot{C}}{J}.$$

(Note: The general technique is to place the highest order derivative on the left side of the equation with all other terms summed on the right side.) Assuming that $\ddot{C}$ is known, it can be integrated twice to obtain $\dot{C}$ and the output variable C. The terms on the right side of the equation are then formed from the diagram and summed as indicated by the equation. The sum, of course, is $\ddot{C}$. This is shown in Figure 2.7, with the integrations shown in transformed (i.e., transfer function) notation. Since the *units* of all terms in Eq. (2.4) are *torque*, a torque summer is an obvious addition to Figure 2.7 as shown, and the obvious inputs to the summer are the developed torque $k_T I$ and the friction torque $f\dot{C}$. The developed torque depends on the current, which is obtained from Eq. (2.5) by observing that the net voltage applied to the armature circuit is the applied voltage less the back emf, i.e.,

$$V(s) - k_B s C = I(R + sL).$$

This is also modeled in detail in Figure 2.7. Note especially that this expanded model includes, explicitly, the torque summing junction. Thus, if a load torque, such as the wind torque on an antenna, must be considered, it is added to the model by introducing another input, T_{load}, to the torque summing junction.

The expanded block diagram is considerably more complex than a simple single input, single output model, but has the advantage of making a torque summing junction available. When external torques need not be considered, the simpler model would be used. Of course, the model of Figure 2.7 may be reduced to a single block with the methods of Sect. 2.4, and the result (for $T_{\text{load}} = 0$) is exactly that of Eq. (2.6a).

2.5 SIGNAL FLOW GRAPHS—MASON'S GAIN RULE

Block diagram models of systems are usually compact. Each block normally represents a piece of equipment, which is modeled mathematically by its transfer function, and the only signals designated are those external to the blocks, i.e., the input and output signals. For some studies, it is advantageous to consider signals that are not explicitly shown on the block diagram. A convenient tool that emphasizes signals (rather than components and transfer functions) is the signal flow graph.[a]

On a signal flow graph, signals are represented as *nodes*, and the flow of a signal between nodes is designated by a *directed line*, as shown in Figure 2.8(a). Each directed line connecting two nodes is called a *branch*, and a transference (or gain) is associated with each branch. These transferences may be simply numbers or they may be transfer functions.

The signal at a node is defined to be the algebraic sum of all incoming signals. In Figure 2.8(a),

$$S_3 = k_2 S_2 + k_6 S_4 + k_8 S_5,$$

where any *signs* are included in the k's. There may be any number of outward directed branches at a node, each carrying the same signal.

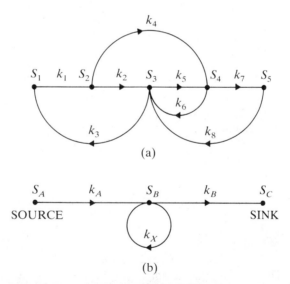

(a)

(b)

FIGURE 2.8 Elements of Signal Flow Graphs: (a) Nodes and Directed Signal Flow where $S_1 \cdots S_5$ are Signals and $k_1 \cdots k_8$ are Branch Transferences (Gains) and (b) Source, Sink, and Self-Loop.

---

[a] For control system studies, block diagrams and signal flow graphs are essentially equivalent. Choosing between them is a matter of personal preference and/or convenience.

A *path* is defined to be any sequence of connected branches such that no branch is traversed more than once. A *forward path* is a path from input to output with no node encountered twice. In Figure 2.8(a), the following may be defined as paths:

Path No. 1: $S_1 \rightarrow S_2 \rightarrow S_3 \rightarrow S_4 \rightarrow S_5$
Path No. 2: $S_1 \rightarrow S_2 \rightarrow S_4 \rightarrow S_5$
Path No. 3: $S_2 \rightarrow S_4 \rightarrow S_3 \rightarrow S_4 \rightarrow S_5$

Note that paths 1 and 2 proceed from input to output with no node included twice, so they would be designated *forward paths*. However, path 3 does not start at the input and uses node S_4 twice and hence, would not be called a forward path.

A *loop* is a path that closes on itself with no branch used twice. Again, in Figure 2.8(a):

Loop No. 1: $S_1 \rightarrow S_2 \rightarrow S_3 \rightarrow S_1$
Loop No. 2: $S_3 \rightarrow S_4 \rightarrow S_3$
Loop No. 3: $S_3 \rightarrow S_4 \rightarrow S_5 \rightarrow S_3$

The *path gain* is the product of all the gains in a path, e.g., for path 1 above:

path gain No. 1 $= k_1 k_2 k_5 k_7$.

The *loop gain* is the product of gains around the loop, e.g., for loop 1 above:

loop gain No. 1 $= k_1 k_2 k_3$

To represent an input (or command or forcing function), a node that has no incoming branches is used. Such a node is called an *input node* or *source node*. In like manner, an output from the system may be represented by a node with no outgoing branches, which is called an *output node* or *sink node*. These are illustrated in Figure 2.8(b).

Occasionally, the signal flow graph includes a branch that starts and terminates on the same node, forming a loop, which is called a *self-loop*, as shown in Figure 2.8(b). Note that for this case the signal at node S_B may be defined as

$$S_B = k_A S_A + k_X S_B$$

$$S_B(1 - k_X) = k_A S_A$$

$$S_B = \frac{k_A S_A}{1 - k_X}.$$

■ **EXAMPLE 2.4**

Consider the dc motor of Example 2.2 with the transfer function as given by Eq. (2.6a). The simplest block diagram for the motor is shown in Figure 2.9(a), and its equivalent signal flow graph is shown in (b). If signals inside the motor are of interest, the block diagram may be expanded as in Figure 2.7, or the equivalent signal flow graph may be constructed as shown in Figure 2.9(c).

A convenient feature of the signal flow method is that the graph is readily constructed for a polynomial and for a ratio of polynomials (i.e., an unfactored transfer function). The resulting signal flow graph may be used with any of the classical analysis methods, but as an additional bonus we can obtain from it the state equations of the system.

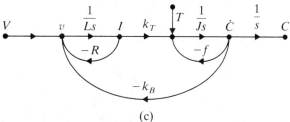

(a)

(b)

(c)

FIGURE 2.9   Signal Flow Graphs for a dc Motor: (a) Transfer Function Block Diagram, (b) Equivalent Signal Flow Graph, and (c) Signal Flow Diagram (Equivalent to Figure 2.7).

 EXAMPLE 2.5

Consider the transfer function:

$$\frac{C(s)}{R(s)} = \frac{10s^2 + 5s + 100}{s^4 + 20s^3 + 45s^2 + 18s + 100}.$$

This may be rewritten:

$$\frac{C(s)}{R(s)} = \frac{10s^2}{\Delta} + \frac{5s}{\Delta} + \frac{100}{\Delta},$$

where

$$\Delta = s^4 + 20s^3 + 45s^2 + 18s + 100.$$

Then

$$C(s) = C_3(s) + C_2(s) + C_1(s)$$

$$= 10s^2\left[\frac{R(s)}{\Delta}\right] + 5s\left[\frac{R(s)}{\Delta}\right] + 100\left[\frac{R(s)}{\Delta}\right].$$

Draw the signal flow graph for $C_1(s)$ as

$$C_1(s)(s^4 + 20s^3 + 45s^2 + 18s + 100) = 100R(s).$$

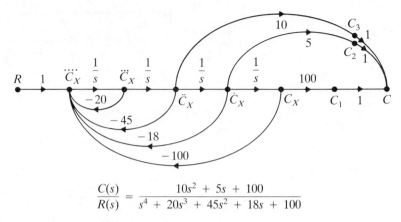

$$\frac{C(s)}{R(s)} = \frac{10s^2 + 5s + 100}{s^4 + 20s^3 + 45s^2 + 18s + 100}$$

FIGURE 2.10 Signal Flow Graph for $C(s)/R(s)$.

Since the system is linear, we divide by 100 and define $C_1(s)/100 = C_X$. Then

$$\dddot{C}_X = R - 20\dddot{C}_X - 45\ddot{C}_X - 18\dot{C}_X - 100C_X.$$

The signal flow graph is shown in Figure 2.10, and C_X is converted to C_1 by multiplying by 100. Then note that C_2 is simply 5 times the derivative of C_X and C_3 is 10 times the second derivative of C_X. Since superposition applies, the appropriate branches are added to define C_2 and C_3, from which C is obtained as shown in Figure 2.10.

By inspection of Figure 2.10, the node signals can be defined as states:

$$X_1 = C_X$$
$$X_2 = \dot{C}_X = \dot{X}_1$$
$$X_3 = \ddot{C}_X = \dot{X}_2$$
$$X_4 = \dddot{C}_X = \dot{X}_3.$$

The state equations are simply the node equations:

$$\dot{X}_1 = X_2$$
$$\dot{X}_2 = X_3$$
$$\dot{X}_3 = X_4$$
$$\dot{X}_4 = R - 20X_4 - 45X_3 - 18X_2 - 100X_1$$

and the output equation is:

$$C = 100X_1 + 5X_2 + 10X_3.$$

Signal flow graphs may be manipulated and reduced in ways very similar to those used with block diagrams. However the input-output transfer function can be obtained very simply by application of <u>Mason's gain rule[b]</u>:

$$G(s) = \frac{\sum\limits_i P_i \Delta i}{\Delta},$$

where

P_i = gain (transfer function) of the i'th forward path

$\Delta = 1.0 - \sum$ all individual loop gains

$+ \sum$ all possible gain products of two nontouching loops

$- \sum$ all possible gain products of three nontouching loops

$+ \cdots$

Δi = the Δ for that part of the graph which does not touch the i'th forward path.

■ EXAMPLE 2.6

To obtain the transfer function $C(s)/R(s)$ for the system of Figure 2.10, note that there are three forward paths and four loops, but all loops are touching. Using the gain rule, $\Delta i = 1.0$ for each forward path,

$$P_1 = \left(\frac{1}{s}\right)\left(\frac{1}{s}\right)\left(\frac{1}{s}\right)\left(\frac{1}{s}\right)(100) = \frac{100}{s^4}$$

$$P_2 = \left(\frac{1}{s}\right)\left(\frac{1}{s}\right)\left(\frac{1}{s}\right)(5) = \frac{5}{s^3}$$

$$P_3 = \left(\frac{1}{s}\right)\left(\frac{1}{s}\right)(10) = \frac{10}{s^2}$$

$$L_1 = \frac{-20}{s}$$

$$L_2 = \frac{-45}{s^2}$$

$$L_3 = \frac{-18}{s^3}$$

$$L_4 = \frac{-100}{s^4}$$

$$\Delta = 1 - \left(\frac{-20}{s} + \frac{-45}{s^2} + \frac{-18}{s^3} + \frac{-100}{s^4}\right)$$

$$= 1 + \left(\frac{20}{s} + \frac{45}{s^2} + \frac{18}{s^3} + \frac{100}{s^4}\right)$$

---

[b] Although developed for signal flow graphs, this gain rule may also be applied directly to block diagrams.

$$\frac{C}{R} = \frac{P_1 + P_2 + P_3}{\Delta} = \frac{\dfrac{100}{s^4} + \dfrac{5}{s^3} + \dfrac{10}{s^2}}{1 + \dfrac{20}{s} + \dfrac{45}{s^2} + \dfrac{18}{s^3} + \dfrac{100}{s^4}}$$

$$= \frac{\dfrac{100 + 5s + 10s^2}{s^4}}{\dfrac{s^4 + 20s^3 + 45s^2 + 18s + 100}{s^4}}$$

$$= \frac{10s^2 + 5s + 100}{s^4 + 20s^3 + 45s^2 + 18s + 100}.$$

■ EXAMPLE 2.7

Figure 2.11 gives the signal flow diagram of a system with three nontouching loops. For this system, there are two paths:

$$P_1 = g_1 g_2 g_3$$
$$P_2 = g_4 g_5 g_6;$$

and four loops:

$$L_1 = -g_1 h_1$$
$$L_2 = -g_2 h_2$$
$$L_3 = -g_3 h_3$$
$$L_4 = -g_5 h_4.$$

Note that L_1 does not touch L_3 and L_4 does not touch L_2 or L_3. Then

$$\Delta = 1 - (L_1 + L_2 + L_3 + L_4) + (+L_1L_3 + L_1L_4 + L_2L_4 + L_3L_4) - L_1L_3L_4.$$

For path P_1, $L_2 = L_3 = 0$ and

$$\Delta_1 = 1 - L_1 - L_4.$$

For path P_2, $L_4 = 0$ and

$$\Delta_2 = 1 - (L_1 + L_2 + L_3) + L_1L_3$$

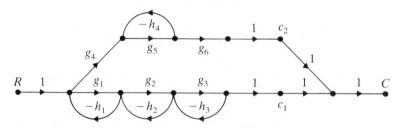

FIGURE 2.11   A Signal Flow Graph with Three Nontouching Loops.

and

$$\frac{C}{R} = \frac{g_1g_2g_3(1 - L_4) + g_4g_5g_6[1 - (L_1 + L_2 + L_3) + L_1L_3]}{1 - (L_1 + L_2 + L_3 + L_4) + (L_1L_3 + L_1L_4 + L_2L_4 + L_3L_4) - L_1L_3L_4}.$$

■ EXAMPLE 2.8

To demonstrate the use of Mason's gain rule with block diagrams, consider the dc Motor of Figure 2.7, letting $T_{load} = 0$. There is one path from V to C:

$$P_1 = \left(\frac{1}{Ls}\right)(k_T)\left(\frac{1}{J}\right)\left(\frac{1}{s}\right)\left(\frac{1}{s}\right) = \frac{k_T}{JL} \cdot \frac{1}{s^3};$$

three loops:

$$L_1 = \frac{-R}{sL}$$

$$L_2 = -\frac{f}{Js}$$

$$L_3 = \frac{-k_Tk_B}{JL} \cdot \frac{1}{s^2};$$

and two non-touching loops:

$$L_1L_2 = \frac{fR}{JL} \cdot \frac{1}{s^2}$$

$$\Delta = 1 - (L_1 + L_2 + L_3) + L_1L_2$$

$$= 1 - \left(\frac{-R}{sL} + \frac{-f}{Js} + \frac{-k_Tk_B}{JL}\frac{1}{s^2}\right) + \left(\frac{fR}{JL} \cdot \frac{1}{s^2}\right)$$

$$\Delta_1 = 1.0$$

$$\frac{C}{V} = \frac{\dfrac{k_T}{JL} \cdot \dfrac{1}{s^3}}{1 + \dfrac{R}{sL} + \dfrac{f}{Js} + \dfrac{k_Tk_B}{JLs^2} + \dfrac{fR}{JL} \cdot \dfrac{1}{s^2}}$$

$$= \frac{\dfrac{k_T}{JLs^3}}{\dfrac{JLs^2 + (JR + fL)s + k_Tk_B + fR}{JLs^2}}$$

$$= \frac{k_T}{s[JLs^2 + (JR + fL)s + fR + k_Tk_B]}.$$

| **2.6** | ACCURACY IN FEEDBACK CONTROL SYSTEMS |

One of the major reasons for choosing to use a feedback control system is a need for accuracy, usually for accuracy in the steady state, occasionally for accuracy during dynamic operation. The need may be for extreme accuracy beyond the capability of the human operator, or of an open-loop control, but it may otherwise be for accuracy over long periods of operation, which may be impractical for a human operator because of the effects of fatigue.

To design a feedback control system that will meet accuracy specifications, we must understand the conditions that are required for accuracy. It must also be understood that the conditions required for accuracy are not necessarily compatible with either stability or required response time. As a result, when a system is adjusted or changed to meet steady-state accuracy requirements, still further design may be required to make the system stable and to provide the desired dynamic response.

Single-loop linear control systems may be represented by a transfer function block diagram as shown in Figure 2.12. Note that the feedback is *unit*; all of the following analysis assumes unity feedback. Since we are interested in accuracy, we shall use the equation for the *error*, because this provides the most direct measure of accuracy. Thus:

$$E(s) = \frac{R(s)}{1 + G(s)}. \qquad (2.7)$$

Equation (2.7) represents the error in the frequency domain, but implies the time variation of the error. If we took the inverse Laplace transform of Eq. (2.7), we would obtain the transient variation in $e(t)$. Since we are interested in the steady-state error, we shall not take the inverse transform but will solve for the steady-state error using the final value theorem of the Laplace transformation. First we need a more explicit expression of $G(s)$. In general, any linear transfer function is a ratio of two polynomials, and to be physically realizable it must have more poles than zeros. If we represent these

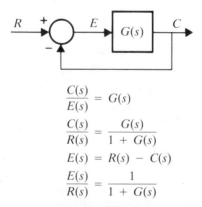

$$\frac{C(s)}{E(s)} = G(s)$$

$$\frac{C(s)}{R(s)} = \frac{G(s)}{1 + G(s)}$$

$$E(s) = R(s) - C(s)$$

$$\frac{E(s)}{R(s)} = \frac{1}{1 + G(s)}$$

FIGURE 2.12   Block Diagram of a Single-Loop Linear Control System.

polynomials in factored form, we obtain

$$G(s) = \frac{K_x \prod_{i=1}^{M} (s\tau_i + 1) \prod_{j=1}^{Q} (a_j s^2 + b_j s + 1)}{s^N \prod_{k=1}^{T} (s\tau_k + 1) \prod_{\ell=1}^{U} (A_\ell s^2 + b_\ell s + 1)}, \tag{2.8}$$

where $N + T + 2U > M + 2Q$; N = number of poles at the origin and is called the *system type number*; and K_x is a gain constant, also called the *error constant* (meaning of x to be forthcoming).

The final value theorem is:

$$\lim_{t \to \infty} f(t) = \lim_{s \to 0} sF(s). \tag{2.9}$$

Paraphrasing in terms of Eq. (2.7), we obtain

$$\lim_{t \to \infty} e(t) = \lim_{s \to 0} sE(s) \triangleq E_{\text{steady state}}. \tag{2.10}$$

Note that the final value theorem requires that the *limit must exist*. For feedback controls, this means that the system must be stable. Initially, we ignore this constraint, and the theorem tells us what is needed to achieve the desired accuracy; if the required accuracy conditions make the system unstable, we simply design a compensator to stabilize the system without altering the accuracy conditions.

To apply the final value theorem, we must know the right side of Eq. (2.7) explicitly, i.e., we must know $R(s)$ and we must know $G(s)$. Interpretation of results may be difficult or even meaningless if R is a complicated function, so it is normal practice to restrict R to the following simple functions:

R	$r(t)$	$R(s)$
Step	$Au(t)$	A/s
Ramp	Bt	B/s^2
Parabola	$Pt^2/2$	P/s^3.

Analysis using these command input signals gives us enough information for practical systems even though actual inputs may differ somewhat.

If the actual transfer function, $G(s)$, is manipulated into the algebraic form of Eq. (2.8), or into the form

$$G(s) = \frac{K_x(Ds^{M+2Q} + \cdots + 1)}{s^N(Fs^{T+2U} + \cdots + 1)}, \tag{2.11}$$

we need only the values of K_x and of N to proceed. Applying the final value theorem to the error equation, we get

$$e(t = \infty) = e_{ss} = \lim_{s \to 0} s \; \frac{R(s)}{1 + \dfrac{K_x(Ds^{M+2Q} + \cdots + 1)}{s^N(Fs^{T+2U} + \cdots + 1)}} \tag{2.12a}$$

$$e_{ss} = \lim_{s \to 0} s \; \frac{R(s)}{1 + \dfrac{K_x}{s^N}}. \tag{2.12b}$$

Note that because the polynomials of Eq. (2.11) were written such that the lowest power term, $s^0 = 1.0$, Eq. (2.12a) reduces to Eq. (2.12b) to simplify the analysis.

We shall analyze three types of systems, i.e.,

$$
\begin{array}{ll}
N = 0 & \text{type zero} \\
N = 1 & \text{type one} \\
N = 2 & \text{type two}
\end{array}
$$

Extension to higher type numbers is straightforward, but systems with higher type number are very rare (possibly nonexistent). For each of these systems, we consider each of the three standard input commands.

Type-Zero System, $N=0$ For the step input $R(s) = A/s$,

$$
e_{ss} = \lim_{s \to 0} s\left(\frac{A/s}{1 + K_p}\right) = \frac{A}{1 + K_p}.
$$

Thus, a type-zero system, when given a step command, will have a steady-state *position* error that is directly proportional to the size of the step A and inversely proportional to the position error constant K_p.

For the ramp input to the type-zero system, $R(s) = B/s^2$,

$$
e_{ss} = \lim_{s \to 0} s\left(\frac{B/s^2}{1 + K_p}\right) = \infty.
$$

The type-zero system cannot follow a ramp input so the error increases continuously. In like manner, a type-zero system cannot follow a parabolic input.

Type-One System, $N=1$ For the step input, $R(s) = A/s$,

$$
e_{ss} = \lim_{s \to 0} s\left(\frac{A/s}{1 + K_v/s}\right) = 0
$$

and the type-one system "follows" the step input in the sense that it reaches steady state with no error.

For the type-one system ramp input, $R(s) = B/s^2$,

$$
e_{ss} = \lim_{s \to 0} s\left(\frac{B/s^2}{1 + K_v/s}\right) = B/K_v.
$$

Thus, the type-one system has a steady-state position error for a ramp input. The amount of error, e_{ss}, is the commanded *velocity* B divided by the velocity error constant K_v. Note that this is a *position* error, the output velocity *exactly* duplicates the commanded velocity, but the output position is not exactly at the desired value.

The parabolic input for the type-one system, $R(s) = P/s^3$,

$$
e_{ss} = \lim_{s \to 0} s\left(\frac{P/s^3}{1 + K_v/s}\right) = \infty.
$$

The type-one system cannot follow the parabolic input.

TABLE 2.3 Steady-State Errors of Feedback Controls (when subjected to input commands)

Input System Type	Step $R(s) = A/s$	Ramp $R(s) = B/s^2$	Parabola $R(s) = P/s^3$
0	$A/1 + K_p$	∞	∞
1	0	B/K_v	∞
2	0	0	P/K_a

$K_p(rad/rad)$ = position error constant
(type-zero system)
$K_v(s^{-1})$ = velocity error constant (type-one system)
$K_a(s^{-2})$ = acceleration error constant
(type-two system)

Type-Two System, $N=2$ For the parabolic input,[c] $R(s) = P/s^3$,

$$e_{ss} = \lim_{s \to 0} s \left(\frac{P/s^3}{1 + K_a/s^2} \right) = P/K_a.$$

Thus, a type-two system *can* follow a parabolic command, but with a finite *position* error in the steady state. The magnitude of the error is P divided by the acceleration error constant K_a. Note that the velocity and acceleration are duplicated exactly at the steady state. These results are conveniently summarized in Table 2.3. The error constants of Table 2.3 are also called:

1. Static error coefficients.
2. Bode gain, because Eq. (2.8), which defines them, is commonly referred to as the "Bode" form of the transfer function.
3. DC gain because each is evaluated by setting $s = 0$ after discarding any poles at the origin.

The accuracy of the control system is affected not only by the type of command signal, but by

1. disturbances introduced at various points in the system, such as load disturbances
2. nonlinearities in the system.

When the system is *linear* but subject to disturbance, we need only invoke the principle of superposition. Compute the error (magnitude and direction) caused by each

---

[c] The type-two system follows step and ramp inputs with zero steady-state error. Proof is left to the student.

disturbance or command and sum them to calculate the total error. This will be illustrated numerically. Certain types of nonlinearities may also contribute to steady-state errors, depending on the location of the nonlinearity in the system. The two types of nonlinearity that commonly introduce errors are Coulomb friction in the motor or load and dead zone in the forward amplification path or in the sensors. These also will be illustrated numerically in Chapter 10.

2.7 ACCURACY IN A TYPE-ZERO SYSTEM

It is instructive to demonstrate the meaning of the mathematical results by presenting a simple physical system and discussing the nature and magnitude of the signals that exist under steady-state conditions. Suitable examples of type-zero, -one, and -two

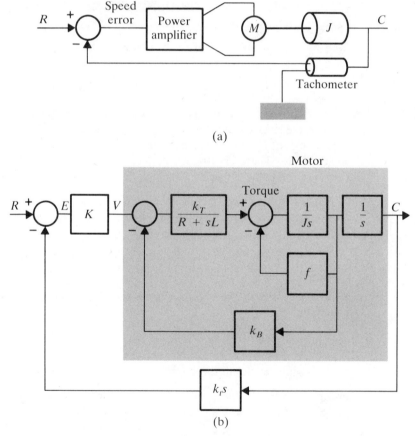

FIGURE 2.13   A Type-Zero System Example—Motor Speed Control: (a) Schematic Diagram and (b) Transfer Function Block Diagram.

systems can be constructed starting with a simple motor speed control as shown in Figure 2.13. The motor itself is modeled by all of the transfer function blocks within the shaded area of Figure 2.13(b). (Simpler models are available; this was chosen because it provides an explicit modeling of the torque summation.) The voltage applied to the motor armature is V, which, of course, cannot be zero if the motor is to run.

Assume that the motor is operating at some *constant* speed at steady state. For this to be true, $V \neq 0$; but $V = K \cdot E$ so $E \neq 0$. Thus, $R \neq$ speed voltage, so the steady-state speed actually achieved is not the commanded speed. We note, however, that actual motor speed depends on V, and $V = K \cdot E$. If K is increased, the speed will increase and E will decrease; a new steady-state condition will be reached at higher speed, increased V, and decreased E. Conceptually, we can make $E \to 0$ as $K \to \infty$.

2.8   ACCURACY IN A TYPE-ONE SYSTEM

The speed control system of Figure 2.13 can be converted to a position control by removing the speed sensor (tachometer) and replacing it with a position sensor as shown in Figure 2.14. If the system is driven with a step input and reaches steady state, the output position is a fixed value, $C = C_{ss}$. Therefore, $\dot{C}_{ss} = \ddot{C}_{ss} = 0$. However, this means that $V_{ss} = 0$; otherwise, the motor would run, changing the value of C. Consequently, $E_{ss} = 0$, since $V_{ss} = K \cdot E_{ss}$ and we conclude that a type-one system has zero steady-state error for a step input.

What happens if this system is commanded to follow a ramp input, $R = Bt$? We note that the rate of change of the input is $dR/dt = B$, a constant. In steady state, then, the servo must have an output velocity $dC/dt = B$; otherwise, we would have $\dot{E} = (\dot{R} - \dot{C}) \neq 0$, i.e., unless $\dot{C} = \dot{R}$, the system is *not in steady state*.

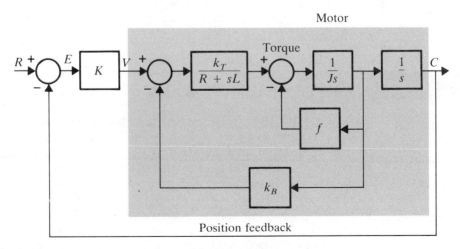

FIGURE 2.14   A Type-One System: a Positioning Servo.

The steady-state speed of a motor depends on the applied voltage. Thus, if $\dot{C} \neq 0$ (i.e., if the motor is running), then $V \neq 0$, and in order to have constant speed, the motor voltage must be some constant value. From Figure 2.14,

$$V = KE$$

and

$$E = R - C$$

and we can conclude as follows: When a type-one servo is driven with a ramp input, the servo exactly duplicates the input *velocity*, but there is a position error. The magnitude of the system error is determined by the forward gain of the servo and is the amount required for the motor to run at the velocity required by the ramp input. This magnitude is

$$E_{ss} = \dot{R}/K_v = \dot{C}_{ss}/K_v,$$

where K_v is the velocity error coefficient. Alternatively,

$$E_{ss} = V_{ss}/K,$$

where V_{ss} is the voltage required at the motor input and K is the dc gain of the amplifiers (see Figure 2.14).

What happens if we try to drive the type-one servo with a parabolic (constant acceleration) command? Clearly, steady-state conditions for the *motor* are a constant speed output caused by a constant voltage input. If the motor speed is to be increased at a constant rate (constant acceleration), then the voltage must be increased at a constant rate, which can only be done if the error changes at a constant rate. Thus, we see that a type-one system *never reaches steady state* when operated with a parabolic input. The final value theorem says that the steady-state value is the value reached as $t \to \infty$, and for the type-one system the result is $E_{ss} = \infty$. This agrees with the physical interpretation; if the system error increases at a constant rate in order to maintain the required acceleration, after a long time, the error would be large and as $t \to \infty$ clearly $E \to \infty$.

2.9   ACCURACY IN A TYPE-TWO SYSTEM

A type-one system has a steady-state error when a ramp input is used. Some applications may require that the steady-state error be reduced to zero. How can this be accomplished? From the physical analysis, we have seen that the steady-state error in the type-one system is due to the nature of the motor: The motor requires a nonzero applied voltage in order to run and in the type-one system we get the required voltage by amplifying the error signal. Since the motor voltage is required, the problem then is to generate the required voltage in such a way that, when steady state is reached, the motor voltage has the required value but the steady-state error is $E_{ss} \equiv 0$.

One solution is to incorporate an *integrator* in the amplifier which generates the motor voltage. This can be done with either a cascade or a parallel arrangement as

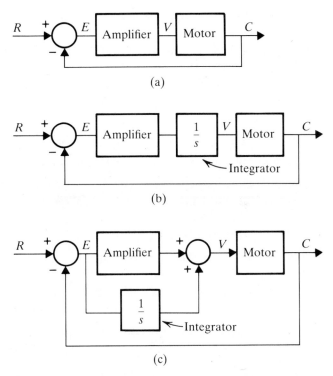

FIGURE 2.15   Conversion of a Type-One System to a Type-Two System: (a) Type-One Feedback Control System; (b) Type-Two Feedback Control System—Cascade Insertion of Integrator; and (c) Type-Two Feedback Control System—Parallel Insertion of Integrator.

shown in Figure 2.15. If we *assume* that steady-state operation has been achieved with a ramp input, the following conditions must exist:

1.   The output velocity is constant *at the command (ramp) value.*
2.   The motor voltage V is constant and is the output of the integrator.
3.   The error is *zero*, otherwise V would not be constant.

Thus, we see that if a type-two system reaches steady state following a ramp input, the steady-state position error will be zero. The insertion of the integrator, however, has an adverse effect on stability, as will be shown, so that additional design based on dynamic analysis is usually required.

What happens if the input to a type-two system is a parabolic (constant acceleration) command? Steady-state operation of the system requires that the output acceleration (in steady state) must be equal to the commanded acceleration. Since the system (in Figure. 2.15) is linear, the voltage applied to the motor must change at a *constant rate.* For the type-two system, the voltage V is the output of an integrator and changes at a constant rate (ramps up) when the integrator input is a constant. The input of the integrator is Figure 2.15 is the error signal. Thus, we conclude that a type-two system will follow a parabolic input, in steady state, with a constant steady-state error, E_{ss}.

The magnitude of this steady-state error may be calculated from the integrator gain and the required rate of change of voltage or, more generally,

$$E_{ss} = \ddot{C}/K_a,$$

where $\ddot{C}$ is the commanded acceleration and K_a is the acceleration error constant.

 EXAMPLE 2.9

The motor of Figures 2.13 and 2.14 has the following parameter values:

$$
\begin{array}{ll}
R = 0.08 \ \Omega & K_T = 0.95 \ \text{N-m/A} \\
L = 0.008 \ \text{H} & K_B = 0.95 \ \text{V/rad/s} \\
J = 0.38 \ \text{kg} \cdot \text{m}^2 & f = 0.001 \ \text{N/rad/s} \\
k_t = 1.0 \ \text{V/rad/s} & K = 150 \ \text{V/V}.
\end{array}
$$

Reducing the block diagram to a single block in the forward path (unity feedback) gives:

$$G(s) = \frac{157.875}{0.00337s^2 + 0.0337s + 1} \ \text{V/V}.$$

This is a type-zero transfer function: $K_p = 157.875$.

If the desired speed is $R = 1200$ rpm (125.6 rad/s), then the speed error is

$$E = \frac{125.6}{1 + K_p} = \frac{125.6}{158.875} = 0.79 \ \text{rad/s}$$

$$= 7.54 \ \text{rpm}.$$

If the system is converted to a type-one position control as in Figure 2.14, the forward path transfer function becomes

$$G(s) = \frac{157.875}{s(0.00337s^2 + 0.0337s + 1)} \ \text{rad/rad}$$

and $K_v = 157.875 \ \text{s}^{-1}$.

The servo is driving an antenna to track a satellite, so $R = 0.05$ deg/s $= 0.0000873$ rad/s. The steady-state error is then

$$E_{ss} = \frac{R}{K_v} = \frac{0.0000873 \ \text{rad/s}}{157.875 \ \text{s}^{-1}} = 0.000000553 \ \text{rad}.$$

2.10 ACCURACY EFFECTS OF LOAD DISTURBANCES AND NONLINEARITIES

The preceding analyses are valid only for *linear* systems and treat only those steady-state errors that arise when the system executes a command. Almost all systems are nonlinear to some extent. The effects of various nonlinearities on the behavior of

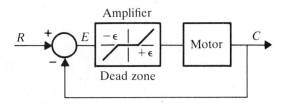

FIGURE 2.16 Dead Zone Nonlinearity in a Servo.

feedback controls involve both static and dynamic conditions and depend not only on the nature of the nonlinearity, but on the location of the nonlinear elements within the system and on the kinds of signals existing in the system. It is important to realize that some nonlinearities affect steady-state errors (as well as dynamic errors). Therefore, a brief discussion will be given of the effects of dead zone and of Coulomb friction on steady-state errors.

2.10A Dead Zone

Many physical devices that transmit signals do not respond to small signals, i.e., if the input amplitude is less than some small value, there will be no output. This is true of many sensors—devices for measuring position, velocity, temperature, acceleration, etc. These devices are said to have a *dead zone*. Amplifiers that are designed to amplify both positive and negative signals often have a dead zone about zero output; i.e., for small input, no output.

Consider Figure 2.16. If the error E is less than the threshold level ϵ there is no output from the amplifier. If this situation exists in steady state, it is clear that we cannot predict the *value* of the steady-state error. We can only guarantee that it is bounded, i.e., $-\epsilon \leq E_{ss} \leq +\epsilon$. If this is not an acceptable condition, then some redesign is in order.

2.10B Coulomb Friction

Another nonlinearity that adversely affects steady-state accuracy is friction. Most of us are familiar with *viscous* friction, which generates a reaction force proportional to relative velocity, but this is a linear effect. In any system where there is relative motion between contacting surfaces, there are several types of friction, all of them nonlinear except the viscous components. The nonlinear effect is variously called *dry friction*, *nonlinear friction*, and *Coulomb friction*. It is, in essence, a drag (reaction) force which opposes motion, but is essentially constant in magnitude regardless of velocity. Examples are common: if you try to *spin* the volume control knob on your radio, you find it will rotate easily when your fingers are in contact, but stops when you release it—it does not spin. The forces holding it are those between the surface of the volume control resistor and the sliding contact; this is Coulomb friction. Another example is an electric motor in which we find Coulomb friction drag due to the rubbing contact between the brushes and the commutator.

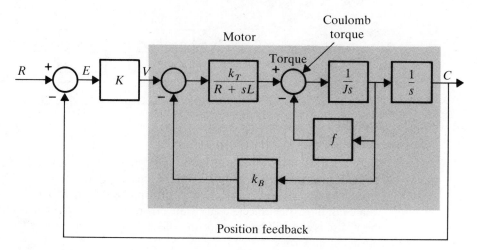

FIGURE 2.17   A Type-One System with Coulomb Friction.

When Coulomb friction is present in a servo system, the torque (or force) that it develops must be included in both dynamic and steady-state analyses. Here we consider only the latter. Figure 2.17 shows a type-one system with the Coulomb torque introduced at the torque summation node. Observe that if the applied torques add to zero there is no net torque to drive the motor inertia. Therefore, it is possible to have the servo at standstill (then $f\dot{C} = 0$) and with a small error such that the drive torque generated by the error is less than the Coulomb friction level, then the system cannot make further corrections so the steady-state condition (in response, say, to a step input) is at standstill but with a small steady-state error.

Note that such errors can be eliminated by converting the type-one system to a type-two system as shown in Figure 2.15.

2.10C   Load Disturbances

Many control systems operate under conditions such that external forces, or torques, or thermal or electrical disturbances may affect accuracy. As long as these disturbances may be treated as signals injected into the system at various points, then the dynamic and static behavior of the linear system may be evaluated by applying the principle of superposition. This means, of course, that any errors introduced by disturbances add algebraically. Computations are not more difficult than finding the error for a normal command signal, but some care is required in interpreting results.

Typical examples of "load disturbances" might be:

1. the force of gravity working on your automobile (Note that it may be neglected on a level road, but slows you down when climbing hills.)

2. wind torques on a radar tracking antenna.

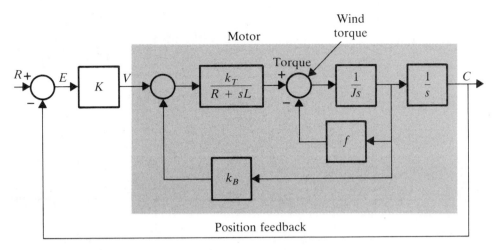

FIGURE 2.18   A Type-One System with Wind (Load) Torque.

Consider the radar tracking antenna in more detail. When the wind blows it tends to rotate the antenna; i.e., it applies a torque to the antenna. Since the servos that drive antennas are essentially the same as the type-one servo we have been discussing, consider Figure 2.18, which shows the system of Figure 2.14 with load torque introduced at the summing junction. (Note that Figure 2.18 is essentially the same as Figure 2.17. However, the Coulomb friction torque is a reaction torque only, whereas the wind torque is an actively applied torque.)

We may apply the principle of superposition to calculate the error due to the input command R and to the wind torque T_W, and add the results to get the total steady-state error. For many conditions, we can also apply the final value theorem to evaluate the error due to the load torque. Note that the quantities to be added are the values of $E = R - C$, since the definition of the position error is the difference between the commanded position R and the output position C.

2.10D   Sensitivity

A system is said to be *sensitive* if its behavior is changed by operating conditions and/or parameter variations. Thus, in a sense, a feedback control system is sensitive to load disturbances. Parameter variations may occur during the life of a given system, perhaps due to temperature or aging; for example, the gain of an amplifier may change and such changes can affect both accuracy and dynamic response. If a small change in parameter value causes an appreciable change in performance (i.e., if the sensitivity is too great), the system may not be satisfactory.

When systems are mass produced (such as magnetic disk memories with production levels of thousands per day), the system parameter values differ because of tolerances. In such cases also, if the sensitivity is too great, the system may not be satisfactory.

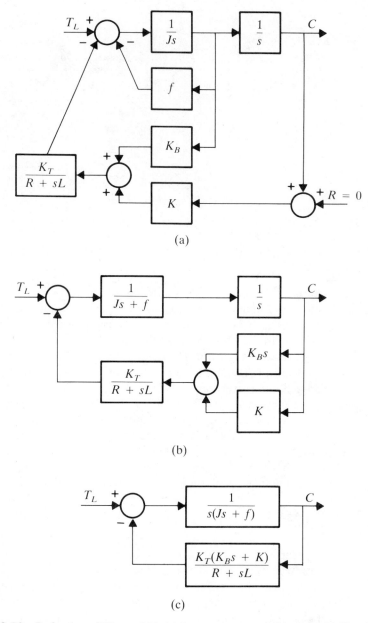

FIGURE 2.19   Reduction of Figure 2.18: (a) Rearrangement of Diagram, (b) First Reduction, and (c) Final Form.

Systems that are not very sensitive to parameter variation or to disturbances are said to be *robust*.

 EXAMPLE 2.10

The system of Figure 2.14 is subjected to a wind load as shown in Figure 2.18. Let $T_L = 1.0$ N-m, effective during tracking at 0.05 deg/s.

Using superposition, with T_L assumed zero, the steady-state error is $E_{ss} = 0.000000553$ rad, as calculated in Example 2.9. Next assume $R = 0$ and use block diagram reduction to obtain the transfer from T_L to C as shown in Figure 2.19. Applying the final value theorem,

$$C_{ss} = \lim_{s \to 0} s \frac{T_L}{s} \frac{C}{T_L} = 5.61 \times 10^{-4}.$$

By inspection of Figure 2.18, when $R = 0$, $E = -C$. Thus, $E_{ss} = -C_{ss}$. The total position error is the algebraic sum of the error due to tracking and the error due to the wind load:

$$E_{ss\,total} = 0.000000553 + 0.000561 = 0.000561553 \text{ rad.}$$

Note that this solution assumes the wind is increasing the error. If the wind reverses direction, the errors subtract.

2.11   DYNAMIC ERROR COEFFICIENTS

For some control problems, the system must accomplish its objective during a transient period. For example, when a radar tracking system is following a maneuvering aircraft, the aircraft motion never reaches a steady-state condition and neither does the error of the radar tracker. In general, it is not possible to design the control system so that it follows with zero error when the command is time varying. We can, of course, simulate the control system and measure such errors if we know and can simulate the time-varying command. Since commands such as the evasive maneuvers of an aircraft cannot be predicted, the use of simulation requires careful interpretation. What we can do, however, is determine bounds on the error and adjust these bounds so that the error of the tracking system will remain within acceptable limits.

If we know the analytic expression for $R(s)$, Eq. (2.7) gives the relationship between error, command, and system characteristics:

$$E(s) = \frac{R(s)}{1 + G(s)} = \frac{1}{1 + G(s)} R(s). \tag{2.7}$$

Since $G(s)$ is, in general, a ratio of two polynomials, these can be manipulated into the form of a ratio of two polynomials. Arranging both polynomials in ascending powers of s and dividing the denominator into the numerator, we obtain

$$E(s) = (K_0 + K_1 s + K_2 s^2 + K_3 s^3 + \cdots)R(s). \tag{2.13}$$

The series in Eq. (2.13) is called the *error series*, and the coefficients $K_0, K_1, K_2, \ldots$ are the dynamic error coefficients.

If we know $R(s)$ to be an analytic expression, we can evaluate Eq. (2.13), but if we have that much information we can also simulate the system and evaluate the error much more easily from the simulation. We would not use the dynamic error coefficients for such a problem. However, if we do not know the explicit command, but can predict its basic characteristics, the error series can be useful.

Expanding Eq. (2.13), we obtain

$$E(s) = K_0 R(s) + K_1 s R(s) + K_2 s^2 R(s) + K_3 s^3 R(s) + \cdots. \tag{2.14a}$$

We next take the inverse transform term by term, but discard the impulses, doublets, triplets, etc., that arise from the multiple derivatives (this is equivalent to saying that we wait until the initial condition transients have damped out and evaluate the error only after the system is operating slowly and smoothly). The result is:

$$e(t) = K_0 R(t) + K_1 \dot{R}(t) + K_2 \ddot{R}(t) + K_3 \dddot{R}(t) + \cdots. \tag{2.14b}$$

If we do not know $R(t)$ explicitly but know the *maximum* expected values of $R(t)$ and its successive derivatives, we evaluate

$$e(t)_{max} = K_0 R(t)_{max} + K_1 \dot{R}(t)_{max} + K_2 \ddot{R}(t)_{max} + \cdots. \tag{2.15}$$

We are guaranteed that the actual error will not exceed the calculated value, because in actual operation it is not possible for all derivatives of $R(t)$ to reach their maximum values simultaneously. Thus, if the value calculated for $e(t)_{max}$ is small enough to be within tolerances, the design is acceptable and, if $e(t)_{max}$ is too large, it is easy to see which terms in the series must be reduced in magnitude. While the series is infinite, it usually converges rapidly so that truncation after only a few terms is sufficiently accurate.

 EXAMPLE 2.11

Consider a simple type-one positioning system for which

$$G(s) = \frac{K_v}{s(s\tau_1 + 1)(s\tau_2 + 1)} = \frac{5}{s(s + 1)(0.1s + 1)}$$

$$1 + G(s) = 1 + \frac{K_v}{s(s\tau_1 + 1)(s\tau_2 + 1)} = \frac{s(s\tau_1 + 1)(s\tau_2 + 1) + K_v}{s(s\tau_1 + 1)(s\tau_2 + 1)}$$

$$\frac{1}{1 + G(s)} = \frac{s(s\tau_1 + 1)(s\tau_2 + 1)}{s(s\tau_1 + 1)(s\tau_2 + 1) + K_v} = \frac{s + (\tau_1 + \tau_2)s^2 + \tau_1\tau_2 s^3}{K_v + s + (\tau_1 + \tau_2)s^2 + \tau_1\tau_2 s^3}.$$

Dividing

$$\frac{1}{1 + G(s)} = \frac{1}{K_v}s + \left(\frac{\tau_1 + \tau_2}{K_v} - \frac{1}{K_v^2}\right)s^2 + \left[\frac{\tau_1 + \tau_2}{K_v} - \frac{2(\tau_1 + \tau_2)}{K_v^2} + \frac{1}{K_v^3}\right]s^3 \cdots,$$

then

$$E(s) = \frac{1}{K_v} sR(s) + \left(\frac{\tau_1 + \tau_2}{K_v} - \frac{1}{K_v^2} \right) s^2 R(s) + \left[\frac{\tau_1 + \tau_2}{K_v} - \frac{2(\tau_1 + \tau_2)}{K_v^2} + \frac{1}{K_v^3} \right] s^3 R(s)$$

from which the dynamic error coefficients are

$$K_0 = 0$$

$$K_1 = \frac{1}{K_v} = \frac{1}{5} = 0.2$$

$$K_2 = \frac{\tau_1 + \tau_2}{K_v} - \frac{1}{K_v^2} = \frac{1.1}{5} - \frac{1}{25} = 0.22 - 0.04 = 0.18$$

$$K_3 = \frac{\tau_1 \tau_2}{K_v} - \frac{2(\tau_1 \tau_2)}{K_v^2} + \frac{1}{K_v^3} = \frac{0.1}{5} - \frac{2.2}{25} + \frac{1}{125} = 0.02 - 0.088 + 0.008 = -0.06.$$

Note especially that the K_0 does not appear in the series, i.e., $K_0 R(s) = 0$; therefore, $K_0 = 0$. This is due to the fact that the system is type one. The series starts with the K_0 term only if the system is type zero; it starts with the K_1 term for a type-one system and with the K_2 term for a type-two system.

Inverting the series to the time domain

$$e(t) = K_1 \dot{r}(t) + K_2 \ddot{r}(t) + K_3 \dddot{r}(t) \cdots$$

$$= 0.2\dot{r}(t) + 0.18\ddot{r}t - 0.06\dddot{r}(t) \cdots.$$

Assume that the input $r(t)$ is slowly varying such that the maximum values of the derivatives are:

$$\dot{r}(t)_{\max} = 0.01; \qquad \ddot{r}(t)_{\max} = 0.012; \qquad \dddot{r}(t)_{\max} = 0.005$$

$$e(t)_{\max} = (0.2)(0.01) + (0.18)(0.012) - (0.06)(0.005)$$

$$= 0.002 + 0.00216 - 0.0003 = 0.00386.$$

This assures that the maximum error during the dynamic operation (but after initial transients have damped out) will not exceed 0.00386. If this is not acceptable, a smaller error may be achieved by altering the values of the error coefficients, which may be done in a variety of ways.

2.12 SUMMARY

The first step in control system analysis is modeling. Transfer functions, block diagrams, signal flow graphs, and Mason's gain rule were developed to aid in the modeling process.

When a suitable model has been obtained, it is first used to study the accuracy of the system in steady state. The final value theorem of the Laplace transform is used to evaluate steady-state errors when the input is a step, a ramp, or a parabola. Accuracy is found to depend on the system gain (error coefficient) and the number of poles at the origin (for the open-loop transfer function); this leads to a classification of systems as type 0, 1, or 2, where the 0, 1, or 2 is the number of poles at the origin.

The effect of load disturbances and nonlinearity on steady-state accuracy was discussed, and the final value theorem is used (together with the superposition theorem) to evaluate steady-state errors due to a constant load disturbance.

It should be noted that analysis cannot proceed to the study of dynamic response until the system type and gain have been established.

BIBLIOGRAPHY

Chestnut, H.; and Mayer, R. W. *Servomechanisms and Regulating Systems Design.* New York: John Wiley and Sons (1953).

Dorf, R. C. *Modern Control Systems.* Reading, Mass.: Addison Wesley (1986).

Gardner, M. F.; and Barnes, J. L. *Transients in Linear Systems.* New York: John Wiley and Sons (1942).

Graybeal, T. D. "*Block Diagram Network Transformation.*" *Elect. Engrg.* 7: 985–90 (1951).

Mason, S. J. "*Feedback Theory—Some Properties of Signal Flow Graphs.*" *Proc. IRE* 41, no. 9 (September 1953).

Thaler, G. J.; and Brown, R. G. *Servomechanism Analysis.* New York: McGraw-Hill Book Co. (1953).

Wylie, C. R., Jr. *Advanced Engineering Mathematics.* New York: McGraw-Hill Book Co. (1960).

PROBLEMS

For each of the following, assume that the component is not loaded by any device connected to its output, and derive the transfer function.

2.1 Basic elementary filters:

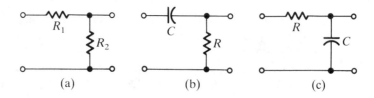

(a)   (b)   (c)

2.2   Servo compensation networks:

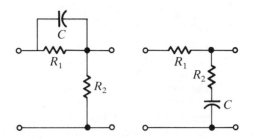

2.3   Resonant filters:

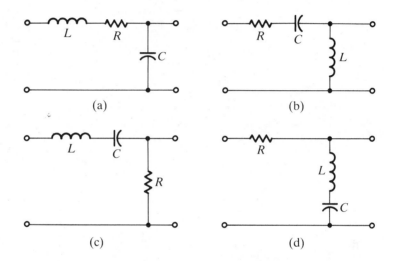

2.4   Write the differential equations, then transform and obtain the transfer function e_o/I.

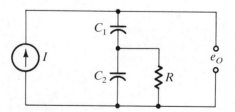

2.5   A dc motor has a fixed field F. A voltage V is applied through an amplifier k. Other symbols:

R = armature resistance
L = armature inductance
e = back emf voltage
k_B = back emf constant
k_T = motor torque, constant
T = torque applied to shaft
J = rotor moment of inertia
f = coefficient of viscous friction of moving parts
θ = angle through which the shaft is rotated.

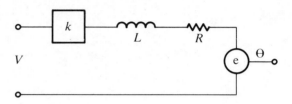

Derive the transfer function of the motor, $\dfrac{\Theta}{V}$.

2.6   The motor of Problem 5 is driven by a *current source* amplifier, i.e.,

The current out of the amplifier is determined by the amplifier design and depends only on the voltage V. Derive the transfer function of the motor, $\dfrac{\Theta}{V}$.

2.7   A dc motor drives a *pointer* and is spring loaded to return the pointer to a reference position:

k_B = back emf constant
k_T = torque constant
k_s = spring constant
J = moment of inertia.

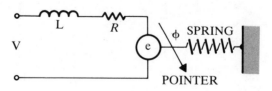

Find the transfer function $\dfrac{\Theta}{V}$.

2.8   An electric dc motor is driving an inertia load J_L through a "springy" shaft with spring constant K_{ML}. Derive the transfer function $\dfrac{\Theta}{V}$.

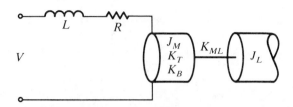

2.9   Reduce each diagram to a single block, determining the resulting equivalent transfer function.

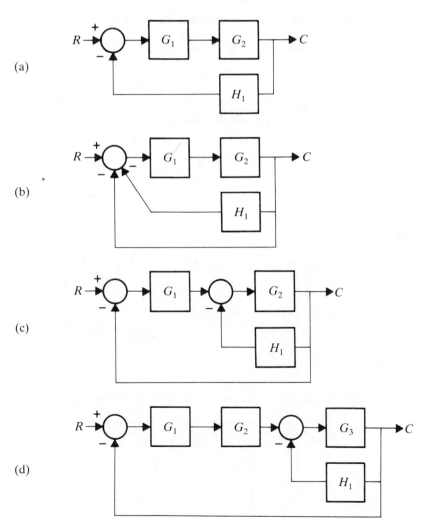

(a)

(b)

(c)

(d)

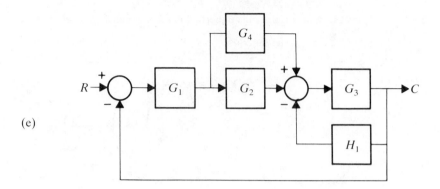

(e)

For each diagram:

2.10   a.   Use block diagram reduction methods to obtain the equivalent transfer function from R to C.

      b.   Check the result of (a) using Mason's gain rule.

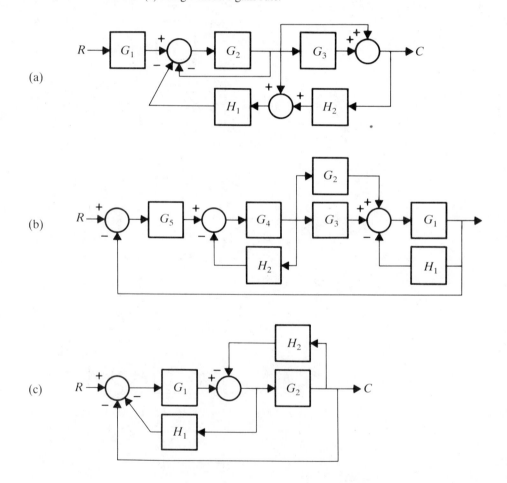

(a)

(b)

(c)

(d)

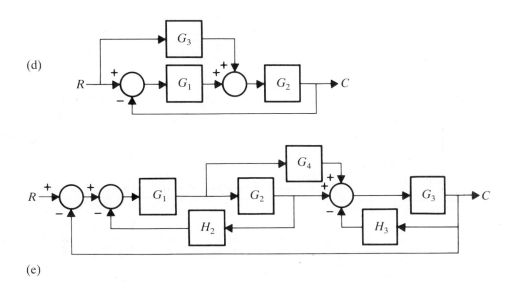

(e)

2.11 a. For each diagram, find the equivalent transfer function.
       b. Use Mason's gain rule to obtain the equivalent transfer function.

(a)

(b)

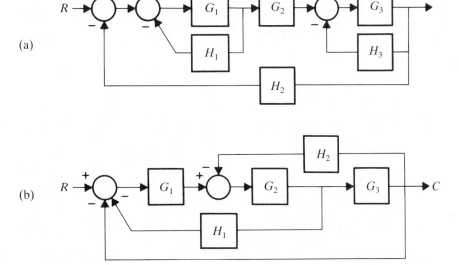

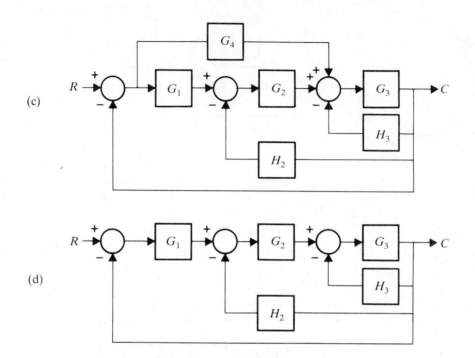

(c)

(d)

2.12   a.   Using block diagram reduction, find the transfer function from each input to the output C.
         b.   Repeat using Mason's gain rule.

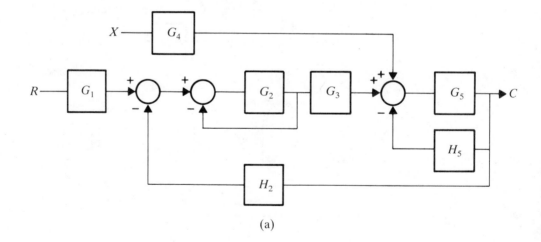

(a)

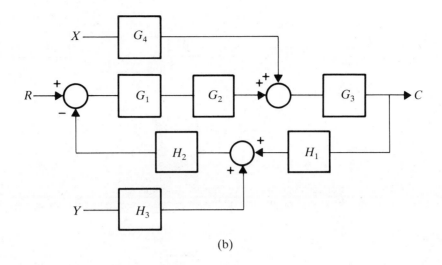

(b)

2.13 a. Derive transfer functions from E to C for each diagram.
  b. How do the minor feedback loops affect the system type number and error coefficient?

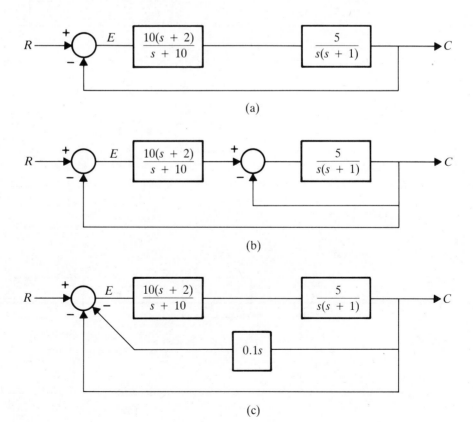

(a)

(b)

(c)

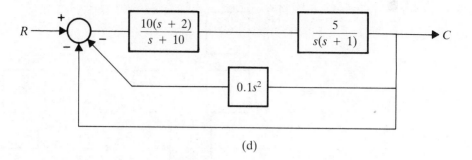

(d)

2.14   a.   For each system, derive transfer functions from E to C.

b.   From the results of (a), determine the system type number and error coefficient for each system.

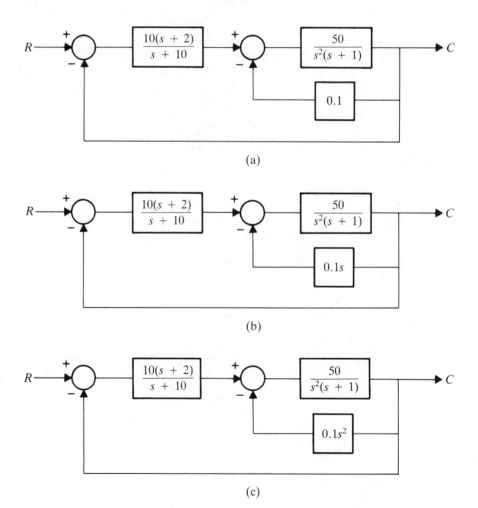

(a)

(b)

(c)

2.15   Obtain open-loop and closed-loop transfer functions for each system.

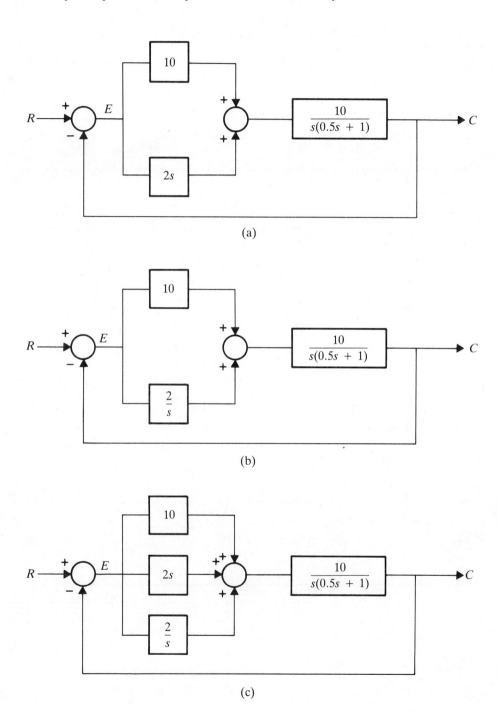

(a)

(b)

(c)

2.16   For each diagram determine:
  a.   number of paths
  b.   number of loops
  c.   number of touching loops
  d.   number of nontouching loops
  e.   transfer function from R to C.

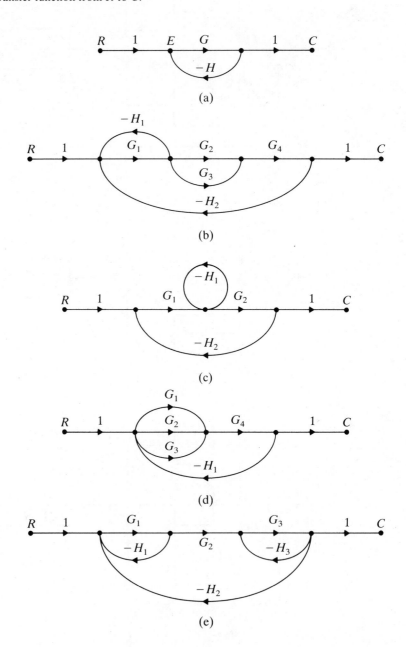

(a)

(b)

(c)

(d)

(e)

2.17 Reduce each diagram using Mason's gain rule.

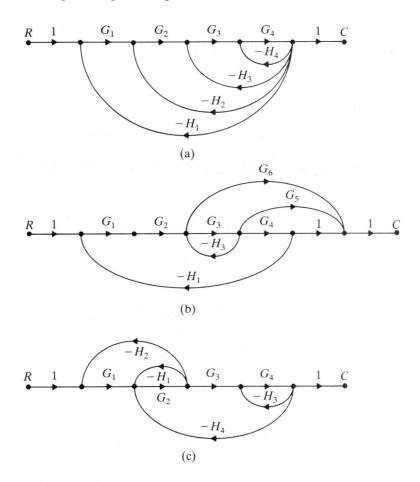

2.18 Redraw the block diagrams of Problem 2.10 as signal flow diagrams.

2.19 Redraw the block diagrams of Problem 2.12 as signal flow diagrams.

2.20 a. Given the following system, R is a unit step, and we assume that the system is stable. What value of K is needed so that $E < 0.1$?

  b. Given the following system, R is a ramp input and $R(t) = 0.1t$. What is the steady-state error?

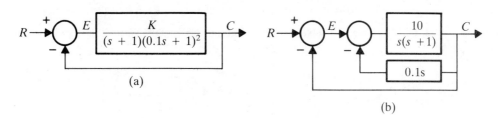

2.21 a. The system is subjected to a ramp input, $R = 0.1t$. What is the steady-state error?
   b. The command input is a constant, $R = 10$. What is the steady-state error?
   c. The command input is $R = 0.2t$. What is the steady-state error?

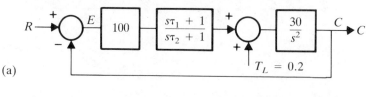

(a)

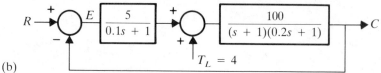

(b)

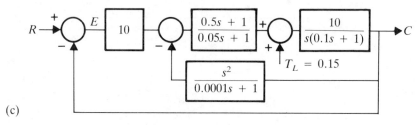

(c)

2.22 For each of the following systems, what is the error coefficient? Give both value and *dimensions*.

a. $$G(s) = \frac{100(0.05s + 1)}{(10s + 1)(0.1s + 1)^2(0.001s + 1)}$$

b. $$G(s) = \frac{50(s + 5)}{s^2}$$

c. $$G(s) = \frac{220}{s(s + 1)(0.001s + 1)}$$

d. $$G(s) = \frac{K(s + z)}{s(s + p)(\tau s + 1)}$$

2.23 A ramp input of 0.1 rad/s is applied to a unity feedback system with forward transfer function:

$$\frac{1500(0.0025s + 1)}{s(0.01s + 1)(0.001s + 1)^2}$$

What is the steady-state error?

2.24 Assume $R(t) = 0.1t$. It is required that $E_{ss} \le 0.005$. What is the required value of K?

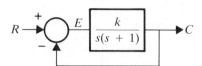

2.25   For each system, evaluate the error coefficient and the steady-state error.

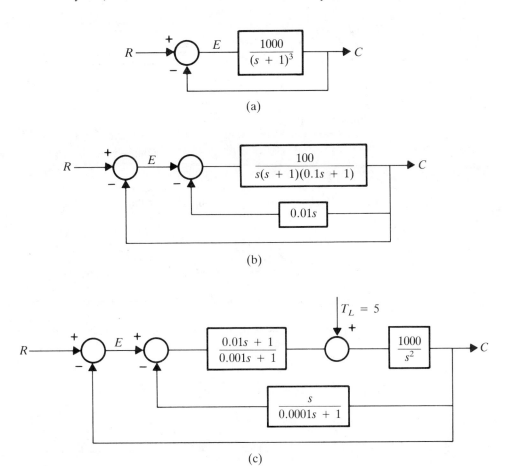

(a)

(b)

(c)

2.26   A dc motor is used in a positioning servo, but the output shaft is restrained by a spring. An approximate block diagram is given. If R is a step of 0.2 units, what is the steady-state error?

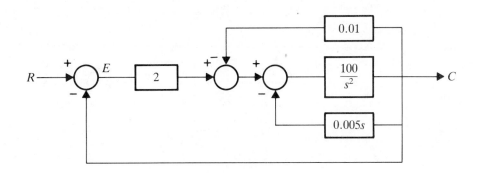

2.27   The block diagram represents a speed regulator.

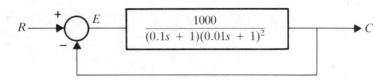

a.   Is the system stable?
b.   Calculate the steady-state value of E if R is a constant and $R = 500$.
c.   What is the system type number, and what is its static error coefficient?

2.28   The block diagram represents a radar antenna drive.
If R is a ramp input $R = (0.1 \text{ rad/s})t$, then what is the steady-state position error? What is the system type number and what is the static error coefficient?

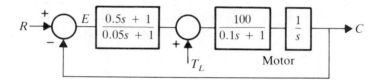

If $R = 0$, the system is at standstill. Assume that the wind blows and applies a load torque $T_L = 0.2$. What is the steady-state error? What can be done to reduce this error? Will such changes affect stability?

2.29   Given:

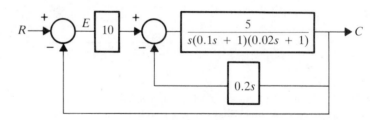

The input is a ramp, $R = 2.0t$. What is the steady-state error?

2.30   Given:

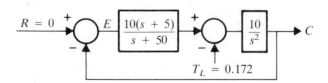

What is the steady-state error?

2.31 Given:

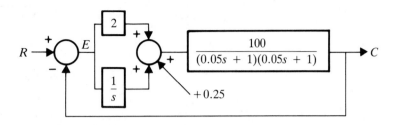

The input is a ramp, $R = 1.5t$. What is the steady-state error?

2.32 Given:

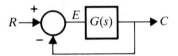

a. The steady-state error for a step input is zero.
b. The characteristic equation of the closed loop system is $s^3 + 4s^2 + 6s + 4 = 0$.
Find $G(s)$. Also find the steady-state error if the input is a unit ramp.

2.33 Given:

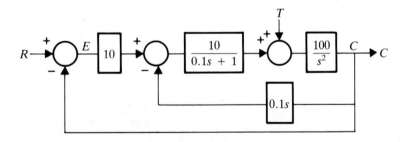

Find the steady-state value of the error E if T is a unit step and $R = 0$.

2.34 The block diagram represents a heat treating oven. The setpoint (desired temperature) is 1000°.
What is the steady-state temperature C?

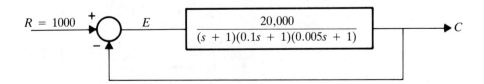

2.35   An instrument servo for controlling position is damped with velocity feedback.

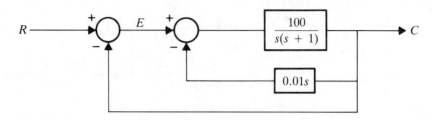

a.   If the input R is a unit step, what is the steady-state error?
b.   If the input R is a ramp: $R = 0.01t$, what is the steady-state error?
c.   What is the system static error coefficient?

2.36   A motor is servo controlled to hoist a weight to a desired position. The load applies a torque of 8 in.-lb to the shaft. The motor torque constant is 3 lb-ft/A. The amplifier gain is 4 A/in. What is the steady-state error when the system comes to rest near the commanded position?

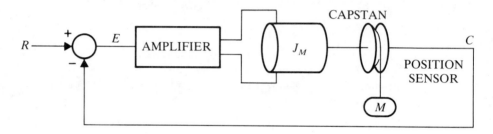

2.37   Given:
Determine the steady-state error if the input is $R = vt$.

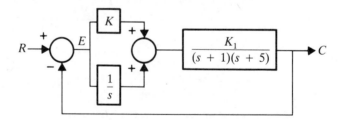

2.38   Given:
If $R = 0$, what is the steady-state error?

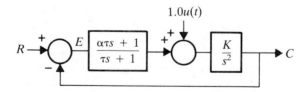

2.39 A tracking servo is given an input command $R = 0.2t$. The system is defined by the following block diagram. Find the steady-state error.

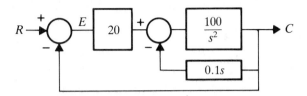

2.40 A positioning system is subject to a constant load torque, $T_L = 1.5$ units. Find the steady-state error.

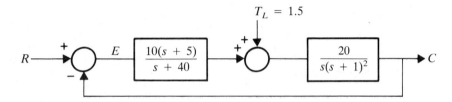

2.41 A positioning system has a command $R = 0.05t$, and a load torque $T_L = -0.6$ units. Find the steady-state error.

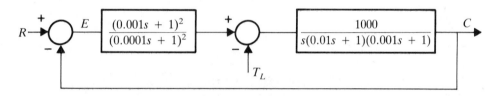

2.42   Given:

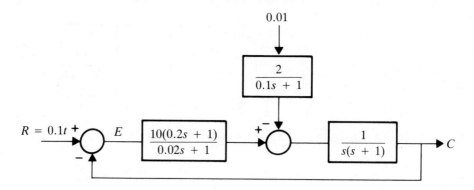

What is the steady-state error?

2.43   An instrument servo consisting of motor, spring-loaded shaft, etc., is shown below. The differential equations are

$$V = iR + L\frac{di}{dt} + K_B \, d\Theta/dt$$

$$K_T i = J\frac{d^2\theta}{dt^2} + K_X\Theta,$$

where

V = voltage in volts
i = current in amperes
R = motor resistance = 1.0 Ω
L = motor inductance = 0.1 H
K_A = amplifier gain = 10 V/V
K_X = spring constant = 0.001 N-m/rad
K_B = back emf constant = 0.01 V/rad/s
K_T = torque constant = 0.01 N-m/A
J = moment of inertia = 0.005 N-m-s^2.

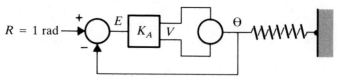

AMPLIFIER   MOTOR   SPRING

If the input is a step of 1.0 rad, what is the steady-state error?

2.44   Given:

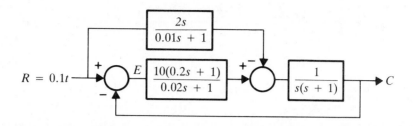

Find the steady-state error.

3 Stability and Dynamic Response

DEFINITIONS AND FUNDAMENTALS

When we say that a system is *stable*, the engineering meaning of this is that a finite duration disturbance causes a response of finite duration, after which the system resumes a steady-state condition. Most control systems, if stable, have a response that includes a few oscillations that damp out. Conversely, when we say that a system is *unstable* we mean that its output readily diverges from an initial value (often with no noticeable disturbance) and the output either changes unidirectionally or oscillates with ever-increasing amplitude.

When the system is linear so that the describing differential equations are linear, the mathematical interpretation of stability is clear. A system is stable if all of the eigenvalues (roots) of the closed-loop equations are negative or have negative real parts. The system is unstable if any eigenvalue is positive or has a real part that is zero or positive.

This becomes clear if we consider the basic mathematical form of the solution to linear differential equations, which is:

$$y(t) = A + Be^{r_1 t} + Ce^{r_2 t} + De^{r_3 t} + \cdots,$$

where the number of terms in the solution is determined by the order of the differential equation and

A, B, C = coefficients determined by the system parameters and the forcing function

r_1, r_2, r_3 = eigenvalues or roots.

63

If all of the r's are negative or have negative real parts, it is clear that all of the exponential terms approach zero as $t \to \infty$. If any of the r's is zero, the corresponding term does *not* change in magnitude and, if any r is positive or has a positive real part, the corresponding exponential term grows without limit. In terms of the complex variable s, the system is stable if all roots are in the left half of the s-plane, but unstable if any roots are on the imaginary axis or in the right half of the s-plane.

For linear systems, then, all stability tests are simply tests to determine whether any root is in the right half of the s plane or on the imaginary axis. There are many such tests available:

1. using state variables or transfer functions, the characteristic polynomial is found
2. a computer is used to find the roots of the polynomials
3. Routh-Hurwitz criterion
4. Nyquist criterion
5. root locus studies.

Evaluating all of the roots by using tests 1 and 2 is certainly a desirable part of any system analysis and gives a direct answer to the question of stability. However, if the analysis is to be followed by changes which are to stabilize the system or otherwise improve its response, the information provided does not guide us; it does not tell us what kind of changes will be helpful, i.e., it does not aid the design effort.

The Routh-Hurwitz criterion is a test for the *existence* of roots in the right half plane. It tells us "yes" or "no," and if "yes" it also tells us how many roots are in the right half plane. It does not tell us where they are and, for most problems, it gives no guidance for design procedures. However, it is a relatively simple manipulation and is easily carried out for low-order systems. For some types of problems, it may be helpful in design.

The Nyquist criterion[1] is a frequency domain test that determines whether there are any roots of the characteristic polynomial in the right half of the s-plane. It is a graphical test based on conformal mapping and complex variable theory. In essence, we plot a curve (or curves), which normally are the open-loop frequency response, and we interpret stability of the system by inspection of these curves. An advantage of the method is that we can develop design procedures based on reshaping these curves.

The root locus method is also a graphical method. Using it, curves are constructed on the s-plane that show the *motion* of all roots as a specified system parameter (such as the gain) is varied. The method permits evaluation of the root location for a specified value of the parameter and, of course, if the curve or curves cross the imaginary axis and go into the right half plane, we can establish the conditions for stability. Again, because of the graphical nature of the method, we can develop design procedures based on reshaping these curves.

It is wise to consider each of these stability tests as a separate tool and to learn to use them all. In practice, the choice of the tool depends on the problem to be solved— the best tool for one problem may not be best for the next problem.

3.2 DETERMINING THE CHARACTERISTIC EQUATION

Consider the system of Figure 3.1. The differential equations of this system are

$$J\ddot{C} + f\dot{C} = k_T i \tag{3.1a}$$

$$V = iR_a + L\dot{i} + k_B\dot{C} \tag{3.1b}$$

$$V = K(R - C) \tag{3.1c}$$

To obtain a single differential equation for the system, first eliminate V and i by substitution, obtaining

$$J\ddot{C} + f\dot{C} = \frac{k_T(KR - KC - L\dot{i} - k_B\dot{C})}{R_a}. \tag{3.2}$$

Differentiate Eq. (3.1a):

$$J\dddot{C} + f\ddot{C} = k_T\dot{i}.$$

Substitute into Eq. (3.2):

$$JR_a\ddot{C} + fR_a\dot{C} = k_T[KR - KC - \frac{L}{K_T}(J\dddot{C} + f\ddot{C}) - k_B\dot{C}].$$

Rearrange:

$$JL\dddot{C} + (JR_a + fL)\ddot{C} + (fR_a + k_B k_T)\dot{C} + Kk_T C = Kk_T R. \tag{3.3}$$

The characteristic equation is the unforced differential equation, and this name is also used for the polynomial obtained by transforming this equation for all initial conditions zero:

$$[JLs^3 + (JR_a + fL)s^2 + (fR_a + k_B k_T)s + Kk_T]C = 0. \tag{3.4}$$

For most systems, it is simpler to transform the individual differential equations first and then combine the resulting algebraic equations. This is often done with block

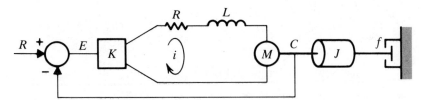

Newton's law: $J\ddot{C} + f\dot{C} = k_T i$
Kirchhoff's law: $V = iR_a + L\dot{i} + k_B\dot{C}$
$V = K(R - C)$

FIGURE 3.1 Differential Equations for a Positioning Servo.

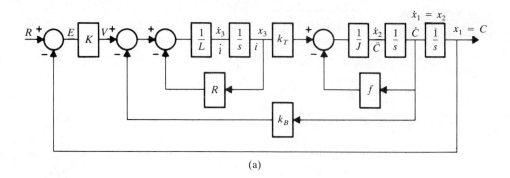

(a)

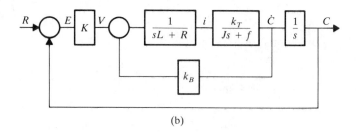

(b)

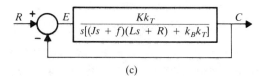

(c)

FIGURE 3.2 Block Diagrams of a Positioning System: (a) Block Diagram from the Differential Equations, (b) Partial Reduction of the Block Diagram, and (c) Transfer Function Form.

diagram manipulations as shown in Figure 3.2. To obtain the characteristic equation note that

$$\frac{C}{R} = \frac{G(s)}{1 + G(s)}$$

as shown in Chapter 2, Figure 2.12. Then

$$[1 + G(s)]C = G(s)R$$

and it is customary to say that

$$1 + G(s) = 0$$

is the characteristic equation.

From Figure 3.2(c):

$$G(s) = \frac{Kk_T}{(s)[(Js + f)(Ls + R_a) + k_Bk_T]} \tag{3.5}$$

and after manipulation

$$
\begin{aligned}
1 + G(s) &= \frac{s[(Js + f)(Ls + R_a) + k_Bk_T] + Kk_T}{s[(Js + f)(Ls + R_a) + k_Bk_T]} \\
&= \frac{JLs^3 + (JR_a + fL)s^2 + (fR_a + k_Bk_T)s + Kk_T}{JLs^3 + (JR_a + fL)s^2 + (fR_a + k_Bk_T)s}. \tag{3.6}
\end{aligned}
$$

Note that the numerator of $1 + G(s)$ is exactly the characteristic polynomial.

3.3   THE ROUTH CRITERION

Once the characteristic equation has been derived, it may be sufficient to test the system for stability, without bothering to find the roots. The Routh criterion is a test for *existence* of roots in the right half plane. It requires that we know the coefficients of the characteristic equation, which are then arranged as rows in an array. Additional rows are formed by simple manipulations, and the stability of the system is then determined by inspection of the *signs* of the left-hand column of the array. We demonstrate the test by examples, without proof.

The general rules for constructing the Routh array are as follows:

1. Given a polynomial of any order, arrange alternate coefficients in two rows, starting the first row with the coefficient of the highest order them. For example:

Polynomial: $a_7s^7 + a_6s^6 + a_5s^5 + a_4s^4 + a_3s^3 + a_2s^2 + a_1s^1 + a_0s^0$

Row 1 (s^7)	a_7	a_5	a_3	a_1
Row 2 (s^6)	a_6	a_4	a_2	a_0

2. Form additional rows according to the following procedure:
    a. The next row is formed using the two preceding rows.
    b. The left column entries of the two preceding rows are used in forming each new entry, as follows:

$$
\begin{array}{cccc}
a_7 & a_5 & a_3 & a_1 \\
a_6 & a_4 & a_2 & a_0 \\
C_1 & C_2 & C_3 & C_4
\end{array}
$$

$$C_1 = \frac{a_6a_5 - a_7a_4}{a_6} \qquad C_2 = \frac{a_6a_3 - a_7a_2}{a_6} \qquad C_3 = \frac{a_6a_1 - a_7a_0}{a_6}$$

■ EXAMPLE 3.1

$$s^7 + s^6 + s^5 + 2s^4 + 3s^3 + 4s^2 + 5s + 6$$

s^7	1	1	3	5
s^6	1	2	4	6
s^5	$\dfrac{(1)(1)-(1)(2)}{1}=-1$	$\dfrac{(1)(3)-(1)(4)}{1}=-1$	$\dfrac{(1)(5)-(1)(6)}{1}=-1$	0
s^4	$\dfrac{(-1)(2)-(1)(-1)}{-1}=1$	$\dfrac{(-1)(4)-(1)(-1)}{-1}=3$	$\dfrac{(-1)(6)-(1)(0)}{-1}=6$	0
s^3	$\dfrac{(1)(-1)-(-1)(3)}{1}=2$	$\dfrac{(1)(-1)-(-1)(6)}{1}=5$	0	0
s^2	$\dfrac{(2)(3)-(1)(5)}{2}=\dfrac{1}{2}$	$+6$	0	0
s^1	$\dfrac{-(0.5)(5)-(2)(6)}{-0.5}=-19$	0	0	0
s^0	$+6$	0	0	0

s^7	1	1	3	5
s^6	1	2	4	6
s^5	-1	-1	-1	0
s^4	$+1$	$+3$	$+6$	0
s^3	$+2$	$+5$	0	0
s^2	$+0.5$	$+6$	0	0
s^1	-19	0	0	0
s^0	$+6$	0	0	0

Interpretation: The *signs* of the terms in the left-hand column are inspected. For the system to be stable (all roots in the left half plane), all of the signs must be the same. In the example, there are two minus signs, so the system is unstable. The *number* of roots in the right half plane is the same as the number of *changes* in sign. In the example, as we inspect the sign in the column progressing downward, the s^6 entry is plus, the s^5 entry is minus (one change in sign), and the s^4 entry is plus (second change in sign); the s^2 entry is minus, and the s^0 entry is plus, giving a total of four sign changes and thus four roots in the right half plane.

In addition to using the Routh criterion as a direct stability test, we may find it convenient to use the test to establish limits on the values of adjustable parameters.

 EXAMPLE 3.2

$$s^3 + 10s^2 + 50s + K = 0. \tag{3.7}$$

What is the maximum value of K if the system is to be stable?

s^3	1	50
s^2	10	K
s^1	$(500 - K)/10$	0
s^0	K	0

From the array, K must be positive. Also, if the s^1 entry is to be positive, it is necessary that $K < 500$. Therefore, the bounds of K for a stable system are $0 < K < 500$. Note that if $K = 500$, the s^1 entry is exactly zero and, if $K > 500$, the s^1 entry is negative. This implies that for $K = 500$ there are two roots *on the imaginary axis*. Thus, the designer knows that he must set the gain to some value less than $K = 500$; probably considerably less than 500 if good damping is to be obtained.

 EXAMPLE 3.3

$$s^3 + 10s^2 + (50 + B)s + K = 0 \tag{3.8}$$

This third-order polynomial contains *two* adjustable variables. It could be the characteristic equation of a positioning system in which K is the gain of the power

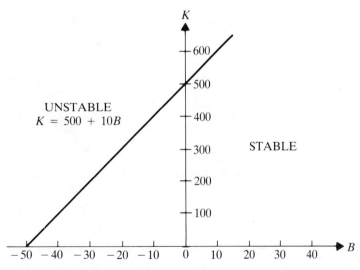

FIGURE 3.3 Stability Boundary Defined by Use of the Routh Criterion.

amplifier and B is the gain of the velocity feedback. The Routh array is:

s^3	1	$50 + B$
s^2	10	K
s^1	$(500 + 10B - K)/10$	0
s^0	K	0

Note that the stability limit is now a function of both K and B. What conditions on B and K provide a stable system? Clearly, K must be positive. Since the s^1 entry must also be positive, it is necessary that $K \leq 500 + 10B$. This equation defines a straight line on a B versus K-plane as shown in Figure 3.3. From Figure 3.3 it is clear that the designer can now raise K above the limit found in the preceding example, i.e., above 500, if he also adjusts B to preserve stability.

It should be noted that some polynomials may give rise to special conditions, i.e.,

1. An entry in the left-hand column may go to zero.
2. An entire row in the array may become zero.

Each of these cases can be resolved rather simply. When an entry in the left-hand column goes to zero (but other entries in that row are nonzero), one simply replaces the zero with a very small number ϵ, completes the array as usual, and determines the signs of the terms in the left-hand column as ϵ goes to zero.

 EXAMPLE 3.4

A proposed control system has $G(s) = K(s + 5)/s^3$ so that the characteristic equation is

$$s^3 + Ks + 5K = 0.$$

The Routh array is

s^3	1	K
s^2	0	$5K$.
s^1	—	—

Inserting ϵ in place of the zero, we obtain

s^3	1	K
s^2	ϵ	$5K$
s^1	$(\epsilon K - 5K)/\epsilon$	0
s^0	$5K$	—

It is clear that K must be positive. But $\epsilon K \ll 5K$, thus the Routh array reduces to

$$
\begin{array}{ccc}
s^3 & 1 & K \\
s^2 & \epsilon & 5K \\
s^1 & -5K/\epsilon & 0 \\
s^0 & 5K & 0
\end{array}
$$

and there are two changes in sign indicating two roots in the right half-plane.

An entire row of the Routh array may go to zero if the polynomial has pure imaginary roots or if it has roots in the right half plane that are mirror images (about the imaginary axis) of roots in the left half plane.

 EXAMPLE 3.5

For the system of our previous Example 3.2, K is set to 500 so that the characteristic equation is

$$s^3 + 10s^2 + 50s + 500 = 0.$$

The Routh array is

$$
\begin{array}{ccc}
s^3 & 1 & 50 \\
s^2 & 10 & 500. \\
s^1 & (500 - 500)/10 = 0 & 0
\end{array}
$$

To modify the row of zeros, we consider the equation of the preceding row, which is called the *auxiliary equation*. For this example, it is

$$10s^2 + 500 = 0.$$

Differentiate this auxiliary equation with respect to s, obtaining

$$\frac{d}{ds}(10s^2 + 500) = 20s + 0.$$

The row of zeros is replaced with this result and the Routh array is then complete:

$$
\begin{array}{ccc}
s^3 & 1 & 50 \\
s^2 & 10 & 500 \\
s^1 & 20 & 0 \\
s^0 & 500 & 0
\end{array}
$$

It is seen that all terms in the first column carry the same sign, so there are no roots in the right half plane.

3.4 SOME RESULTS FROM THEORY OF EQUATIONS

It is well known that the roots of a polynomial are functions of the coefficients and vice versa. The relationships are easily demonstrated for a monic[a] polynomial. For example, given a polynomial

$$s^4 + As^3 + Bs^2 + Cs + D, \tag{3.9}$$

assume that we know the roots and they are r_1, r_2, r_3, r_4. Then the polynomial may be rewritten in factored form:

$$(s + r_1)(s + r_2)(s + r_3)(s + r_4).$$

If we expand this latter by multiplying we get

$$s^4 + (r_1 + r_2 + r_3 + r_4)s^3 + (r_1r_2 + r_1r_3 + r_1r_4 + r_2r_3 + r_2r_4 + r_3r_4)s^2$$
$$+ (r_1r_2r_3 + r_1r_2r_4 + r_2r_3r_4 + r_1r_3r_4)s + r_1r_2r_3r_4. \tag{3.10}$$

From this we can readily define all of the root-coefficient relations for any order polynomial:

$$s^N + \sum_{i=1}^{N} r_i s^{N-1} + \sum_{\substack{i=1 \\ j=1 \\ i \neq j}}^{N} r_i r_j s^{N-2} + \sum_{\substack{i=1 \\ j=1 \\ k=1 \\ i \neq j \neq k}}^{N} r_i r_j r_k s^{N-3} + \cdots + r_1 r_2 r_3 \cdots r_N s^0. \tag{3.11}$$

It is clear that no coefficient can be zero or negative unless there is at least one root in the right half plane (at least one negative value for some r). This provides an inspection test—we inspect the terms of the polynomial and if any term is missing this means the coefficient is zero and, thus, the system is unstable. In like manner, if any coefficient is negative, i.e., if there is a change in sign, the system is unstable. Unfortunately, the converse is not true; when all terms are present and all signs are positive, there may still be roots in the right half plane, so a stability test is needed.

It is also helpful to note that the coefficient of the $N - 1$ power term is the *sum* of *all* the roots. Many practical adjustments and changes in system design do not change the value of the coefficients of the higher power terms, and this permits us to predict some performance changes. For example, when we increase the forward gain in a feedback control, this increases the value of the coefficient of the s^0 term, which is the *product* of the roots. Clearly, the product of the roots cannot be increased unless some of the roots get larger. However, if the sum of the roots is unchanged, then for most systems other roots must get smaller. This provides the broad picture that as we increase the gain we can expect some roots to move left, others right. If the gain increase is too great, some of the roots moving to the right will move into the right half plane, and the system becomes unstable. This approach to analysis can be applied when the

---

[a] A monic polynomial is one for which the coefficient of the highest power term is 1.0.

adjustment used is other than the gain. It does not result in numerical answers, but it does indicate what we can expect and thus helps guide both analysis and design.

 EXAMPLE 3.6

Consider a simple second-order control system with $G(s) = K/s(s + 5)$ and unity feedback. The characteristic equation is

$$s^2 + 5s + K = 0.$$

The roots of this characteristic equation are given by

$$r_1, r_2 = \frac{-5 \mp \sqrt{25 - 4K}}{2}$$

and it is clear that the roots move as K is changed:

K	r_1, r_2
2	$-0.438, -4.562$
3	$-0.697, -4.303$
4	$-1.0, -4.0$
10	$-2.5 \mp j1.936$
20	$-2.5 \mp j3.708$
100	$-2.5 \mp j9.682$

 EXAMPLE 3.7

In the system of Example 3.6, the gain is set to $K = 100$. System behavior is too oscillatory, so velocity feedback is used to provide damping. The characteristic equation becomes:

$$s^2 + (5 + 100k)s + 100 = 0,$$

where k is the gain of the velocity feedback. The roots of this characteristic equation are given by

$$r_1, r_2 = \frac{-(5 + 100k) \mp \sqrt{(5 + 100k)^2 - 400}}{2}$$

and it is clear that the roots move as k is varied:

k	r_1, r_2
0.001	$-2.55 \mp j9.669$
0.01	$-3 \pm j9.539$
0.1	$-7.5 \mp j6.614$
1.	$-0.961, -104.04$

3.5   CAUCHY'S PRINCIPLE OF ARGUMENT AND THE NYQUIST CRITERION

A very useful stabilty criterion was developed by H. Nyquist in 1932 at Bell Laboratories.[1] At that time, H. S. Black was developing the negative feedback amplifier for use in telephone repeater circuits. At times these amplifiers became oscillators and no one undestood why. Nyquist was asked to study the problem and as a result developed his stability test. It should be clearly understood that this criterion is simply a means of determining whether the characteristic polynomial has any roots in the right half plane.

The Nyquist criterion uses some basic results in complex variable theory. The technique is to draw a curve that is a map of a closed contour on the *s*-plane, and this map is interpreted using Cauchy's principle of argument. In simple terms, the principle of argument states: if you have a polynomial (or a ratio of two polynomials) and you choose to map a closed contour on the complex plane using your polynomial (or ratio) as a mapping function, it is possible to determine whether any singularities (poles or zeros) of the mapping function lie inside of the chosen contour. This is done by observing the change in the argument (angle) of the mapping function as the mapping point moves completely around the contour. The *net* number of singular points inside of the contour is equal to the

$$\frac{\text{(net angle change)}}{2\pi}$$

of the mapping function. This is most easily done by drawing the specific curves (maps), so the method is basically a graphical method.

To interpret stability using the Nyquist criterion, recall that the characteristic equation of a feedback control system is:

$$1 + GH(s) = 0. \tag{3.12}$$

The system is stable if all of the roots of this equation are in the left half plane. To test for roots in the right half plane, we choose a closed mapping contour on the *s*-plane which encloses the entire right half plane, as shown in Figure 3.4(a). This contour consists of the imaginary axis of the *s*-plane ($s = 0 + j\omega$, $-\infty < \omega < +\infty$) and a closing semicircle of infinite radius. If the system being tested has poles of $GH(s)$ *on* the imaginary axis, it is customary to modify the contour [as shown in Figure 3.4(b)] excluding these poles from the interior of the contour.

This contour is mapped using $GH(s)$ as a mapping function:

1.   Substitute values of $s = 0 + j\omega$ into $GH(s)$ and plot on polar coordinates.[b] The result is the open-loop frequency response of the system.

2.   Construct the mirror image of this curve in the real axis. This gives the map of

---

[b] In modern practice, this would be done with a computer program or by using the Bode diagram (see Chapter 5) as a nomograph.

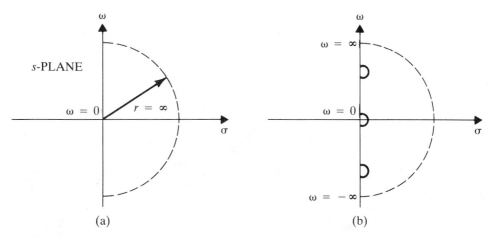

FIGURE 3.4 The Nyquist Mapping Contour: (a) Basic Contour Enclosing Right Half Plane (by Convention Mapping Direction is Clockwise) and (b) Modified Contour to Exclude any Poles on the Imaginary Axis.

the negative imaginary axis, since all values of $-\omega$ are conjugates to the corresponding value of $+\omega$.

3. All points on the infinite semicircle of Figure 3.4 map to the origin of the GH plane; $|s| = \infty$ for all of these points and, therefore, $|GH| = 0$ because GH has more poles than zeros for any physical system.

If the system is type zero, the above procedure results in a closed curve. If the type number is greater than zero, a closure must be constructed as explained later.

Figure 3.5 is a polar Nyquist plot of $GH(s)$ for a type zero system, which is minimum phase [all poles of $GH(s)$ are in the left half plane]. The relationship between $GH(s)$ and the characteristic equation $1 + GH(s)$ is shown graphically in Figure 3.6, from which it is clear that the origin of coordinates for $1 + GH(s)$ is the $-1 + j0$ point on the polar plot of $GH(s)$, and the angle change pertinent to Cauchy's principle of argument is the angle of the vector $1 + GH(s)$ as shown in Figure 3.6. Returning to Figure 3.5 and drawing the $1 + GH$ vector to some chosen frequency, ω_1, we interpret the angle change of this vector by letting its tip trace the entire contour, ending at ω_1. (Either direction of rotation may be chosen, but it is conventional to let the tip of the vector move in the direction of increasing ω.) We evaluate the *net* change in the angle of $1 + GH(s)$. For the plot of Figure 3.5(a), there is no net angle change. This means that there are no zeros of $1 + GH(s)$ inside the mapping contour and, of course, this means no roots of the characteristic equation in the right half plane, so the system is stable.

To demonstrate the Nyquist conditions for an unstable system, the gain of the system of Figure 3.5 is raised. This increases the length of every $GH(s)$ vector by a constant factor, and the result is the plot of Figure 3.5(b). Observe that the "shape" of the curve is unchanged but the $-1 + j0$ point [origin of the $1 + GH(s)$ coordinate system] is now *inside* the closed curve. When we repeat the procedure of tracing the

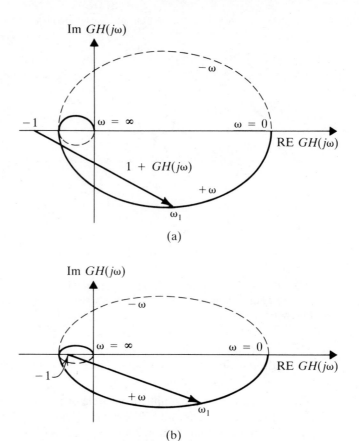

FIGURE 3.5 Nyquist Polar Plot for a Type-Zero System: (a) Stable System and (b) Unstable System.

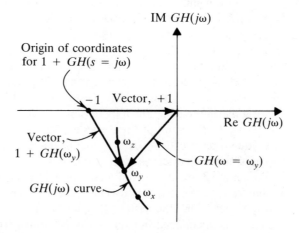

FIGURE 3.6 Converting the Map of $GH(s = j\omega)$ to a Map of $1 + GH(s = j\omega)$.

curve with the tip of the $1 + GH$ vector, it is seen that the vector makes two complete clockwise rotations, changing the angle of $1 + GH(\omega_1)$ by 4π radians. This indicates two zeros of $1 + GH(s)$ inside the mapping contour, thus two roots in the right half plane, and the system is unstable.

In practice, one merely observes that if the -1 point is inside the curve the system is unstable. For specific cases where "inside the curve" is not easily seen by inspection, counting rotations of the $1 + GH$ vector is recommended. Additional examples of type-zero systems are given in Figure 3.7.

When the system is type one or higher, the loop transfer function $GH(s)$ has one or more poles at the origin of the s-plane. Since this point lies on the chosen mapping contour (on the ω-axis), the magnitude of $GH(j0)$ would go to infinity. Thus, the plot would close through infinity but the interpretation of the change in the angle of $1 + GH(s)$ is not clear. We therefore choose to alter the mapping contour to simplify interpretation of stability. As shown in Figure 3.8(a), the contour is changed by detouring around the poles at the origin, excluding these poles from the inside. The detour is chosen as a semicircle of infinitesimal radius with center at the origin.

To demonstrate by example, the Nyquist plot for a simple type-one system is given in Figure 3.8(b) where

$$GH(j\omega) = \frac{K}{j\omega(j\omega\tau + 1)}.$$

Start at point A on the mapping contour, where ω is small so $\omega \ll 1$ and $GH(j\omega) = K/j\omega$. Then $|GH(\omega = A)| = \infty$ and $\angle GH(\omega = A) \approx -90°$ so the point A plots as shown in Figure 3.8(b). The mapping point moves up the $+\omega$-axis to $\omega = \infty, |GH| \to 0$, $\angle GH \to -180°$ resulting in the curve in Figure 3.8(b). For all points on the infinite radius semicircle, $|s| \approx \infty$, so $|GH(s)| = 0$, and all points map at the origin of the $GH(j\omega)$ plane. When the mapping point moves up the negative ω axis from $\omega = -\infty$ to $\omega = B$, the curve obtained on the $GH(j\omega)$ plane is the mirror image of that obtained for the $+\omega$ axis, as also shown in Figure 3.8(b). To complete the traverse of the mapping contour, the mapping point must go around the infinitesimal semicircle from B to C to A. For all points on this semicircle, $|s| = \epsilon$, a small number, and

$$|GH(\epsilon)| = \frac{K}{\epsilon}, \left|\frac{K}{\epsilon}\right| \approx \infty.$$

Thus, the infinitesimal semicircle on the s-plane will map as an infinite radius semicircle on the $GH(j\omega)$ plane. This large semicircle starts at B in Figure 3.8(b) when the mapping point is at B in Figure 3.8(a). As the mapping point proceeds from B to C to A, the tracing vector (vector from the pole at the origin to the mapping point) rotates *counterclockwise* through 180°. In the mapping function,

$$GH(s) = \frac{K}{s(s/P + 1)}.$$

This angle change is the change in $\angle s$, which is in the denominator of $GH(s)$ and, therefore, the $\angle GH(s)$ varies through 180°, but clockwise. Thus, the map of the

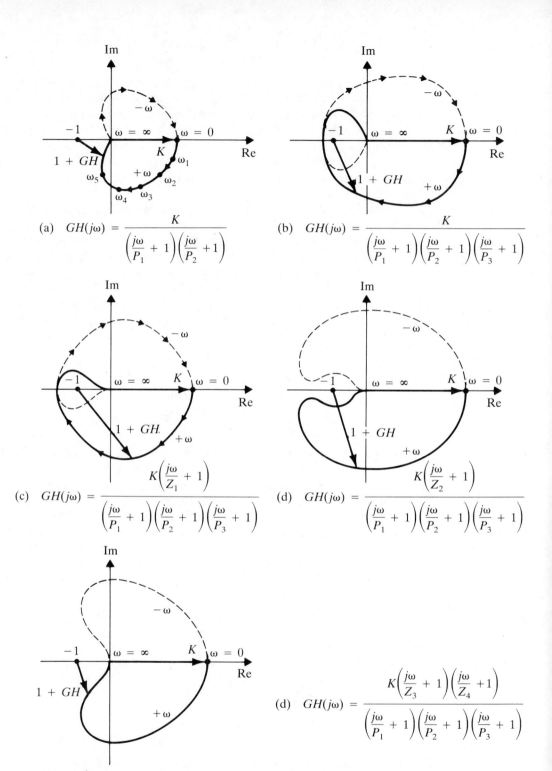

(a) $GH(j\omega) = \dfrac{K}{\left(\dfrac{j\omega}{P_1} + 1\right)\left(\dfrac{j\omega}{P_2} + 1\right)}$

(b) $GH(j\omega) = \dfrac{K}{\left(\dfrac{j\omega}{P_1} + 1\right)\left(\dfrac{j\omega}{P_2} + 1\right)\left(\dfrac{j\omega}{P_3} + 1\right)}$

(c) $GH(j\omega) = \dfrac{K\left(\dfrac{j\omega}{Z_1} + 1\right)}{\left(\dfrac{j\omega}{P_1} + 1\right)\left(\dfrac{j\omega}{P_2} + 1\right)\left(\dfrac{j\omega}{P_3} + 1\right)}$

(d) $GH(j\omega) = \dfrac{K\left(\dfrac{j\omega}{Z_2} + 1\right)}{\left(\dfrac{j\omega}{P_1} + 1\right)\left(\dfrac{j\omega}{P_2} + 1\right)\left(\dfrac{j\omega}{P_3} + 1\right)}$

(d) $GH(j\omega) = \dfrac{K\left(\dfrac{j\omega}{Z_3} + 1\right)\left(\dfrac{j\omega}{Z_4} + 1\right)}{\left(\dfrac{j\omega}{P_1} + 1\right)\left(\dfrac{j\omega}{P_2} + 1\right)\left(\dfrac{j\omega}{P_3} + 1\right)}$

FIGURE 3.7 Nyquist Polar Plots of Type-Zero Systems.

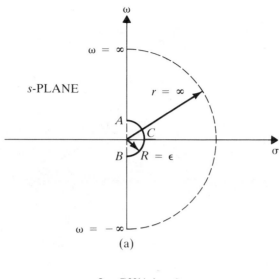

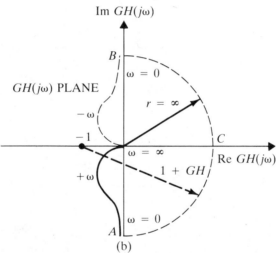

FIGURE 3.8   Mapping Contour for a Type-One System and Example Nyquist Plot for a Type-One System: (a) Modification of a Mapping Contour for a Pole at the Origin and (b) Nyquist Plot for $GH(j\omega) = \dfrac{K}{j\omega(j\omega\tau + 1)}$.

infinitesimal semicircle is the infinite semicircle in Figure 3.8(b), which proceeds *clockwise* from *B* to *C* to *A*.

Stability of type-one systems is determined in exactly the same way as it was for type-zero systems. The $1 + GH$ vector is added to Figure 3.8(b). The number of net revolutions of this vector is determined as the tip of the $1 + GH$ vector traverses the

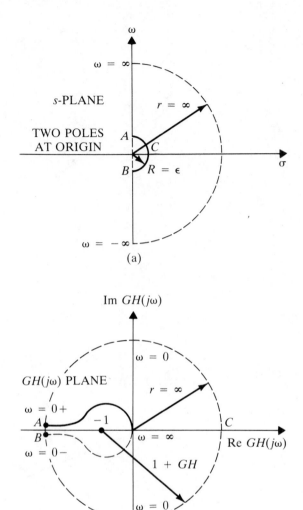

FIGURE 3.9 Mapping Contour for a Type-Two System and Example Nyquist Plot for a Type-Two System: (a) Modification of a Mapping Contour for Two Poles at the Origin and (b) Nyquist Plot for $GH(j\omega) = \dfrac{K}{(j\omega)^2(j\omega\tau + 1)}$.

contour and this is interpreted as before. For the system of Figure 3.8(b), there are no net rotations and no poles in the right half plane, so the system is stable.

For type-two systems, the loop transfer function has two poles at the origin. The second pole at the origin modifies the Nyquist plot as will be shown. Consider the mapping contour of Figure 3.9(a), which is exactly the same as that of Figure 3.8(a), but note that there are now *two* poles at the origin. Figure 3.9(b) shows a sample map for a type-two transfer function.

To trace out the map of Figure 3.9(b), the mapping point starts at point A in Figure 3.9(a) and this point maps to A in Figure 3.9(b). The *angle* of this initial point is $-180°$ because $GH(\omega = A) \approx K/(j\omega)^2$ and each pole at the origin contributes $-90°$. Mapping of the $+\omega$ axis, infinite semicircle, and $-\omega$ axis proceeds as previously discussed for the type-one system. When the map has been completed to point B, the mapping point must next go around the infinitesimal semicircle. Again the tracing vector rotates through $180°$ counterclockwise, but this counts *twice* in the $GH(j\omega)$ function because there are two poles at the origin. Thus, $\angle GH$ changes by $360°$ *clockwise*. The map of Figure 3.9(b) thus proceeds clockwise from point B and traces out a complete circle ending at point A. This closes the map and stability interpretation proceeds as usual. In the case of Figure 3.9(b), the $1 + GH$ vector makes two clockwise rotations, indicating two roots in the right half s-plane, so the system is unstable.

 EXAMPLE 3.8   NYQUIST PLOT FOR A TYPE-ZERO SYSTEM

Let the system have unity feedback and

$$G(j\omega) = \frac{25}{(j\omega + 1)(0.1j\omega + 1)(0.05j\omega + 1)}.$$

Figure 3.10(a) shows the Nyquist plot for $+\omega$. The student may add the $-\omega$ curve, which is the mirror image of the $+\omega$ curve in the horizontal axis. By inspection, the system appears to be stable, but the scale of the plot does not give much resolution in the vicinity of the -1 point. Figure 3.10(b) expands the plot near the origin.

 EXAMPLE 3.9   NYQUIST PLOT FOR A TYPE-ONE SYSTEM

Let the system have unity feedback and

$$G(j\omega) = \frac{50(0.2j\omega + 1)}{j\omega(j\omega + 1)(0.02j\omega + 1)}.$$

Figure 3.11(a) shows a large-scale plot and Figure 3.11(b) expands the plot near the origin. When there are poles at the origin, the lowest frequency points are very far from the origin and cannot be included when the plot scale is reasonable. However, once the plot is known in the vicinity of the origin and the -1 point, the rest of the plot can be sketched and the closing circles are added.

 EXAMPLE 3.10   NYQUIST PLOT FOR A TYPE-TWO SYSTEM

Let the system have unity feedback and

$$G(j\omega) = \frac{500(0.2j\omega + 1)(0.1j\omega + 1)}{(j\omega)^2(j\omega + 1)(0.02j\omega + 1)}.$$

Figure 3.12 shows the plot near the origin. Again, the lowest frequency points cannot be plotted. However, that portion can be sketched and the closure added.

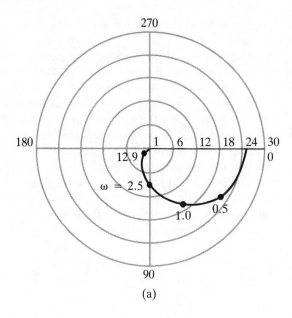

(a)

Gain Margin (dB) = 3.21		
Phase Margin (deg) = 9.54		
Mag	Phase	Frequency
0.5	−190.0	18.065
1.0	−170.5	12.867
1.5	−157.9	10.368
2.0	−149.2	8.886
3.0	−134.1	6.731

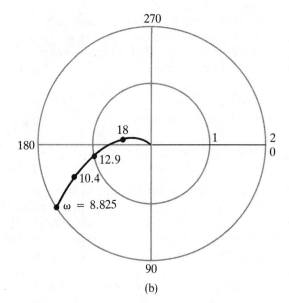

(b)

Gain Margin (dB) = 3.05		
Phase Margin (deg) = 9.51		
Mag	Phase	Frequency
0.5	−189.9	18.050
1.0	−170.5	12.875
1.5	−158.1	10.400
2.0	−148.8	8.825
3.0	−132.9	6.575

FIGURE 3.10   Example 3.8:

$$\text{a.} \quad G(s) = \frac{25}{(s + 1)(0.1s + 1)(0.05s + 1)}$$

$$\text{b.} \quad G(s) = \frac{25}{(s + 1)(0.1s + 1)(0.05s + 1)}.$$

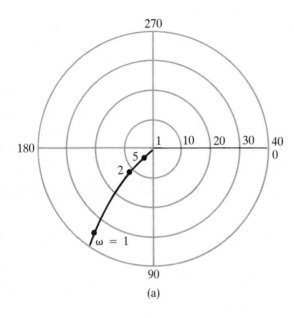

Phase Margin (deg) = 58.25		
Mag	Phase	Frequency
0.5	−122.7	19.387
1.0	−121.8	10.777
1.5	−124.1	7.898
2.0	−126.6	6.310
3.0	−129.9	4.786

(a)

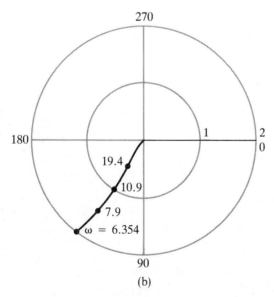

Phase Margin (deg) = 58.30		
Mag	Phase	Frequency
0.5	−122.8	19.541
1.0	−121.7	10.895
1.5	−124.2	7.837
2.0	−126.5	6.354

(b)

FIGURE 3.11   Example 3.9:

$$\text{a.}\quad G(s) = \frac{50(0.2s + 1)}{s(s + 1)(0.02s + 1)}$$

$$\text{b.}\quad G(s) = \frac{50(0.2s + 1)}{s(s + 1)(0.02s + 1)}.$$

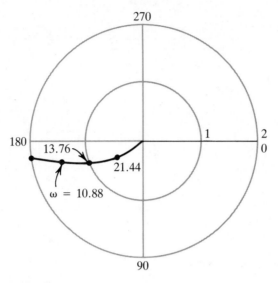

FIGURE 3.12   Example 3.10:

$$G(s) = \frac{500(0.1s + 1)(0.2s + 1)}{s^2(s + 1)(0.02s + 1)}.$$

Phase Margin (deg) = 22.79		
Mag	Phase	Frequency
0.5	−148.7	21.440
1.0	−157.2	13.760
1.5	−164.3	10.880

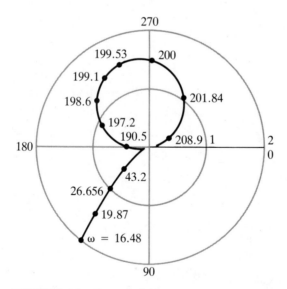

FIGURE 3.13   Example 3.11:

$$G(s) = \frac{50(0.05s + 1)}{s(0.1s + 1)(0.02s + 1)\left[\left(\dfrac{s}{200}\right)^2 + \dfrac{0.02s}{200} + 1\right]}.$$

Gain Margin (dB) = 8.85		
Phase Margin (deg) = 45.46		
Mag	Phase	Frequency
0.5	−142.9	43.192
1.0	−134.5	26.656
1.5	−130.3	19.872
2.0	−127.6	16.480
3.0	−123.1	12.240

EXAMPLE 3.11 NYQUIST PLOT FOR A TYPE-ONE SYSTEM WITH LIGHTLY DAMPED COMPLEX POLES

Let the system have unity feedback and

$$G(j\omega) = \frac{50(0.05j\omega + 1)}{j\omega(0.1j\omega + 1)(0.02j\omega + 1)\left[\left(\dfrac{j\omega}{200}\right)^2 + \dfrac{0.02j\omega}{200} + 1\right]}.$$

Figure 3.13 shows the portion of the Nyquist plot near the origin, and the effect of the complex poles is seen to be very significant. For this case, the system is stable, but high-frequency oscillations are to be expected.

3.6 NYQUIST INTERPRETATION OF NONMINIMUM PHASE SYSTEMS

The mapping function actually used by the Nyquist criterion is $1 + GH(s)$. Since $GH(s)$ is a ratio of two polynomials, then $1 + GH(s)$ is also a ratio, and two sets of singular points must be considered—the zeros (roots) of the numerator and the zeros of the denominator (which are the poles of the function). If any of the poles are in the right half of the s-plane, the system is *nonminimum phase*. When there are poles inside of the mapping contour, they too contribute to the rotations of the $1 + GH$ vector. Using the standard mapping direction (i.e., ω increasing) each *root* inside the contour causes a net angle change of $+2\pi$, but each *pole* inside the contour causes a net angle change of -2π. In applying the Nyquist test, what one actually observes is an angle change equal to $\theta = 2\pi(\#Z_R - \#P_R)$ where P_R and Z_R are, respectively, the poles and zeros in the right half plane. However, if there are *any zeros* in the right half plane, the system is unstable. Therefore, in applying the Nyquist criterion to any system, if there are any net clockwise rotations of the $1 + GH$ vector, the system is surely unstable, but if there are no net rotations, or some counterclockwise rotations, the number of right half plane poles of $1 + GH(s)$ must be known before stability can be predicted.

Fortunately, the nonminimum phase condition is easily detected. Note that the poles of $1 + GH$ are identical to the poles of GH. When the loop transfer function is in factored form, we determine right half plane poles by inspection. If the denominator is not factored, application of the Routh test usually supplies the number of right half plane poles.

 EXAMPLE 3.12 NYQUIST PLOTS OF NONMINIMUM PHASE SYSTEMS

A type-zero system has unity feedback and

$$G(j\omega) = \frac{10}{\left(\dfrac{j\omega}{5} + 1\right)\left(\dfrac{j\omega}{10} - 1\right)}.$$

Figure 3.14(a) shows the Nyquist plot. Clearly, there will be *one* clockwise rotation of

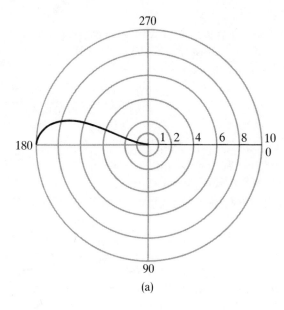

Mag	Phase	Frequency
0.5	−188.6	31.623
1.0	−192.0	21.135
1.5	−194.2	16.788
2.0	−195.8	14.125
3.0	−198.1	10.593

(a)

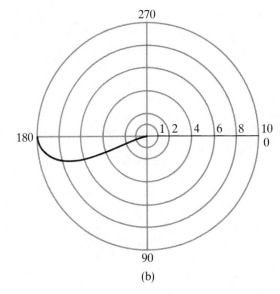

Mag	Phase	Frequency
0.5	−122.7	19.387
1.0	−121.8	10.777
1.5	−124.1	7.898
2.0	−126.6	6.310
3.0	−129.9	4.786

(b)

FIGURE 3.14   Example 3.12, Nonminimum Phase Systems:

a. $G(s) = \dfrac{50}{(s + 5)(s - 10)}$

b. $G(s) = \dfrac{500}{(s + 10)(s - 5)}$

c. $G(s) = \dfrac{60}{s(s - 5)}$.

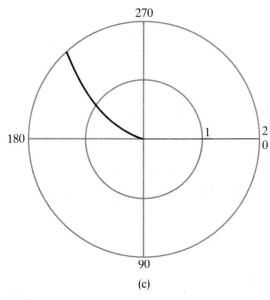

Mag	Phase	Frequency
0.5	− 205.3	10.593
1.0	− 215.5	6.998
1.5	− 222.3	5.495
2.0	− 227.2	4.624
3.0	− 235.9	3.388

(c)

FIGURE 3.14   (*Continued*)

the $1 + G(s)$ vector signifying at least one root in the right half plane. Because there is one pole in the right half plane, there are actually *two* roots there.

A type-zero system has unity feedback and

$$G(j\omega) = \frac{10}{\left(\dfrac{j\omega}{10} + 1\right)\left(\dfrac{j\omega}{5} - 1\right)}.$$

Figure 3.14(b) shows the plot. The $1 + G(s)$ vector makes one counterclockwise rotation and, since there is one pole in the right half plane, the system is stable.

A type-one system has unity feedback and

$$G(j\omega) = \frac{10}{j\omega\left(\dfrac{j\omega}{5} - 1\right)}.$$

Figure 3.14(c) shows the computer-generated plot. If the negative frequency half and closing semicircle are added as in the sketch of Figure 3.8b the $1 + G(s)$ vector is seen to make *one* clockwise rotation, so there are *two* roots in the right half plane.

3.7   RELATIVE STABILITY: GAIN MARGIN AND PHASE MARGIN

The Nyquist stability criterion as derived and discussed in previous sections is basically only a test for *existence* of roots in the right half of the *s*-plane. If this were the limit of its capability, it would be no more useful than the Routh test, but fortunately we can

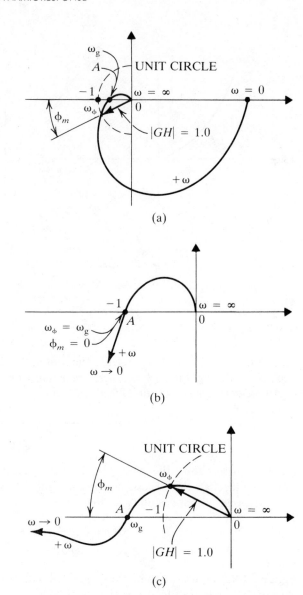

FIGURE 3.15   Graphical Definition of Phase and Gain Margins:

a.   Nyquist Plot for

$$GH(j\omega) = \frac{K}{\left(\dfrac{j\omega}{P_1} + 1\right)\left(\dfrac{j\omega}{P_2} + 1\right)\left(\dfrac{j\omega}{P_3} + 1\right)}$$

$$\left. \begin{array}{l} \text{Gain Margin} = \dfrac{1}{OA} > 1.0 \\[2mm] \text{Phase Margin} = \phi_m \end{array} \right\} \quad \text{Stable}$$

extend our interpretation to define *relative* stability. This is possible because of the graphical nature of the Nyquist criterion.

Once the engineer is aware that he can predict stability with a given method, it is natural to ask "Can we predict <u>how</u> stable? or how unstable?" After considering a number of specific cases, it is clear that when the map of $+\omega$ crosses the negative real axis only once, the system is stable if the crossing point is between -1 and the origin; the system is unstable if the crossing point is more negative than -1; and, if the map passes *through* the -1 point, it is clear that the system is at the limit of stability, which we readily associate, correctly, with roots *on* the imaginary axis of the *s*-plane. It is then a small step to associate "how stable" and "how unstable," i.e., the relative stability, with the distance from the curve to the -1 point. The association (at this point in the discussion) is relative: if the system is stable, we expect damping of the transient to increase as the distance from the -1 point to the curve increases. In like manner, if the system is unstable, we would expect the amplitude of oscillations to increase more rapidly as the distance from the -1 point to the curve increases.

To quantify this "distance," two measures have been defined and found to be useful. They are *gain margin* and *phase margin*. Figure 3.15 defines them graphically on the Nyquist plot. First we consider gain margin.

On each of the plots [Figures 3.15(a), (b), (c)], the point at which the polar plot for $+\omega$ crosses the negative real axis is marked A, and the frequency at that point is designated ω_g. In each case, that point is the tip of the vector $GH(j\omega_g) = OA$. For Figure 3.15(a), the vector OA is short, its tip does not reach to the critical -1 point, so the system is *stable*. We can increase the length of this GH vector (of <u>any</u> GH vector) by an increase in gain. If we were to increase the gain by the factor $1/OA$, the length of the vector would be $GH(j\omega_g) = OA(1/OA) = 1.0$. The length of the vector would thus be

---

FIGURE 3.15   (*Continued*)

  b.   Nyquist Plot for

$$GH(j\omega) = \frac{K}{j\omega\left(\dfrac{j\omega}{P_1} + 1\right)\left(\dfrac{j\omega}{P_2} + 1\right)}$$

$$\left.\begin{array}{l} \text{Gain Margin} = \dfrac{1}{OA} \equiv 1.0 \\[2mm] \text{Phase Margin} = \phi_m = 0. \end{array}\right\} \quad \begin{array}{l}\text{System is at}\\ \text{stability limit}\end{array}$$

  c.   Nyquist Plot for

$$GH(j\omega) = \frac{K\left(\dfrac{j\omega}{Z} + 1\right)}{(j\omega)^2\left(\dfrac{j\omega}{P_1} + 1\right)\left(\dfrac{j\omega}{P_2} + 1\right)}$$

$$\left.\begin{array}{l} \text{Gain Margin} = \dfrac{1}{OA} < 1.0 \\[2mm] \text{Phase Margin} = -\phi_m \end{array}\right\} \quad \text{Unstable}$$

one, its tip would lie on the -1 point [so the $GH(j\omega)$ curve would go through the critical point], and the system would be at the limit of stability. We therefore call the multiplying factor $1/OA$ the *gain margin*, because it is the factor by which the gain must be changed to put the system at the stability limit. In Figure 3.15(b), the $GH(j\omega)$ curve goes *through* the critical point, so $OA = 1.0$ and also the gain margin $1/OA = 1.0$; in Figure 3.15(c), the system is unstable, so $OA > 1.0$ and $1/OA < 1.0$.

In all three cases, however, the <u>definition</u> of gain margin is the same: gain margin is the factor by which we change the gain such that the vector with $\angle GH(\omega_g) = -\pi$ has a length of one unit.

The numerical value obtained for the gain margin of a system specifies how much the gain must be changed to put the system at the stability limit. Thus, it certainly is a measure of "How stable?" or "'How unstable?" the system is. In a graphical sense, it is also a measure of the distance from the critical point $(-1.)$ to *one point* on the $GH(j\omega)$ curve. Gain margin also has direct physical interpretation: gain is a quantity we can change directly.

Another measure of distance from the $GH(j\omega)$ curve to the critical point is called *phase margin*. Note that for almost any $GH(j\omega)$, there will exist at least one vector which is exactly one unit long. This vector is easy to locate on the Nyquist plot—we can always draw a circle with center at the origin and unit radius; the intersection of this circle with the $GH(j\omega)$ curve defines the unit vector and its frequency, ω_ϕ. This is demonstrated in Figures 3.15(a), (b), and (c). Note also that the tip of this vector can be placed on the critical (-1.0) point by rotating the vector through the angle ϕ, which is defined to be the phase margin. This is clearly another measure of the distance from the critical point to the $GH(j\omega)$ curve. It is not associated with any adjustment that is readily available in the physical system—there is no knob we can turn to rotate a $GH(j\omega)$ vector!

These two measures of relative stability have proven very useful in both analysis and design, and are probably used more widely than any other criteria in the design of feedback controls. The engineer, of course, wants a more *quantitative* measure of relative stability than has been provided thus far. Development of quantitative meaning for phase margin is included in the next chapter.

 EXAMPLE 3.13   PHASE MARGIN AND GAIN MARGIN

The Nyquist plot of Figure 3.10(b) is reproduced as Figure 3.16(a). The phase margin is marked, as is the gain margin:

$$\phi_m = 9.51°$$

$$GM = 3.05 \text{ dB.}$$

From these values the system is seen to be stable.

The Nyquist plot of Figure 3.11(b) is reproduced as Figure 3.16(b), with phase margin marked:

$$\phi_m = 58.3°$$

$$GM = \text{not defined.}$$

This system is stable.

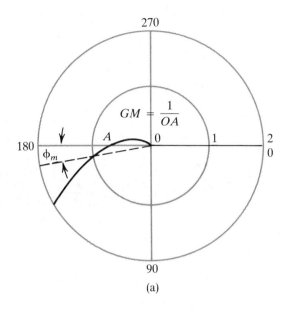

Gain Margin (dB) = 3.05		
Phase Margin (deg) = 9.51		
Mag	Phase	Frequency
0.5	−189.9	18.050
1.0	−170.5	12.875
1.5	−158.1	10.400
2.0	−148.8	8.825
3.0	−132.9	6.575

(a)

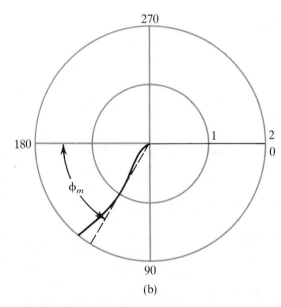

Phase Margin (deg) = 58.30		
Mag	Phase	Frequency
0.5	−122.8	19.541
1.0	−121.7	10.895
1.5	−124.2	7.837
2.0	−126.5	6.354

(b)

FIGURE 3.16   Example 3.13:

$$(a) \quad G(s) = \frac{25}{(s+1)(0.1s+1)(0.05s+1)}$$

$$(b) \quad G(s) = \frac{50(0.2s+1)}{s(s+1)(0.02s+1)}$$

$$(c) \quad G(s) = \frac{500(0.1s+1)(0.2s+1)}{s^2(s+1)(0.02s+1)}$$

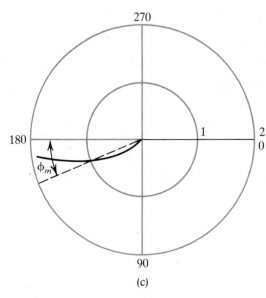

Phase Margin (deg) = 22.79		
Mag	Phase	Frequency
0.5	− 148.7	21.440
1.0	− 157.2	13.760
1.5	− 164.3	10.880

(c)

FIGURE 3.16   (*Continued*)

The Nyquist plot of Figure 3.12 is reproduced as Figure 3.16(c). Again, the phase margin is marked:

$$\phi_m = 22.79°$$

$$GM = \text{not defined.}$$

This type-two system is also stable.

| **3.8** | SOME CONSTRAINTS AND CAUTIONS |

Phase margin and gain margin are measures of *relative stability* that we develop from the basic absolute stability criterion, the Nyquist criterion. They are therefore subject to a major constraint arising from this criterion:

> Phase margin and gain margin are valid measures of relative stability only if $GH(j\omega)$ has *no poles* in the *right half* plane.

It should also be observed that the sketches of Figure 3.15 each cross the negative real axis once only and also cross the unit circle only once. This is not a necessary condition; for systems with numerous zeros and poles, the Nyquist plot may cross the negative real axis several times, or it may cross the unit circle more than once. For such

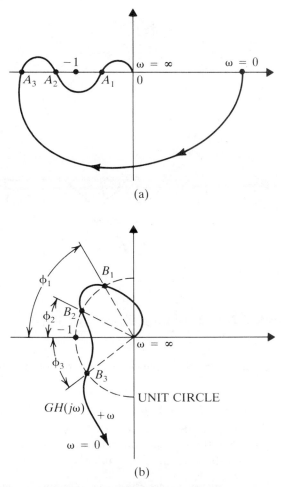

FIGURE 3.17   Nyquist Plots Requiring Careful Interpretation of Gain Margin and Phase Margin: (a) Nyquist Plot that Crosses the Negative Real Axis Three Times, where the Gain Margins are: $1/OA_1$, $1/OA_2$, $1/OA_3$; and (b) Nyquist Plot that Crosses the Unit Circle Several Times, where the Phase Margins are: ϕ_1, ϕ_2, ϕ_3.

curves, several values of gain margin and phase margin are defined, as shown in Figure 3.17. Caution is suggested when interpreting the stability of such systems.

3.9   CLOSED-LOOP FREQUENCY RESPONSE

The feedback control system is a closed-loop system. Using the Nyquist criterion, we predict its stability using plots of the open-loop transfer function, but there are a variety of circumstances under which we need to know how to convert the open-loop

Definition: $M = \left| \dfrac{C(j\omega)}{R(j\omega)} \right| = \left| \dfrac{G(j\omega)}{1 + G(j\omega)} \right|$

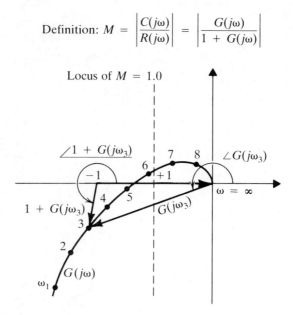

FIGURE 3.18   Polar Relationships Between Open- and Closed-Loop Frequency Responses.

frequency response to the closed-loop response, and vice versa. For example:

1. Occasionally systems are given a periodic command and we are interested in steady-state response.

2. When testing systems, closed-loop frequency measurements are often easier to make than open-loop measurements, but we wish to obtain the open-loop transfer function for use in analysis and design.

The polar plot shows the basic relationship between open-loop and closed-loop frequency response as in Figure 3.18. For some transfer function with plot $G(j\omega)$ as shown, a vector from the origin to a point on this curve such as ω_3 is the vector $G(j\omega_3)$ as shown; a vector from the -1 point to the origin is a vector $+1$, and the sum of these two vectors is the vector $1 + G(j\omega_3)$ as shown. One point on the closed-loop curve, then, can be calculated by measuring the lengths and angles of these vectors:

$$\left| \frac{C}{R}(j\omega_3) \right| = \frac{|G(j\omega_3)|}{|1 + G(j\omega_3)|} \qquad \textbf{(3.13a)}$$

$$\angle \frac{C}{R}(j\omega_3) = \angle G(j\omega_3) - \angle[1 + G(j\omega_3)]. \qquad \textbf{(3.13b)}$$

Obviously, this calculation can be repeated at many points and the entire closed-loop frequency response can be calculated in this manner. It is easier to prepare nomographs so that the results can be read off without additional calculations.

Consider the graphical configuration of Figure 3.18 as a problem in geometry and trigonometry, without regard to frequency response. The lines for $G(j\omega_3)$ and $1 + G(j\omega_3)$ are simply vectors which have lengths and angles associated with them when drawn as shown from the origin and from the -1 point to a common point with coordinates associated with ω_3. By definition

$$\left| \frac{G(j\omega)}{1 + G(j\omega)} \right| = M, \text{ a number.} \tag{3.14}$$

Are there any other points in the plane such that the same value of M would be associated with vectors drawn to them? In like manner, we can define

$$\angle G(j\omega_3) - \angle [1 + G(j\omega_3)] = N, \text{ an angle.} \tag{3.15}$$

Again, are there any other points in the plane such that the value of N would be the same at those points? The answer to both questions is *yes*. An obvious example is that the perpendicular bisector of the segment 0 to -1 is a locus of $M = 1.0$ as shown in Figure 3.18.

Curves of constant M on the polar plane can be shown to be circles with center at

$$X = \frac{-M^2}{M^2 - 1}$$
$$Y = 0 \tag{3.16}$$

and radius

$$r = \frac{M}{M^2 - 1}. \tag{3.17}$$

Curves of constant N are also circles with center at

$$X = -\frac{1}{2}$$
$$Y = +\frac{1}{2N} \tag{3.18}$$

and radius

$$r = \sqrt{\frac{1}{4} + \left(\frac{1}{2N} \right)^2}. \tag{3.19}$$

A family of constant M circles is shown in Figure 3.19(a) and a family of constant N circles in Figure 3.19(b).

The use of these circles in constructing the closed-loop response from the open loop is straightforward but seldom used because Nyquist analysis is normally carried out on the Bode diagram and it is more efficient to use a nomograph compatible with the Bode data. The nomograph used is the Nichol's chart, which is discussed in Chapter 5.

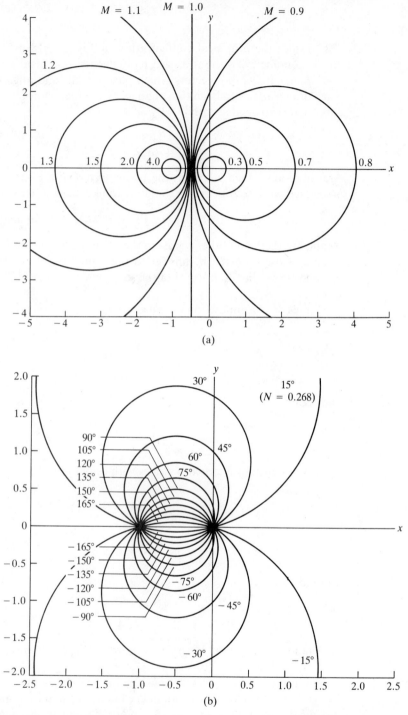

FIGURE 3.19 (a) Constant Magnitude Circles *M* and (b) Constant Phase Circles *N* on the Nyquist Polar Plane.

3.10   ROOT MOTION ON THE *s*-PLANE AND LOCUS OF ROOTS

In previous sections, we have seen that the stability of a control system is determined by the location of the roots of its characteristic equation on the *s*-plane. It is also clear that root locations can be changed by changing the value of one or more parameters in the characteristic equation. In particular, the discussion of gain margin indicated that roots could be moved from the left half plane to the right half plane (or vice versa) by changing the value of the gain.

If the value of a given parameter, for example, the gain *K*, is varied over some range, then for every value of *K* in that range the characteristic polynomial can be factored (by computer, of course) and the root values plotted as points on the *s*-plane. The points thus plotted form continuous curves on the *s*-plane and are called, variously, a *root locus* or a *locus of roots*. If the curves are entirely in the left half plane, then the system is stable under all conditions. For most systems, the root locus will cross the ω-axis so that it is partly in the left half plane and partly in the right half plane. Such curves can be very helpful in analysis and design. Clearly, the limit of stability is defined by the intersections of the root locus with the imaginary axis, and the nature of the transient response is determined by specific points which are the locations of the roots.

Most large computer centers have programs designed to find the roots of polynomials, and many have special programs to calculate and plot the root loci. This is obviously the easiest way to obtain the desired curves. Use of the curves in both analysis and design is expedited and the results improved if the engineer understands how the root locus of the closed loop is related to the poles, zeros, and gain of the open-loop transfer function. The details of these relationships are presented in Chapter 6.

3.11   SUMMARY

A linear system is unstable if any roots of its characteristic equation are in the right half of the *s*-plane (i.e., have positive real parts). A direct procedure is to use the computer to factor the characteristic polynomial. Since knowledge of root locations does not help design efforts very much, other methods of stability analysis are needed and useful. The Routh test uses the coefficients of the characteristic equation to determine whether a given system is stable or unstable and also permits definition of stability limits in terms of permissible values of one or more parameters. The Nyquist test uses the frequency response of the open loop to predict absolute stability of the closed-loop system. It also permits estimation of relative stability using phase margin and gain margin.

REFERENCES

1. Nyquist, H. "Regeneration Theory." Bell System Tech. J. 11 (January 1932).

BIBLIOGRAPHY

Churchill, R. V.; and Brown, J. W. *Complex Variables and Applications*. New York: McGraw-Hill Book Co. (1984).

Dorf, R. C. *Modern Control Systems*. Reading, Mass.: Addison Wesley (1986).

Eveleigh, V. W. *Introduction to Control System Design*. New York: McGraw-Hill Book Co. (1972).

James, H. M.; Nichols, N. B.; and Phillips, R. S. *Theory of Servomechanisms*. New York: McGraw-Hill Book Co. (1947).

MacDuffee, C. C. *Theory of Equations*. New York: John Wiley and Sons (1954).

Routh, E. J. *Dynamics of a System of Rigid Bodies*. London: MacMillan and Co., Ltd. (1905).

Thaler, G. J.; and Brown, R. G. *Analysis and Design of Feedback Control Systems*. New York: McGraw-Hill Book Co. (1961).

PROBLEMS

3.1   Check the stability of each of the following systems using
a.   the Routh criterion
b.   the Nyquist polar plot.

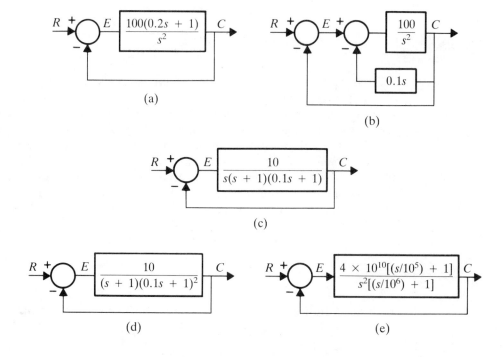

(a)

(b)

(c)

(d)

(e)

3.2 The following polar Nyquist plots are sketches of the map of the positive imaginary axis of the *s*-plane. None of the *G(s)* functions have poles in the right half plane.

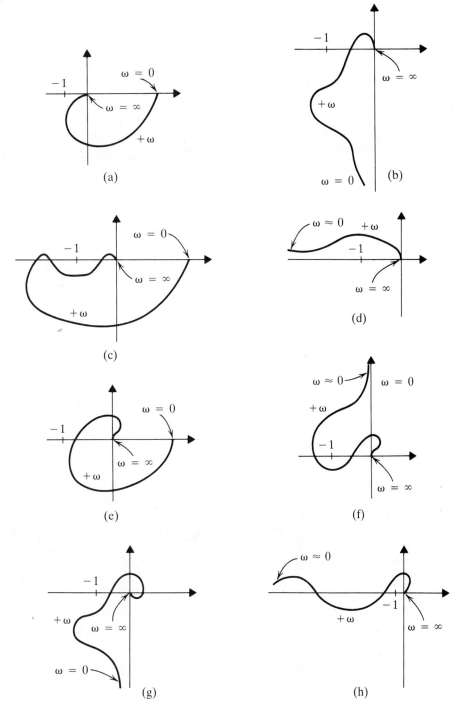

    a.   Complete each plot—i.e., add the map of the negative imaginary axis and any required closing circular arcs.

    b.   Is the system stable?

    c.   What is the phase margin? (Indicate on the plot.)

    d.   What is the gain margin? (Indicate on the plot.)

    e.   What is the system type number?

3.3   Use the Nyquist criterion to determine the stability of the following system. *SHOW ALL WORK.*

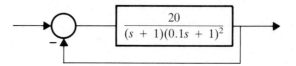

3.4   Determine the range of K values for which the following system is stable.

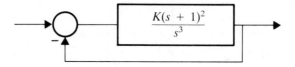

3.5   a.   Is the following system stable? (Use Routh)

        b.   Draw the Nyquist plot. Is the system stable? What are the phase and gain margins?

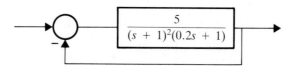

3.6   For each of the following Nyquist polar plots, determine the number of roots in the right half plane. None of the systems has any poles in the right half plane.

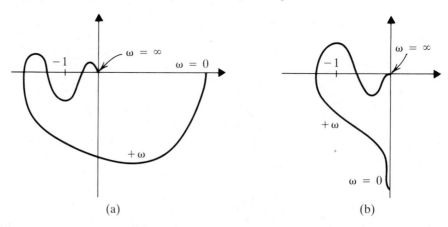

        (a)   (b)

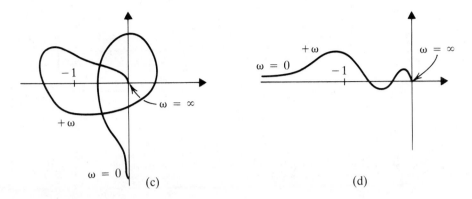

(c)   (d)

3.7 a. Complete the Nyquist plot for the single-loop, unity feedback system shown below.
   b. Is the system stable?
   c. Indicate the phase margin on the plot. Give an approximate value in degrees.
   d. From Chapter 4, what are values for ζ, M_{pt}, $M_{p\omega}$?
   e. Indicate the gain margin on the plot. Give an approximate numerical value.

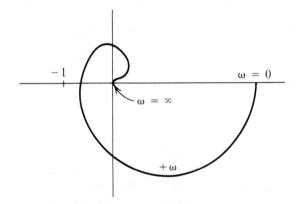

3.8 a. Complete the Nyquist plot for the following system.
   b. How many roots of the closed-loop system are in the right half plane?

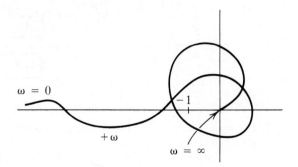

3.9 Given:

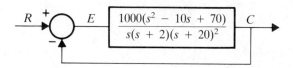

Is the system stable?

3.10 For the following, the polar plot given is a map of the $+\omega$-axis only.

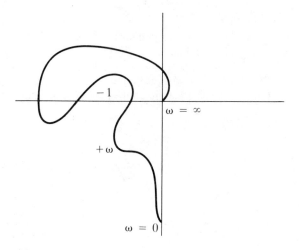

a. Complete the Nyquist plot.
b. Is the system stable? (All poles are in the left half plane.)
c. Define the phase margin(s) on the plot by adding appropriate lines and symbols.
d. Define the gain margins(s) in like manner.

3.11 Given:

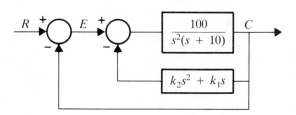

a. The outer loop is unstable. Find the relationship between k_1 and k_2 to put the system at the limit of stability.
b. If only velocity feedback is used ($k_2 = 0$), what value of k_1 is required at the stability limit?

3.12   Given:

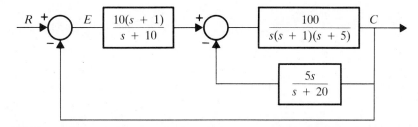

Is the system stable?

NOTE: The Nyquist plots of Problems 3.13 and 3.14 have been drawn to an expanded scale. In answering you may, if you wish, redraw (sketch) these plots to a more convenient scale.

3.13   A single-loop, unity feedback control system has a transfer function

$$G(s) = \frac{8(0.25s + 1)(0.125s + 1)}{s(2.5s + 1)(1.25s + 1)(0.02s + 1)}.$$

The Nyquist plot for $+\omega$ is given.

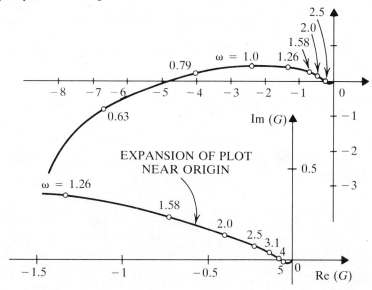

a.   Complete the plot. Is the system stable? If not, how many roots are in the right half plane?
b.   Evaluate the phase margin from the plot.
c.   Evaluate the gain margin from the plot.

3.14   Another single-loop, unity feedback control system has a transfer function

$$G(s) = \frac{0.1s(s + 1)^3}{s^2(0.1s + 1)^4}.$$

The Nyquist plot for $+\omega$ is given.

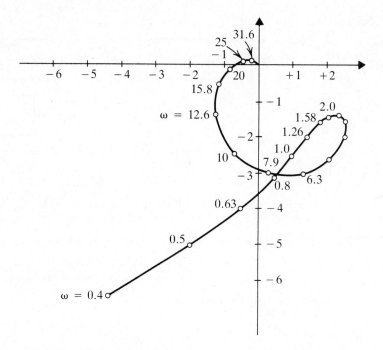

a. Complete the plot. Is the system stable? If not, how many roots are in the right half plane?
b. Evaluate the phase margin from the plot.
c. Evaluate the gain margin from the plot.

3.15   Given

$$G(s) = \frac{50(s + 1)}{(s - 1)^2},$$

is the system stable?

3.16   Given

$$G(s) = \frac{20\left[\left(\frac{s}{5}\right)^2 + 0.64s + 1\right]}{s(0.1s + 1)\left[\left(\frac{s}{5}\right)^2 - 0.04s + 1\right]},$$

is the system stable?

3.17   Given

$$G(s) = \frac{3(s - 1)}{(0.5s - 1)(0.1s + 1)^2},$$

is the system stable?

3.18   Given

$$G(s) = \frac{3(s - 1)}{(0.5s + 1)(0.1s + 1)^2},$$

is the system stable?

4 Performance Characteristics of a Second-Order System

The dynamic behavior of a linear system is determined by the roots of the characteristic equation. We have seen that the system is unstable if any roots are in the right half of the *s*-plane. We would like to correlate the dynamic behavior of a *stable* system with the locations of its roots in the left half of the *s*-plane. More specifically, we would like also to correlate the dynamic behavior of a system with its phase margin and gain margin, since those quantities will be readily available whenever we check stability using the Nyquist criterion.

Since the feedback control systems we may analyze and design are quite varied in nature and in complexity, we cannot expect to find simple correlations that quantitatively evaluate the contributions of a large number of roots. What we hope to do is define rather simple criteria that provide a reasonably accurate evaluation of those features in the dynamic behavior which are most important for control systems.

What features of control system response are most important? The system must be stable, of course, and for a stable system we are almost always concerned with the nature of any transient oscillation, such as the amount of overshoot (if any) and its frequency. We are *always* interested in the *duration* of the transient (usually called *settling time*) because in many applications we must wait until the transient is over before proceeding with the next step in the operation of the system.

The nature of any oscillations depends on where the roots are in the *s*-plane, and the duration of the transient depends on how long these roots contribute to the response; which is another way of saying that the duration of the transient also depends on the location of the roots! Therefore, we need to know more about the *root distribution* that we encounter with feedback controls.

If we examine the *root patterns* on the *s*-plane for a variety of feedback control systems that are operating satisfactorily, we find two general categories:

1. Most regulators and a few position controls have *all real roots*.

2. Some regulators and almost all position and tracking controls have a pair of complex roots relatively close to the origin, and the remaining roots are usually real and to the left of the complex pair.

Of course, all of the roots are in the left half plane (the systems are stable) and, of course, there are exceptions to these categories. However, the percentage of systems that fits the categories is quite high, and it is useful to define performance criteria for at least one of them, category 2. The resulting quantitative criteria are in common use. When the engineer encounters an exception, i.e., a system whose root pattern does not fit the category, he must recognize that his problem is a special case. Fortunately, special cases are rare.

4.2   ANALYSIS OF PERFORMANCE FROM ROOT PATTERNS: DOMINANCE

Consider the general solution of a linear differential equation when the specific disturbance is a step input. The output response is described by

$$C(t) = A_0 + A_1 e^{r_1 t} + A_2 e^{r_2 t} + A_3 e^{r_3 t} + \cdots. \tag{4.1}$$

For stable systems, the values of all of the *r*'s are negative real, or complex with negative real part. Thus, as *t* increases, all of the exponential terms decrease.

When the root pattern fits category 1, i.e., all roots real, and if $|r_1| < |r_2| < |r_3| < \cdots$, and $A_1 > A_2 > A_3 > \cdots$, which is usually true, then in a relatively short time the higher numbered terms approach zero and so contribute nothing to $C(t)$. In almost all cases, we find that essentially all of the response is due to one or two (r_1, r_2) of the roots, and these roots are said to be *dominant*. For systems in category 1, the dominant roots are the two smallest real roots.

When the root pattern fits category 2, i.e., when a pair of complex roots exists and these complex roots are near the origin, it is almost always true that the contributions from all other roots disappear quickly. Then the complex roots are said to be dominant.

It is reasonably clear, just from these considerations, that the transient performance of most systems is largely due to two roots, since the contributions of additional roots are almost entirely negligible. Therefore, we use a second-order system as a standard model. We choose a type-one system for convenience and analyze it completely[a] for both transient (step) response and frequency response. Having these results, we compare them to establish *quantitative* performance criteria. In particular, we are able to define a quantitative relationship between the phase margin and the damping of the step response.

---

[a] This is a reasonable undertaking because the second-order system is simple, having only two parameters.

THE SECOND-ORDER SYSTEM: TIME RESPONSE

Figure 4.1 gives the basic transfer function block diagram for the second-order type-one system and explains the notation used. For the closed-loop system[b],

$$\frac{C(s)}{R(s)} = \frac{\omega_n^2}{s^2 + 2\zeta\omega_n s + \omega_n^2} = \frac{\omega_n^2}{(s + \zeta\omega_n + j\omega_n\sqrt{1 - \zeta^2})(s + \zeta\omega_n - j\omega_n\sqrt{1 - \zeta^2})} \quad \text{(4.2)}$$

and for a unit step input, $R(s) = 1/s$. Thus,

$$C(s) = \frac{\omega_n^2}{s(s + \zeta\omega_n + j\omega_n\sqrt{1 - \zeta^2})(s + \zeta\omega_n - j\omega_n\sqrt{1 - \zeta^2})}. \quad \text{(4.3)}$$

Taking the inverse transform,

$$C(\omega_n t) = 1 - \frac{e^{-\zeta\omega_n t}}{\sqrt{1 - \zeta^2}} \sin\left(\omega_n\sqrt{1 - \zeta^2}\, t + \tan^{-1}\frac{\sqrt{1 - \zeta^2}}{\zeta}\right). \quad \text{(4.4)}$$

A family of curves for the step response is given in Figure 4.2. To find the maximum overshoot of the step response, we differentiate $C(\omega_n t)$ with respect to $\omega_n t$ and equate

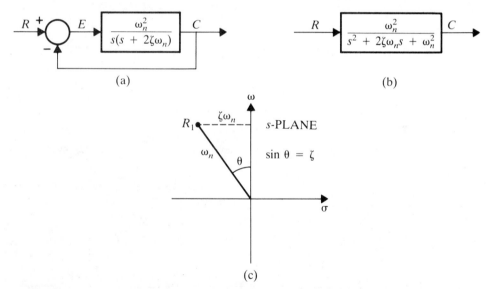

(a)　　　　　　　　　　　　　　　　(b)

(c)

FIGURE 4.1   The Second-Order System: (a) Block Diagram Showing Open-Loop Transfer Function, (b) Closed-Loop Diagram Showing Characteristic Function, and (c) s-Plane Plot Showing Definitions of ζ and ω_n.

---

[b] The standard model is an all pole (no zeros) model. If zeros are present, the residues at the roots are altered, i.e., the magnitudes of the coefficients in Eq. (4.1) are affected. Damping, settling time, etc., are not.

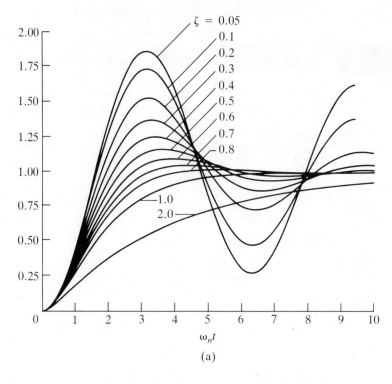

(a)

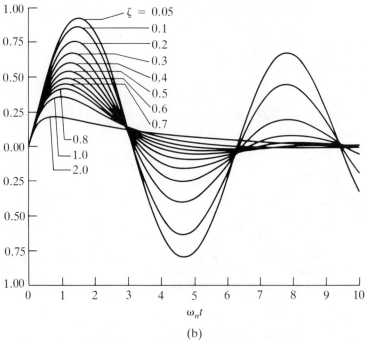

(b)

FIGURE 4.2 Step Response of a Second-Order System for (a) Position Versus Time and (b) Velocity Versus Time.

to zero. Omitting the intermediate steps, one determines first the *time* at which the maximum overshoot occurs:

$$\omega_n t_p = \frac{\pi}{\sqrt{1 - \zeta^2}}. \tag{4.5}$$

Substituting Eq. (4.5) into (4.4):

$$C(\omega_n t)_{max} = 1 - \frac{\exp\left(\dfrac{-\pi\zeta}{\sqrt{1 - \zeta^2}}\right)}{\sqrt{1 - \zeta^2}} \sin\left(\pi + \tan^{-1}\frac{\sqrt{1 - \zeta^2}}{\zeta}\right), \tag{4.6}$$

which reduces to:

$$C(\omega_n t)_{max} = 1 + \exp\left(\frac{-\pi\zeta}{\sqrt{1 - \zeta^2}}\right) \triangleq M_{pt}, \tag{4.7}$$

where

$M =$ *ratio* of output over input
subscript, $p =$ peak value
subscript, $t =$ *time response*
subscript, $\omega =$ *frequency response.*

It is important to note that, for this second-order system, the maximum overshoot for a step input depends only on the value of ζ and is independent of the value of ω_n. Since the definition of ζ from Figure 4.1 is

$$\zeta \triangleq \sin \theta, \tag{4.8}$$

it is clear that ζ defines a radial line on the *s*-plane and, therefore, complex roots anywhere on this line produce the same percentage overshoot, regardless of the value of ω_n (distance from origin).

From Eq. (4.4) the envelope of the transient response is

$$e^{-\zeta\omega_n t} = e^{-t/\tau} \tag{4.9}$$

and clearly the time constant is

$$\tau = \frac{1}{\zeta\omega_n}. \tag{4.10}$$

Defining[c] the duration of the transient (or settling time) to be four time constants,

$$T_s = \frac{4}{\zeta\omega_n}. \tag{4.10a}$$

Also, from Eq. (4.4), the transient oscillating frequency is

$$\omega_t = \omega_n\sqrt{1 - \zeta^2}. \tag{4.11}$$

---

[c] Calculation of ω_n is possible for the second-order system, but is not easily evaluated for higher order systems.

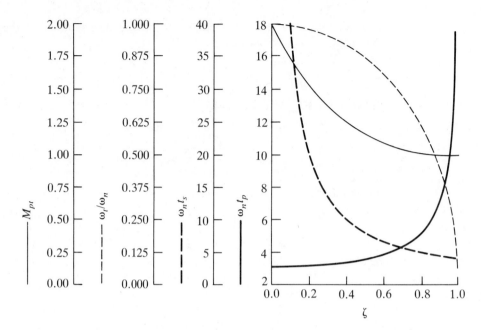

FIGURE 4.3   Time Response Characteristics of a Second-Order System.

Equations (4.5) through (4.11) define the important characteristics of the *transient response* in terms of the root location as defined by ζ and ω_n. Plots of the transient characteristics are shown in Figure 4.3.

The curves of Figure 4.3 summarize the dynamic step response characteristics available with one pair of complex roots. They are useful in both analysis and design. Since design specifications place bounds on performance, in particular on maximum overshoot (M_{pt}) and settling time ($4/\zeta\omega_n$), one can enter the curves of Figure 4.3 with the specified M_{pt}, evaluating ζ. Then, knowing ζ, the curves yield a value for $4/\zeta\omega_n$ and thus the required ω_n can be calculated and, if desired, the peak time t_p can be evaluated. In practice, step testing is common and values obtained from the step response can be used in conjunction with Figure 4.3 to extend the test data and to check the assumed dominance of a complex pair of roots. The curves of Figure 4.3 are most valuable, however, when used in *conjunction with* the frequency response curves developed in the next section.

4.4   THE SECOND-ORDER SYSTEM: FREQUENCY RESPONSE

The next step is to derive the important characteristics of the frequency response. To do this, assume a sinusoidal input of adjustable frequency, ω:

$$\frac{C}{R}(j\omega) = \frac{\omega_n^2}{(j\omega)^2 + 2\zeta\omega_n j\omega + \omega_n^2} = \frac{\omega_n^2}{(\omega_n^2 - \omega^2) + j2\zeta\omega_n\omega}. \tag{4.12}$$

Families of curves for Eq. (4.12) are given in Figure 4.4. The *magnitude* of the closed-loop response is

$$\left|\frac{C}{R}(j\omega)\right| = \frac{\omega_n^2}{[(\omega_n^2 - \omega^2)^2 + (2\zeta\omega_n\omega)^2]^{1/2}}.\qquad(4.13)$$

To evaluate the height of the resonance peak, which is the maximum value of Eq. (4.13), differentiate Eq. (4.13) with respect to ω and set equal to zero. After manipulation we obtain the value of ω at which the resonance peak ω_r occurs. The result is:

$$\omega_r = \omega_n\sqrt{1 - 2\zeta^2}.\qquad(4.14)$$

Substituting Eq. (4.14) into (4.13):

$$\left.\left|\frac{C}{R}(j\omega)\right|\right|_{max} \triangleq M_{p\omega} = \frac{1}{2\zeta\sqrt{1 - \zeta^2}}.\qquad(4.15)$$

Clearly, the height of the resonance peak is a function of ζ only, as was the amount of overshoot to a step input.

One other characteristic of the closed-loop frequency response is of interest— the *bandwidth*. The normal definition of bandwidth is *the range of frequencies over which the magnitude of the response to a unit amplitude input exceeds 70.7%*. In mathematical notation,

$$\left|\frac{C}{R}(\omega = \omega_b)\right| = \frac{1}{\sqrt{2}} = \frac{\omega_n^2}{[(\omega_n^2 - \omega_b^2)^2 + 4\zeta^2\omega_n^2\omega_b^2]^{1/2}}.\qquad(4.16)$$

Solving for ω_b, the result is

$$\omega_b = \omega_n\sqrt{1 - 2\zeta^2 + (2 - 4\zeta^2 + 4\zeta^4)^{1/2}}.\qquad(4.17)$$

Of course, only the positive value of ω_b is used. Note that because the system is type one it passes zero frequency without attenuation and thus Eq. (4.17) defines the upper frequency limit of the bandwidth. The lower frequency limit is $\omega = 0$.

Finally, and perhaps most important for purposes of analysis and design, we wish to know the phase margin of the system. [We cannot define gain margin because $|G(\omega)| \to 0$ as $\angle G(\omega) \to -180°$.] By definition

$$\phi_m = 180° - \angle G(j\omega)|_{|G(j\omega)| = 1.0}.\qquad(4.18)$$

Since

$$G(j\omega) = \frac{\omega_n^2}{j\omega(j\omega + 2\zeta\omega_n)},\qquad(4.19)$$

$$|G(j\omega)| = 1.0 = \frac{\omega_n^2}{[(-\omega^2)^2 + (2\zeta\omega\omega_n)^2]^{1/2}},\qquad(4.20)$$

from which

$$\omega_\phi = \omega_n\sqrt{-2\zeta^2 + (4\zeta^4 + 1)^{1/2}}.\qquad(4.21)$$

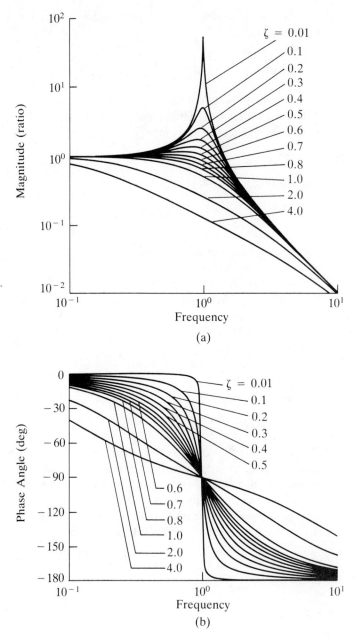

FIGURE 4.4 Frequency Response of a Second-Order System: (a) Magnitude (dB) Versus $\log \omega$ and (b) Phase Angle (deg) Versus $\log \omega$.

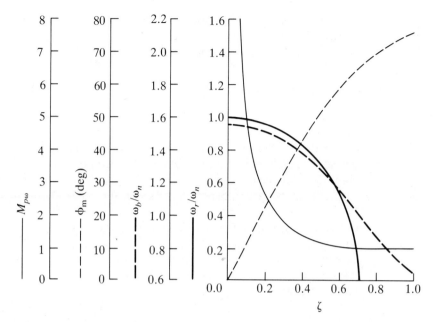

FIGURE 4.5   Frequency Response Characteristics of a Second-Order System.

Now, in general, the angle of G for the second-order type-one system is

$$\angle G = -\frac{\pi}{2} - \tan^{-1}\frac{\omega}{2\zeta\omega_n} = -\tan^{-1}\frac{2\zeta\omega_n}{\omega}. \tag{4.22}$$

So the phase margin is

$$\phi_m = +\tan^{-1}\frac{2\zeta}{\sqrt{-2\zeta^2 + (4\zeta^4 + 1)^{1/2}}}. \tag{4.23}$$

Clearly, the phase margin is a function of ζ only. These frequency response characteristics are plotted in Figure 4.5.

The curves of Figure 4.5 are exceptionally useful. They permit us to evaluate the essential dynamic parameter ζ from either the open-loop frequency response using the phase margin ϕ_m or from the closed-loop frequency response using the height of the resonance peak $M_{p\omega}$. We may then proceed to predict the time response or we may proceed to use these results to guide design efforts.

4.5   USES AND CAUTIONS

These second-order curves are used in various ways with low-order and high-order systems. First of all, inspection of the transient response curves of Figure 4.2 and/or the frequency response curves of Figure 4.4 permits intelligent choice of the desired damping. This provides a value for ζ. Then, if we can specify the desired (or permissible) duration of the transient, we can calculate a value for ω_n, i.e., we can interpret these curves to determine a suitable location for a pair of dominant roots.

If we wish to *design* the system using frequency response methods, the value of ζ, determined as in the preceding paragraph, is used to enter Figure 4.5 and find the phase margin needed to provide that amount of damping. Figure 4.5 also supplies the required dimensionless bandwidth, ω_b/ω_n, and the desired value of ω_n is used to determine the bandwidth ω_b. The frequency response curves are then reshaped to obtain the desired phase margin and bandwidth. The procedures for doing this are developed in Chapters 5 and 8. Other interpretations of Figures 4.3 and 4.5 are often convenient for the designer.

It should be remembered that use of the second-order system as a standard for reference is based on the assumption of *dominance*. We are assuming that the contributions of all *additional roots* may be neglected. There will certainly be cases where this assumption is not valid. It is not practical to develop rules to define the exceptions (though an experienced analyst can often tell by inspection of the frequency response curves), so the transient response should be determined, usually by simulation, as a check.

4.6   EXAMPLES

The preceding analysis of the second-order system has provided tools for estimating control system behavior, which have been used successfully on higher order systems, because the dynamic behavior of most systems is dominated by one pair of complex roots. It should be clear that the estimates obtained will be precisely correct only for a second-order system without zeros. The examples in this section are intended to demonstrate:

1. procedures for estimating performance and the degree of accuracy of the estimates
2. some of the ways in which estimates may deviate from the correct performance.

A computer-aided design program is used to obtain accurate answers for comparison with the estimates. (Computer-generated frequency response curves are in logarithmic coordinates for convenience.)

 EXAMPLE 4.1

Consider a second-order two-pole system for which the plant transfer function is

$$G(s) = \frac{100}{s\left(\dfrac{s}{40} + 1\right)}.$$

Figure 4.6(a) shows the Nyquist plot; the phase margin, $\phi_m = 34.5°$; and the gain crossover frequency, $\omega_x = 57$ rad/s. Entering Figure 4.5 with the phase margin, we

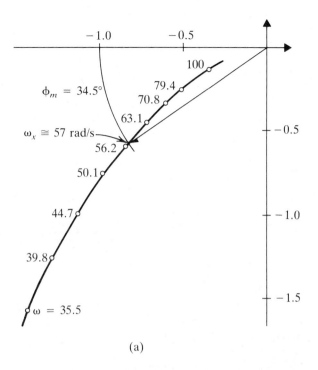

(a)

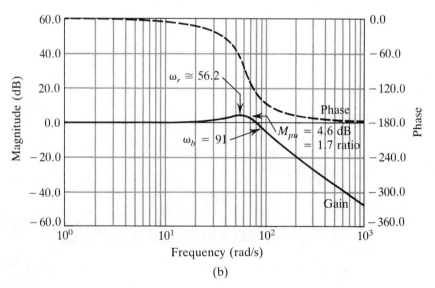

(b)

FIGURE 4.6 (a) Nyquist Plot for

$$G(s) = \frac{100}{s\left(\dfrac{s}{40+1}\right)};$$

(b) Closed-Loop Response for System of 4.6(a); and (c) Step Response for System of 4.6(b).

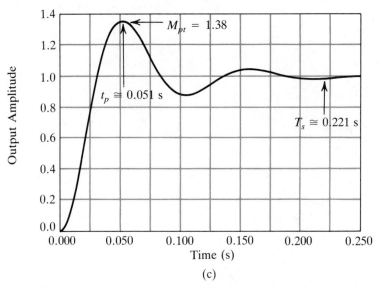

FIGURE 4.6   (*Continued*)

TABLE 4.1   Performance of a Second-Order System

$$G(s) = \frac{100}{s\left(\dfrac{s}{40} + 1\right)}$$

Estimates		Computer-Calculated Values
	Closed loop roots are at $s = -20 \mp j60$	
	$\zeta = 0.31 \qquad \phi_m = 34.5° \qquad \omega_x = 57 \text{ rad/s}$	
$\omega_n = \omega_x = 57^a$	if $\omega_n \triangleq 63.25$	$\sqrt{4000} = 63.25^a$
$M_{p\omega} = 1.8$		1.7^b
$M_{pt} = 1.37$		1.38^c
$\dfrac{\omega_r}{\omega_n} = 0.9; \ \omega_r = 51.3$	56.9	56.2^b
$\dfrac{\omega_b}{\omega_n} = 1.45; \ \omega_b = 82.65$	91.7	91^b
$\omega_n t_p = 3.35; \ t_p = 0.059$	0.053	0.051^c
$\omega_n T_s = 13; \ T_s = 0.228$	0.205	0.221^c

[a] Calculation of ω_n is possible for the second-order system, but is not easily evaluated for higher order systems.
[b] See Figure 4.6(b).
[c] See Figure 4.6(c).

obtain $\zeta = 0.31$ and then, from Figures 4.3 and 4.5, we obtain performance estimates which are compared with calculated results in Table 4.1.

Observe from Table 4.1 that M_{pt} and $M_{p\omega}$ are estimated quite accurately. Other quantities suffer somewhat because use of the gain crossover frequency as an estimate for ω_n introduces some error.

 EXAMPLE 4.2

Consider next a third-order system for which

$$G(s) = \frac{100}{s\left(\dfrac{s}{40} + 1\right)\left(\dfrac{s}{800} + 1\right)}.$$

This system differs from the second-order system in that a remote pole (at -800) has been included, but the error coefficient has not been changed.

Since the pole is far out in the left half of the s-plane, one expects that the root introduced by this pole will also be remote and will contribute little to system response; thus, the phase margin and gain crossover frequency should not be changed appreciably, the complex root pair should be dominant, and the estimates of per-

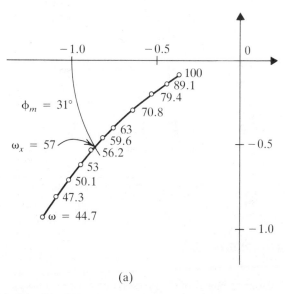

(a)

FIGURE 4.7 (a) Nyquist Plot for

$$G(s) = \frac{100}{s\left(\dfrac{s}{40} + 1\right)\left(\dfrac{s}{800} + 1\right)};$$

(b) Closed-Loop Response for System of 4.7(a); and (c) Step Response for System of 4.7(b).

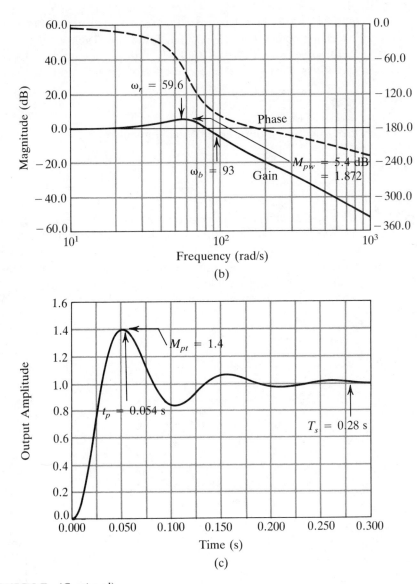

FIGURE 4.7   (*Continued*)

formance based on Figures 4.3 and 4.5 should be essentially unchanged by introduction of the pole and should be valid for the third-order system.

Figure 4.7(a) shows the Nyquist plot with ϕ_m and ω_x marked. Using Figures 4.3 and 4.5 again, the performance estimates are evaluated and listed in Table 4.2. To check the accuracy of these estimates, Figure 4.7(b) gives the closed-loop frequency response and Figure 4.7(c) gives the step response. From these we obtain the calculated values listed in Table 4.2.

TABLE 4.2 Performance of a Third-Order System

$$G(s) = \frac{100}{s\left(\dfrac{s}{40} + 1\right)\left(\dfrac{s}{800} + 1\right)}$$

Estimates	Computer-Calculated Values
Closed loop roots are at $s - 17.4 \mp j60.6;\ -850.2$	
$\zeta = 0.26 \qquad \phi_m = 31° \qquad \omega_x = 57$ rad/s	
$\omega_n = \omega_x = 57$	
$M_{p\omega} = 1.8$	$M_{p\omega} = 1.872$
$M_{pt} = 1.44$	$M_{pt} = 1.4$
$\dfrac{\omega_r}{\omega_n} = 0.93;\ \omega_r = 53$	$\omega_r = 59.6$
$\dfrac{\omega_b}{\omega_n} = 1.49;\ \omega_b = 84.9$	$\omega_b = 93$
$\omega_n t_p = 3.253;\ t_p = 0.057$	$t_p = 0.054$
$\omega_n T_s = 15.38;\ T_s = 0.27$	$T_s = 0.28$

Table 4.2 shows that the second-order estimates predict response of the third-order system quite accurately, and comparison with Table 4.1 shows that the performance of the third-order system is essentially the same as that of the second-order system, i.e., the remote pole contributes very little to the response.

 EXAMPLE 4.3

To illustrate some effects of a transfer function zero on the performance estimates, consider a system with

$$G(s) = \frac{1600\left(\dfrac{s}{150} + 1\right)}{s\left(\dfrac{s}{40} + 1\right)\left(\dfrac{s}{800} + 1\right)}.$$

Note that the poles of the preceding example have been retained, but a zero has been introduced at $s = -150$, and the gain has been increased. The performance of this system cannot be compared with that of the preceding illustrations but the estimates from Figures 4.3 and 4.5 can be compared with computer calculated results. Figure 4.8(a) shows the Nyquist plot and indicates the phase margin and gain cross-over frequency. Using these values and Figures 4.3 and 4.5, performance estimates are obtained and listed in Table 4.3. Figure 4.8(b) gives the closed-loop frequency response and Figure 4.8(c) gives the step response. Again, the appropriate values are listed in Table 4.3.

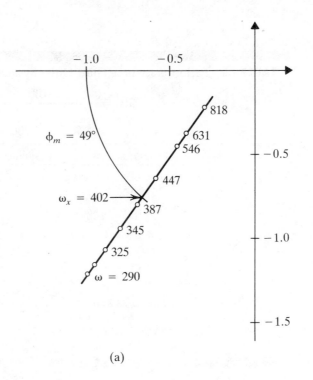

(a)

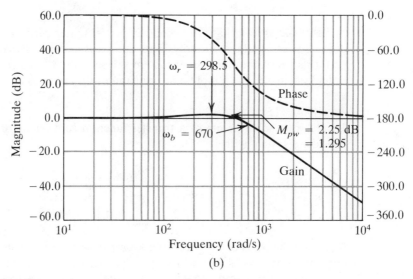

(b)

FIGURE 4.8 (a) Nyquist Plot for

$$G(s) = \frac{1600\left(\dfrac{s}{150} + 1\right)}{s\left(\dfrac{s}{40} + 1\right)\left(\dfrac{s}{800} + 1\right)};$$

(b) Closed-Loop Response for System of 4.8(a); and (c) Step Response for System of 4.8(b).

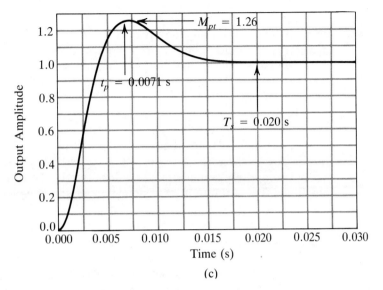

FIGURE 4.8   (*Continued*)

TABLE 4.3   Performance of a Third-Order System with Zero

$$G(s) = \frac{1600\left(\dfrac{s}{150} + 1\right)}{s\left(\dfrac{s}{40} + 1\right)\left(\dfrac{s}{800} + 1\right)}$$

Estimates	Computer-Calculated Values
Closed loop roots are at $s = -214;\ -313 \mp j376$	
$\zeta = 0.45 \qquad \phi_m = 49° \qquad \omega_x = 402$	
$\omega_n = \omega_x = 402$	
$M_{p\omega} = 1.3$	$M_{p\omega} = 1.295$
$M_{pt} = 1.2$	$M_{pt} = 1.26$
$\dfrac{\omega_r}{\omega_n} = 0.771;\ \omega_r = 310$	$\omega_r = 298.5$
$\dfrac{\omega_b}{\omega_n} = 1.32;\ \omega_b = 531$	$\omega_b = 670$
$\omega_n t_p = 3.52;\ t_p = 0.00875$	$t_p = 0.0071$
$\omega_n T_s = 8.88;\ T_s = 0.0221$	$T_s = 0.020$

Table 4.3 shows that the second-order predictions are quite close to the calculated values and, thus, the presence of a zero in the transfer function does not necessarily introduce much error. There are some observable variations in the step response that should be understood. The step response of Figure 4.8(c) does not exhibit oscillations. There is an initial overshoot, then the response apparently decays asymptotically to the steady-state value. With $\zeta = 0.45$, one expects several oscillations before reaching steady state. Closed-loop roots at $-313 \mp j376$ reinforce this expectation, yet the oscillations are not apparent. Note that there is a real root at $s = -214$, which is smaller than the real part of the complex roots. This real root location is due to the zero at $s = -150$. Because of the real root, the complex roots are not completely dominant; the contribution of the real root, though not large in amplitude, has a duration exceeding that of the complex root contribution. This affects the system response in two ways:

1. M_{pt} is increased slightly. This condition is often caused by a zero in the system transfer function.

2. After the peak overshoot has been reached, the complex root contributions oscillate about the exponential decay of the real root contribution. In this example, the variation is hard to detect in Figure 4.8(c).

■ EXAMPLE 4.4

Many systems have open-loop dynamics which includes a pair of lightly damped high-frequency complex poles (these may be due to, for example, a mechanical resonance). The contribution of such poles to the open-loop and/or closed-loop frequency response is readily observable.

The criteria developed from the second-order system predict only the behavior of a dominant pair of complex roots, so the transient contribution of such a secondary pair of roots is not available. In general, the phase margin and gain crossover frequency correctly predict the behavior of the dominant root pair, and the high-frequency poles do not contribute significantly. This is not always true, however; the high-frequency poles may superimpose an oscillation on the response predicted by the second-order relationships. Consider the system with

$$G(s) = \frac{1600\left(\dfrac{s}{150} + 1\right)}{s\left(\dfrac{s}{40} + 1\right)\left(\dfrac{s}{800} + 1\right)\left[\left(\dfrac{s}{2000}\right)^2 + \dfrac{0.1\,s}{2000} + 1\right]}.$$

Figure 4.9(a) gives the Nyquist plot with phase margin and gain crossover marked. These are used to predict the dominant mode behavior of the system. Note that the values are very nearly the same as for Figure 4.8(a), so the behavior of this system should be nearly the same as that of the previous example. The effect of the complex poles is quite apparent on the Nyquist plot but simple procedures for predicting the quantitative effect are not readily available (except computer-aided analysis, of course).

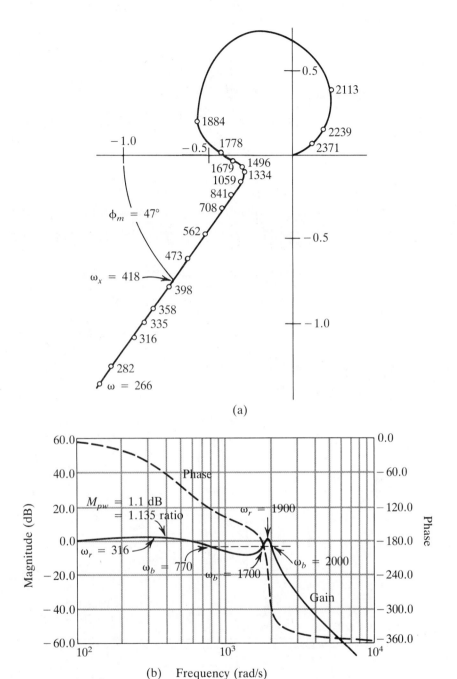

(a)

(b)   Frequency (rad/s)

FIGURE 4.9   (a) Nyquist Plot for

$$G(s) = \frac{1600\left(\dfrac{s}{150} + 1\right)}{s\left(\dfrac{s}{40} + 1\right)\left(\dfrac{s}{800} + 1\right)\left[\left(\dfrac{s}{2000} + 1\right)^2 + \dfrac{0.1s}{2000} + 1\right]};$$

(b) Closed-Loop Response for System of 4.9(a); (c) Step Response for System of 4.9(b); and
(d) Comparison of Step Response of Figure 4.9(c) with that of Figure 4.8(c).

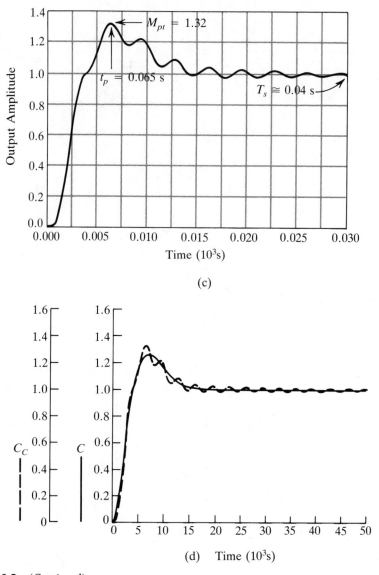

(c)

(d)   Time (10^3s)

FIGURE 4.9   (*Continued*)

Table 4.4 lists the predictions from the second-order model and also gives the results of computer analysis. The predictions are essentially the same as those of Table 4.3, as expected. The deviations from the calculated results are readily explained by observations of the step response of Figure 4.9(c): there is a high-frequency oscillation of long duration modifying the transient. This oscillation changes the

TABLE 4.4 Performance of a Fifth-Order System with Zero and Lightly Damped Complex Poles

$$G(s) = \frac{1600\left(\dfrac{s}{150} + 1\right)}{s\left(\dfrac{s}{40} + 1\right)\left(\dfrac{s}{800} + 1\right)\left[\left(\dfrac{s}{2000}\right)^2 + \dfrac{0.1s}{2000} + 1\right]}$$

Estimates	Computer-Calculated Values
Closed-loop roots[a] are at $s = -213.94; -337 \mp j385; -767 \mp j1911$	
$\zeta = 0.43 \qquad \phi_m = 47° \qquad \omega_x = 418$	
$\omega_n = \omega_x = 418$	
$M_{p\omega} = 1.2$	$M_{p\omega} = 1.135$
$M_{pt} = 1.21$	$M_{pt} = 1.3$
$\dfrac{\omega_r}{\omega_n} = 0.79; \omega_r = 330$	$\omega_r = 316; 1900$
$\dfrac{\omega_b}{\omega_n} = 1.35; \omega_b = 564$	$\omega_b = 770; \begin{Bmatrix} 1700 \\ 2000 \end{Bmatrix}$
$\omega_n t_p = 3.5; t_p = 0.0084$	$t_p = 0.0065$
$\omega_n T_s = 9.3; T_s = 0.0222$	$T_s = 0.04$

[a] Note that the roots are not significantly different from those listed in Table 4.3 except for the additional complex pair.

values of M_{pt}, t_p, and T_s. However, this does not prevent the second-order methods from correctly predicting the response of the lower frequency dominant roots. Figure 4.9(d) shows the step response of Figure 4.8(c) with that of Figure 4.9(c) superimposed. It is clear that the high-frequency ripple due to the complex poles is oscillating about the response of Figure 4.8(c) as a baseline. The baseline has been predicted correctly by the second-order approximation, which, unfortunately, has no mechanism for including the additional effect.

In summary , we may note that the basic dynamic response of the example systems has been well predicted by using the phase margin and gain crossover frequency in conjunction with the performance curves of the second-order system.

The question arises: Of what use are these methods if a computer with appropriate software is available? If only analysis is of interest, the methods are clearly obsolete. But when design is to follow (the usual case), phase margin, gain crossover frequency, and the second-order results provide valuable guidance as discussed in Chapter 8.

4.7 SUMMARY

The step response and frequency response of a second-order type-one linear system are analyzed in detail, with all response characteristics related to the damping ratio ζ.

Because the dynamic response of most control systems is dominated by a pair of complex roots, the characteristics of a second-order system can be used to estimate (predict) the behavior of higher order systems. Knowledge of any one characteristic, such as maximum overshoot to a step input, permits estimation of the value of ζ, which then permits prediction of other response characteristics. The phase margin of a system is found to be a convenient parameter.

BIBLIOGRAPHY

Chu, Y. "Correlation Between Frequency and Transient Response of Feedback Control Systems." *AIEE Trans. Appl. Ind.* Part II (May 1953).

PROBLEMS

4.1 A second-order servo has unity feedback and an open-loop transfer function:

$$G(s) = \frac{500}{s(s + 15)}.$$

   a. Draw a block diagram for the closed-loop system.
   b. What is the characteristic equation of the closed loop?
   c. What are the numerical values of ζ and ω_n?
   d. Sketch the transient response to a unit step input.
   e. From your sketch, and from other available information, estimate the time from start of transient to maximum overshoot.
   f. What is the settling time of the system?
   g. If the system is subjected to a ramp input of 0.5 rad/s, what is the steady-state error?

4.2 The open-loop frequency response of a unity feedback control system shows a phase margin of 40°. What can be predicted about its step response?

4.3 For the system of Problem 4.2, the closed-loop frequency response shows a resonance peak of $M_{p\omega} = 1.5$. What can be predicted about its step response?

4.4 For the system of Problems 4.2 and 4.3, the resonant frequency is observed to be $\omega_r = 4.7$ rad/s. What can be predicted about the step response?

4.5 It is desired that the step response of a system have a peak overshoot not exceeding $M_{pt} = 1.35$. How much phase margin is expected in the open-loop frequency response?

4.6 The settling time of a system must be less than 0.1 s. Interpret this in terms of specified values for the locations of dominant complex roots (i.e., ζ and ω_n). Also interpret in terms of desired values for open-loop frequency response characteristics (i.e., ϕ_m, ω_x, ω_ϕ).

4.7 Performance specifications for a system require that the maximum overshoot to a step input must not exceed 1.5 (50% overshoot) and that the settling time should be approximately 0.25 s. Interpret these requirements in terms of desired ζ and ω_n. Also interpret them in terms of phase margin, closed-loop resonance frequency, and closed-loop bandwidth.

4.8 The open-loop frequency response of a system shows $\phi_m = 20°$ and $\omega_x = 5000$ rad/s. Using the second-order curves, predict the characteristics of the closed-loop step response.

4.9 The closed-loop frequency response of a system shows $M_{p\omega} = 2.0$ and $\omega_r = 1.7$ rad/s. Predict the characteristics of the closed-loop step response.

5 Frequency Response Analysis

INTRODUCTION

To use the Nyquist criterion to analyze system stability, and to extend this analysis to evaluate transient response using phase margin, gain margin, and bandwidth, the Nyquist plot must be obtained. If the system equations are known, complete with numerical values, there are a variety of ways to do this. If the equations and/or parameter values are *not* known, but the physical system or its components exist and are available for test, we can obtain the required Nyquist plot by measuring the frequency response of the system or components. This is a very valuable feature. When the hardware is available for test, frequency response measurements are usually inexpensive and easy to make, while parameter evaluation by other means may be very difficult. Furthermore, it provides a ready experimental verification when the curves are calculated from parameter values.

Methods of calculating the frequency response (transfer function) curves from the known expression for $G(j\omega)$ are readily available:

1. FORTRAN programs (or BASIC, etc.) (Such programs[a] may be based on transfer functions or may use matrix manipulations based on the state equations.)
2. programs for hand-held calculators
3. step-by-step calculations with nonprogrammable calculators
4. graphical construction of the Bode diagram using asymptotes and templates.

---

[a] When the computer is used, it is probably simplest to calculate the magnitude and angle of $G(j\omega)$, then convert the magnitude into decibels for the Bode plot. Conversion to the real part and imaginary part of the number would be carried out if the polar Nyquist plot is desired.

To relate transfer functions to curve shapes and to develop techniques for interpretation of the plots, we proceed with development of graphical methods.

5.2 THE BODE DIAGRAM: MAGNITUDE VERSUS ω

Given a transfer function $G(s = j\omega)$, we wish to calculate values of $G(j\omega)$ for various ω and plot them to permit interpretation of the Nyquist criterion. For each value of ω, $G(j\omega)$ is a complex number, which may be evaluated in either rectangular or polar coordinates. For longhand (graphical) computation, Bode has developed a method that uses the polar coordinate form. We can express

$$G(j\omega) = |G(j\omega)| \angle G(j\omega) \tag{5.1}$$

and plot two curves, $\angle G(j\omega)$ versus ω and $|G(j\omega)|$ *in decibels* versus ω. The advantage of this from a computational point of view is that the magnitude curve can be constructed rather than calculated, the construction consisting of drawing some straight line segments of predetermined slope on semilogarithmic graph paper. In like manner, the angle curve can be constructed using straight line segments. Either or both curves are easily refined if greater accuracy is desired. The use of the straight line provides a very quick computational procedure and, as an added advantage, little refinement of the curves is needed for analysis and design purposes.

The graphical construction technique is a straightforward, rigorous interpretation of the transfer function. In general, we have

$$G(j\omega) = \frac{K_x \prod_{i=0}^{M} (j\omega\tau_i + 1) \prod_{k=0}^{Q} [A_k(j\omega)^2 + B_k j\omega + 1]}{(j\omega)^N \prod_{\ell=0}^{U} (j\omega\tau_\ell + 1) \prod_{m=0}^{V} [a_m(j\omega)^2 + b_m j\omega + 1]}. \tag{5.2}$$

Interpreting as in Eq. (5.1) and expressing the magnitude in decibels[b] gives:

$$20\log_{10}|G(j\omega)| = 20\log_{10}|K_x| + \sum_{i=0}^{M} 20\log_{10}|j\omega\tau_i + 1|$$

$$+ \sum_{k=0}^{Q} 20\log_{10}|A_k(j\omega)^2 + B_k j\omega + 1| - \begin{cases} 20\log_{10}|(j\omega)^N| \\ \text{or} \\ 20N\log_{10}|j\omega| \end{cases}$$

$$- \sum_{\ell=0}^{U} 20\log_{10}|j\omega\tau_\ell + 1| - \sum_{m=0}^{V} 20\log_{10}|a_m(j\omega)^2 + b_m j\omega + 1|$$

$$\tag{5.3}$$

---

[b] Use of the decibel measure is well established in the literature. Use of logarithms to base ϵ would be equally valid, but there does not seem to be any reason to make the change.

and the angle equation is

$$\angle G(j\omega) = \sum_{i=0}^{M} \tan^{-1}\omega\tau_i + \sum_{k=0}^{Q} \tan^{-1}\frac{\omega B_k}{1 - A_k\omega^2} - N\left(\frac{\pi}{2}\right)$$

$$- \sum_{\ell=0}^{U} \tan^{-1}\omega\tau_\ell - \sum_{m=0}^{V} \tan^{-1}\frac{\omega b_m}{1 - a_m\omega^2}. \tag{5.4}$$

In plotting curves defined by Eqs. (5.3) and (5.4), ω is plotted as the abscissa to a logarithmic scale; magnitude is in decibels and angle in degrees as the ordinates to linear scales. Note that in converting the magnitude to decibels, multiplication and division as defined by Eq. (5.2) are converted to addition and subtraction, which is advantageous.

To develop a simple plotting construction for the magnitude curve of Eq. (5.3), we first consider the plot of each of the kinds of terms in Eq. (5.3):

1. Since $20\log_{10}|K_x|$ is the logarithm of a constant, it too is a constant independent of the value of ω and plots as a horizontal straight line as shown in Figure 5.1(a).

2. Since $|j\omega\tau + 1| = \sqrt{\omega^2\tau^2 + 1}$, we observe that if $\omega\tau \ll 1$, the $\omega\tau$ can be neglected and $|j\omega\tau + 1| = 1.0$. Thus, at low frequencies, $20\log_{10}|j\omega\tau + 1| = 0\,\text{dB}$. At higher frequencies $\omega\tau \gg 1$, the 1 can be neglected and $|j\omega\tau + 1| = |\omega\tau|$ so $20\log_{10}|\omega\tau|$ plots as a straight line with a *slope* of $+20\,\text{dB/decade}$. This is shown in Figure 5.1(b). Note that the two lines intersect at $\omega = 1/\tau$, which is called the *break frequency* or *corner frequency*, because the curve *breaks* or *goes around the corner* at that frequency.

3. For the low frequencies of $20\log_{10}|A(j\omega)^2 + Bj\omega + 1|$ both $A(j\omega)^2$ and $Bj\omega$ may be neglected and the term reduces to 1.0 so at low ω, $20\log_{10}|(Aj\omega)^2 + Bj\omega + 1|$ has a magnitude of 0 dB. For higher frequencies, the $A(j\omega)^2$ term dominates, we neglect $Bj\omega + 1$, and $20\log_{10}|A\omega^2|$ plots as a straight line with slope of $+40\,\text{dB/decade}$. The two straight lines intersect at $\omega = \dfrac{1}{\sqrt{A}}$, which is also called a corner frequency. This is shown in Figure 5.1(c). The term $Bj\omega$ is the *damping term* and determines the deviation of the true curve from the asymptotes. Figure 5.1(c) shows only one example of a "true" curve. See Figure 4.4(a) for a typical family.

4. The term $-20N\log_{10}|j\omega|$ is exactly a straight line with negative slope. The numerical value of the slope is $20N\,\text{dB/decade}$, where N is the number of poles at the origin, i.e., the type number of the system. This is shown in Figure 5.1(d).

5. The terms $-20\log_{10}|j\omega\tau + 1|$ and $-20\log_{10}|a(j\omega)^2 + bj\omega + 1|$ differ from those in items 2 and 3 above only in the minus sign. This gives a negative slope to the high-frequency asymptote as shown in Figures 5.1(e) and (f).

The magnitude (or gain) plot of $G(j\omega)$ is defined by Eq. (5.3). The individual plots of Figure 5.1 are possible components of that gain curve. One procedure for obtaining the desired curve is to plot each component separately and at each value of ω add the

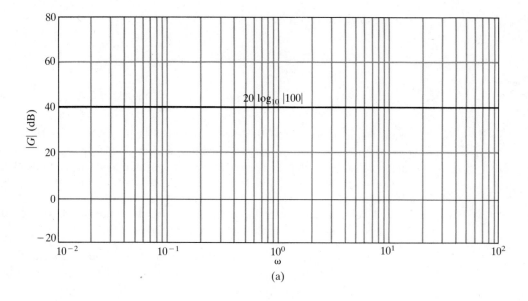

(a)

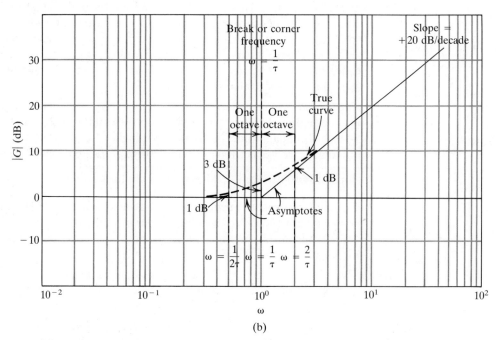

(b)

FIGURE 5.1 Bode Diagram for (a) a Constant, $G(s) = 100$; (b) for a Real Zero, $G(s) = s + 1$, (c) for Complex Zeros, $G(s) = s^2 + 0.6s + 1$; (d) for Poles at the Origin, $G(s) = 1/s^N$ where $N = 1, 2, 3$; (e) for a Real Pole, $G(s) = 1/(s + 1)$; and (f) for Complex Poles, $G(s) = 1/(s^2 + 0.6s + 1)$.

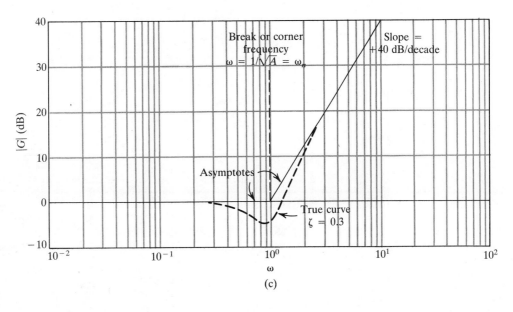

(c)

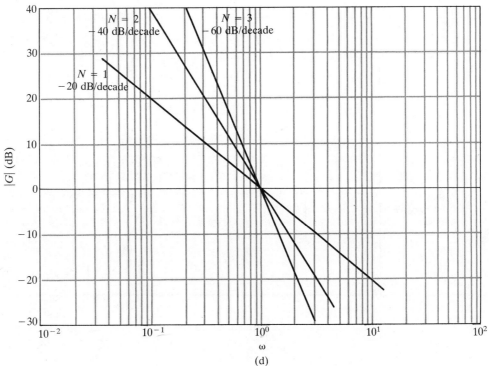

(d)

FIGURE 5.1   (*Continued*)

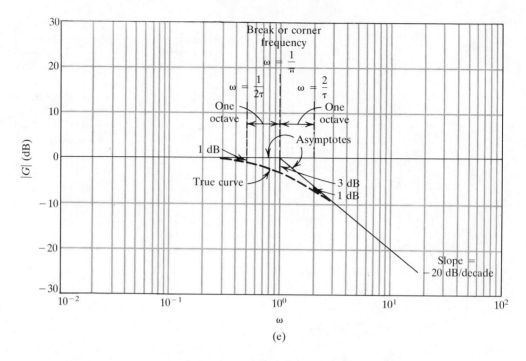

(e)

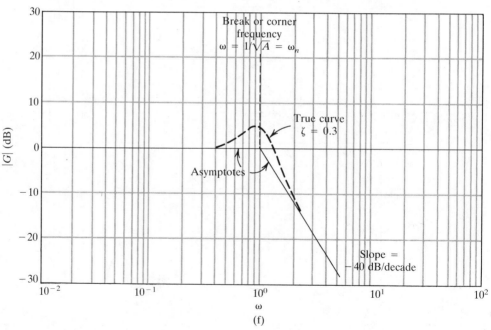

(f)

FIGURE 5.1 (*Continued*)

ordinates. An alternative (and much simpler) procedure is based on the following observations:

1. The straight line asymptotes represent the component terms with reasonable accuracy.
2. The permissible *slopes* of these asymptotes are restricted to multiples of 20 dB/decade.
3. The asymptotes change slope only at corner frequencies.

The resulting procedure is:

1. Prepare the semilog graph sheet, choosing scales and marking all corner frequencies as shown in Figure 5.2(a).
2. Construct the lowest frequency asymptote to start the plot.
3. Complete the asymptotic plot by extending the asymptote to higher frequencies while changing the slope at each corner frequency. This is demonstrated by Figure 5.2(b).
4. If necessary (or desired), obtain the *true* curve by correcting at appropriate frequencies.

To construct the lowest frequency asymptote, as specified in step 2 above, we return to Eq. (5.3) and observe that at low frequencies all terms of the form $20\log_{10}|j\omega\tau + 1|$ or $20\log_{10}|A(j\omega)^2 + Bj\omega + 1|$ reduce to the form $20\log_{10}|1.0|$, which evaluates to 0.0 dB. Thus, at low frequencies, Eq. (5.3) reduces to

$$20\log_{10}|G(j\omega)| = +20\log_{10}|K_x| - 20N\log_{10}|j\omega|. \tag{5.5a}$$

Rearranging

$$20\log_{10}|G(j\omega)| = -20N\log_{10}|j\omega| + 20\log_{10}|K_x|, \tag{5.5b}$$

which is seen to be the equation of a straight line in the form

$$y = mx + b \tag{5.5c}$$

This straight line is the low-frequency asymptote, and its slope is seen to be $-20N$ dB/decade, where N is the number of poles at the origin. Thus, the initial slope depends only on the system type number:

Type Number	Slope of Lowest Frequency Asymptote
Zero	0.0 dB/decade
One	-20 dB/decade
Two	-40 dB/decade

To draw in this asymptote precisely, we need to evaluate one point on the line. We can do this by choosing any value of ω, substituting in Eq. (5.5b), and evaluating the ordinate at that ω. For a type-zero system, of course, $N = 0$ and we must evaluate $20\log_{10}|K_p|$. For type-one and type-two systems, observe that if we choose $\omega = 1.0$,

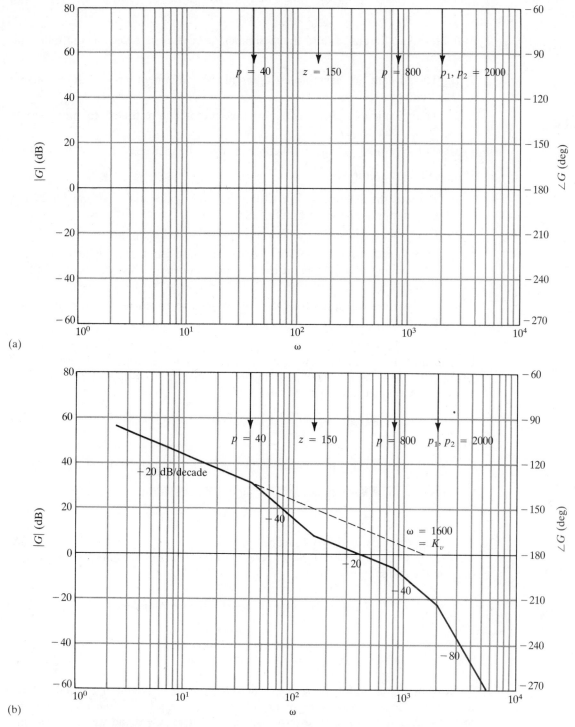

FIGURE 5.2 Bode Diagram: (a) Marking Corner Frequencies and (b) Construction Asymptotes for:

$$G(j\omega) = \frac{1600[(j\omega/150) + 1]}{j\omega[(j\omega/40) + 1][(j\omega/800) + 1][(j\omega/2000)^2 + 2\zeta j\omega/2000 + 1]}.$$

then $\log_{10}|1.0| = 0$, so in each case the magnitude of the desired point is just $20 \log_{10}|K|$, plotted at $\omega = 1$. Alternatively, for type-one and type-two systems, the slope of the initial asymptote is negative, so the line (extended) must cross the 0-dB axis at some value of ω. At this point, the ordinate is zero. Then the right side of Eq. (5.5b) must evaluate to zero, which provides the value of ω at the crossover point:

Type	ω at Crossing of 0-dB Axis
One	$\omega = K_v$
Two	$\omega^2 = K_a;\ \omega = \sqrt{K_a}$

Applying these procedures to construct the Bode asymptotic magnitude diagram of Figure 5.2(b), observe that:

$$G(j\omega) = \frac{1600\left(\dfrac{j\omega}{150} + 1\right)}{j\omega\left(\dfrac{j\omega}{40} + 1\right)\left(\dfrac{j\omega}{800} + 1\right)\left[\left(\dfrac{j\omega}{2000}\right)^2 + \dfrac{2\zeta j\omega}{2000} + 1\right]}.$$

The transfer function is type one, so the lowest frequency asymptote must have a slope of -20 dB/decade. The error coefficient is $K_v = 1600$. To construct the lowest frequency asymptote, we locate $\omega = K_v = 1600$ on the 0-dB axis, draw a straight line with slope of -20 dB/decade through this point and extend this line to low frequencies. The complete diagram is now constructed. Starting with this low-frequency line as $|G(j\omega)|$ (in decibels), we proceed along it to higher frequencies, encountering the lowest frequency corner at $\omega = p = 40$ where the line slope is made more negative by 20 dB/decade. At $\omega = 150$, the slope changes by $+20$ dB/decade and for $150 < \omega < 800$ the slope is again -20 dB/decade. At $\omega = 800$, the slope changes by -20 dB/decade so, for $800 < \omega < 2000$, the line slope is -40 dB/decade. At $\omega = 2000$, the corner frequency represents the ω_n of the complex pole (quadratic) factor, so the slope changes by -40 dB/decade. This completes the straight line (or asymptotic) representation of the Bode magnitude curve. Note that for the complex poles the value of ζ was not defined. It is not needed for the asymptotic curve, only for the true curve.

5.3   THE BODE DIAGRAM: PHASE ANGLE VERSUS ω

Equation (5.4) indicates that the angle of the transfer function $\angle G(j\omega)$ is to be obtained by finding the angle of each *factor* and adding these angles. Each of the individual factors contributes an angle that varies as an arctan function of ω (except the poles at the origin, each of which contributes $-90°$ at all ω).

There are two "short cuts," either of which may be used to construct the curve of the phase angle versus ω. The first of these uses templates (or master curves) to draw a curve of angle versus ω *for each factor*. We then construct the *total* angle curve by

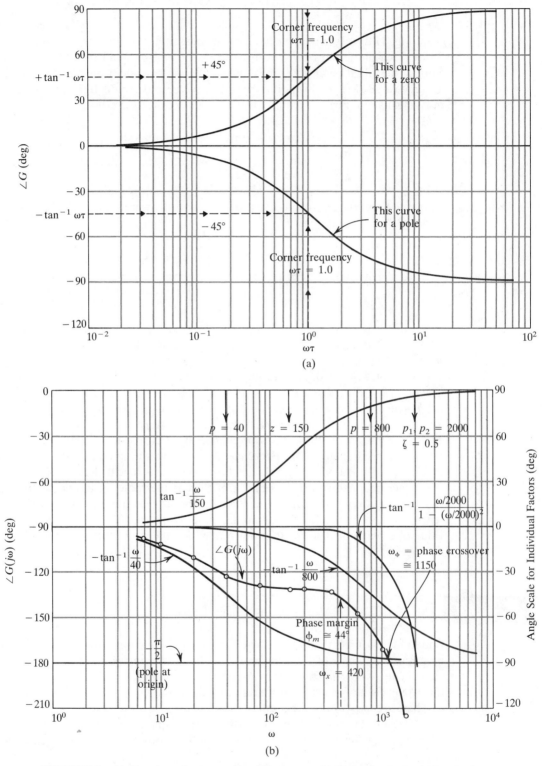

FIGURE 5.3 (a) Templates for ($\mp \tan^{-1} \omega\tau$) Versus $\omega\tau$; (b) Phase Angle Curve for the System of Figure 5.2; and (c) Approximation of the Phase Curve for a Real Pole or Zero.

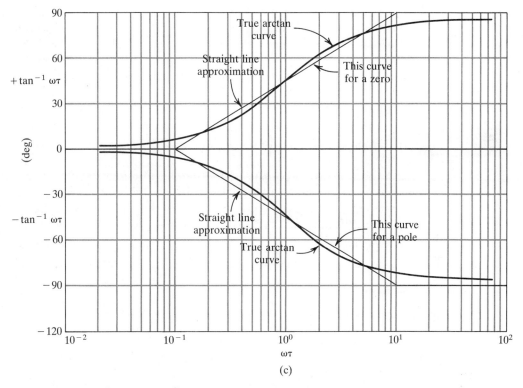

(c)

FIGURE 5.3 (*Continued*)

adding ordinates at selected values of ω, plotting the points obtained, and drawing a smooth curve through the calculated points.

A template may be made for the function $\tan^{-1} \omega\tau$ as shown in Figure 5.3(a). One would usually mount on cardboard and cut out along the edge (or make it of sheet plastic). The broken lines marking the 45° angle and the corner frequency should be marked on the template to assist in alignment. Note that the angle scale (degrees/inch) may be chosen arbitrarily; the ω-scale is set by the choice of graph paper. A template can be used only with semilog paper in which the scale (inches/decade) is the same as that used in making the template.

When complex poles or zeros occur in the transfer function, the specific arctan curve depends on the value of ζ. A *family* of templates can be made, or master curves such as those shown in Figure. 4.4(b) are used to plot the angle of the quadratic factor. Figure 5.3(b) shows the generation of the phase angle versus ω curve from the template curves for the $G(j\omega)$ function of Figure 5.2. Calculated points on the curve for $G(j\omega)$ are shown, together with the resulting curve.

The second shortcut method approximates the arctan curve with straight lines. Two different techniques 1,2 have been suggested; here only the easiest to remember is presented, as shown in Figure 5.3(c). The total excursion of the arctan function for a single time constant factor is 90°. We approximate this function by assuming a straight line relationship on the semilog coordinates such that the 90° change is made over a two-decade frequency range with the corner frequency at the middle of this range. The maximum error is 6°. For a quadratic factor, the total phase excursion is 180°. A straight line approximation distributing this phase over two decades is also used, but the maximum error depends on the value of ζ and can be quite large for small ζ.

5.4   SYSTEM STABILITY ANALYSIS FROM THE BODE DIAGRAM

The Nyquist criterion permits us to interpret absolute stability from the $GH(j\omega)$ function and, for systems with no poles of $GH(s)$ in the right half plane, we can interpret the criterion using phase margin and gain margin. Once the Bode diagram has been drawn we can evaluate both phase and gain margins directly[c] from the plots:

1. To evaluate gain margin, refer to the curve for $\angle G(j\omega)$ and determine the frequency at which the phase is $-180°$. This frequency is called the *phase crossover* frequency. For the system of Figure 5.2(b),

$$\omega_\phi = \text{phase crossover frequency} = 1150$$

as determined from Figure 5.3(b). This is the frequency of the polar vector with $-180°$ phase angle. The length of this vector is determined by reference to the Bode magnitude curve at that frequency. From Figure 5.2(b), at $\omega = 1150$, $|G(1150)| = -12$ dB. Thus, the gain must be raised 12 dB to put the system at the stability limit. Thus the gain margin is:

$$\text{gain margin (decibels)} = +12$$
$$\text{gain margin (ratio)} = 4.$$

2. To evaluate the phase margin, refer to the curve for $|G(j\omega)|$(dB) and determine the frequency at which the gain is 0 dB. This frequency is called the *gain crossover* frequency. For the system of Figure 5.2(b),

$$\omega_x = \text{gain crossover frequency} = 420$$

as determined from Figure 5.2(b). This is the frequency at which the polar vector has a length of one unit. From Figure 5.3(b), the $\angle G(420) = -136°$, so that unit vector would have to be rotated an additional 44° to place the system at the stability limit. Thus, the phase margin $= +44°$.

The preceding discussion has evaluated stability using two separate curve sheets: Figure 5.2(b) for the Bode magnitude curve and Figure 5.3(b) for the Bode phase curve.

---

[c] Alternatively, we can plot the polar curve, reading coordinates of points from the Bode plot, i.e., using the Bode plot for a nomograph. If $G(j\omega)$ has poles in the right half plane, we cannot use phase margin and gain margin so use of the polar plot is recommended. However, for the usual problem, interpretation is made directly from the Bode diagram.

It is convenient to plot *both curves* on the same sheet. Note that both have the same abscissa variable and scale, but the two ordinates are completely independent, so the ordinate scales are chosen as convenient. It is recommended, however, that the horizontal axis for $|G|(\text{dB}) = 0$ dB and that the horizontal line for $\angle G = -180°$ be made the same line. This simplifies interpretation of phase margin and gain margin as shown in Figure 5.4. On this composite diagram,

1. locate the gain crossover and read the phase margin
2. locate the phase crossover and read the gain margin.

When the phase curve is *above* the gain crossover, the phase margin is positive and the system is stable. In like manner, when the gain curve is below the phase crossover, the gain margin is positive, which also indicates a stable system.

Use of a computer to calculate and plot the Bode diagram can provide more accurate curves and requires less time and effort. Construction of the lowest frequency asymptote can be used as a partial check of the computer results. Figure 5.4(a) shows

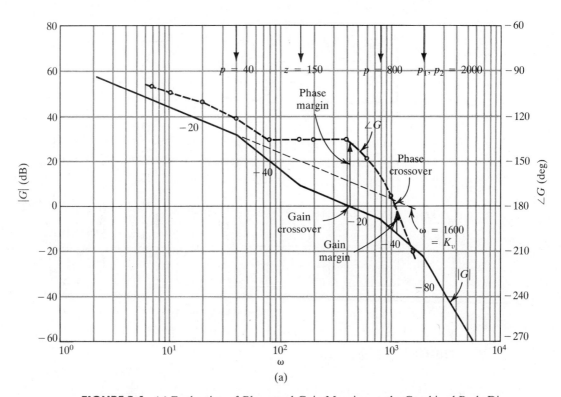

(a)

FIGURE 5.4   (a) Evaluation of Phase and Gain Margins on the Combined Bode Diagram, (b) Addition of Lowest Frequency Asymptote; (c) Detection of Plotting Error.

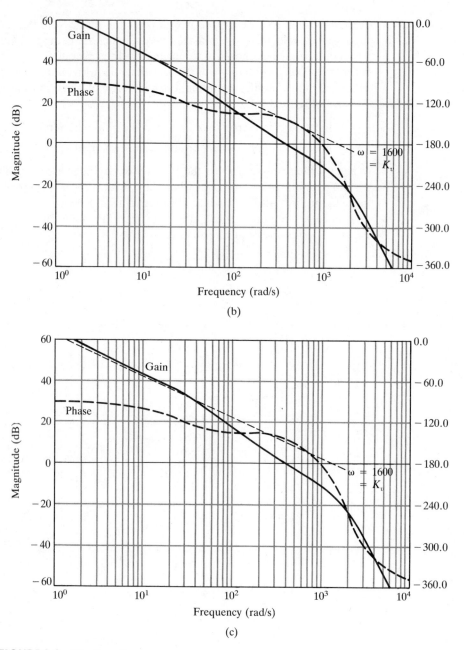

(b)

(c)

FIGURE 5.4   (*Continued*)

the graphically constructed plot for

$$G(j\omega) = \frac{1600\left(\dfrac{j\omega}{150} + 1\right)}{j\omega\left(\dfrac{j\omega}{40} + 1\right)\left(\dfrac{j\omega}{800} + 1\right)\left[\left(\dfrac{j\omega}{2000}\right)^2 + \dfrac{2\zeta j\omega}{2000} + 1\right]}.$$

Figure 5.4(b) shows the computer-generated plot of the same transfer function. The lowest frequency asymptote has been added to Figure 5.4(b) by locating the point $\omega = 1600$ and drawing in the straight line with slope of -20 dB/decade. Since this line is clearly asymptotic to the computer-generated curve, we conclude that the gain constant is correct. If the gain had been entered incorrectly, e.g., 2000, the Bode diagram would appear as on Figure 5.4(c), and addition of the asymptote for $K = 1600$ discloses the error.

Checking the computer drawn plot for other types of errors is desirable, but the graphical construction procedures are not very useful for such purposes. Marking the corner frequencies on the Bode plot is of some assistance; the magnitude curve should change slope at a corner frequency, and the rate of change of phase due to the pole or zero is maximum. Thus, inspection of the magnitude and phase curves at the specified corners should indicate whether the pole or zero was correctly entered in the program.

5.5 SYSTEM TRANSIENT ANALYSIS FROM THE BODE DIAGRAM

Phase margin and gain margin are measures of the separation (distance) of the $G(j\omega)$ curve from the critical (-1.0) point. When the system is stable, they are a measure of the amount of damping to be expected. Assuming dominance of a pair of complex roots, the transient (step) response of the closed-loop control system can be estimated quantitatively using the results from second-order system analysis as given in Chapter 4. For the sytem of Figure 5.2, using these second-order correlations,

Phase margin $= 44°$, $\zeta = 0.4$ (Figure 4.5)

Maximum overshoot to a step input $= M_{pt} = 1.26$ (Figure 4.3)

Ratio of bandwidth to natural frequency $= \dfrac{\omega_b}{\omega_n} = 1.37$

From a master family of transient response curves, such as Figure 4.2, the step response for $\zeta = 0.4$ has a substantial first overshoot, a non-negligible first undershoot, and a small second overshoot. Additional oscillations, while theoretically present, are probably negligible. Thus, we expect a step response with two overshoots and one undershoot, the first overshoot (from $M_{pt} = 1.26$) exceeding the desired steady-state value by about 26%.

We would like to know the settling time also:

$$T_s = \frac{4}{\zeta\omega_n}, \tag{5.6}$$

but the value of ω_n is not known. A value for ω_n can be obtained from ω_b/ω_n if we can evaluate ω_b, or from $\omega_r/\omega_n = \sqrt{1 - 2\zeta^2}$ if we can evaluate ω_r. To find a quick (but approximate) way to evaluate ω_n, note that

$$\left|\frac{C}{R}\right| = \frac{|G|}{|1 + GH|}, \tag{5.7a}$$

$$= \frac{|G|}{|1 + G|} \text{ for } H = 1.0. \tag{5.7b}$$

Clearly, if $|G| \gg 1$, we can neglect the 1 in the denominator and

$$\left|\frac{C}{R}\right| = \frac{|G|}{|G|} = 1.0, \tag{5.8a}$$

while if $|G| \ll 1$,

$$\left|\frac{C}{R}\right| = \frac{|G|}{1.0} = |G|. \tag{5.8b}$$

By applying this to any plot of $|G|(j\omega)|$ as in Figure 5.5 and relaxing the restrictions from *much greater than* to simply *greater than* (for simplicity in interpretation), observe that for all frequencies lower than the gain crossover frequency ω_x, $|G|$ $dB > 0$ dB; $|G| > 1$, and for $\omega < \omega_x$,

$$\left|\frac{C}{R}\right| = 1.0. \tag{5.9a}$$

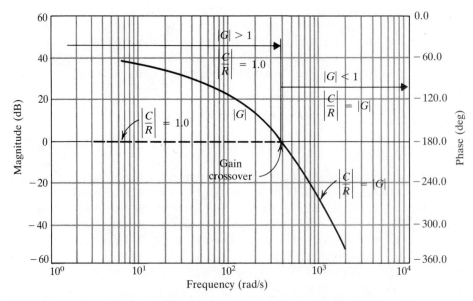

FIGURE 5.5 Construction of Closed-Loop Frequency Response from $|G|$.

In like manner, for $\omega > \omega_x$,

$$\left| \frac{C}{R} \right| = |G|. \tag{5.9b}$$

The closed-loop frequency response determined in this manner is shown by the dotted lines in Figure 5.5. The gain crossover frequency for the open-loop transfer function curve is the corner frequency for the closed-loop frequency response. Now it is clear that the closed-loop bandwidth frequency ω_b will always be greater than the open-loop gain crossover, and clearly the resonant frequency ω_r (if the dominant roots are complex) will be less than ω_b. It is therefore reasonable to estimate that the gain crossover frequency is approximately ω_n.

$$\omega_n = \omega_x.$$

For the system of Figure 5.2 then, if $\zeta = 0.4$ and $\omega_n = \omega_x = 420$, an estimate for the settling time is $T_s = 4/\zeta\omega_n \simeq 4/(0.4)(420) = 0.0238$ s.

Note that the value for gain margin, 12 dB, was not used at all in obtaining quantitative estimates for the response. There are no comparable criteria using gain margin because the gain margin of a second-order system is undefined (or infinite). If,

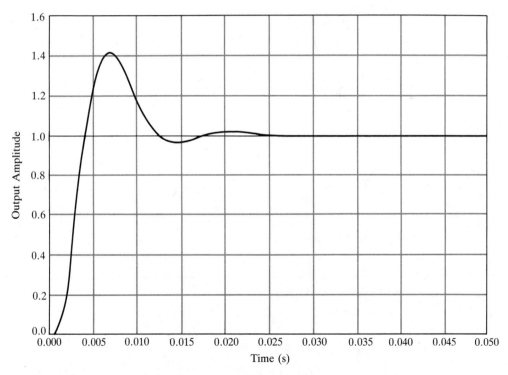

FIGURE 5.6 Step response of System of Figure 5.4(a)

however, we consider the gain margin of a second-order system to be *infinite*, then it seems reasonable to assume that the larger the gain margin (for a high-order system) the more nearly the system response will approximate that of a second-order system. On this basis, we conclude that if the gain margin is *large*, the predictions based on second-order dominance will be accurate. Experience has shown that if the gain margin is 10 dB or greater the second-order predictions are reasonably good. Simulation results in Figure 5.6 show that the 420 rad/s component of the response behaves as predicted. However, there is a high-frequency component due to the complex poles that persists for a longer interval. For this, an additional compensation would be required.

 EXAMPLE 5.1

Draw the Bode diagram for

$$G(s)\frac{25}{(s + 1)(0.1s + 1)(0.05s + 1)}.$$

The diagram is shown in Figure 5.7(a). The procedure follows:

1. The term $G(s)$ has three poles (no zeros) at 1, 10, and 20. They are marked at the appropriate values of ω.

2. The system is type zero, so the low-frequency asymptote has a slope of 0 dB/decade, and is plotted at $|G| = 20\log_{10}|25| = 27.96$ dB.

3. Asymptote slope changes to -20 dB/decade at the first corner frequency, $\omega = 1$; then to -40 dB/decade at $\omega = 10$; and to -60 dB/decade at $\omega = 20$.

4. The magnitude curve is corrected over a portion of the frequency range to obtain a more accurate determination of the gain crossover frequency, which is $\omega_x = 12.6$.

FIGURE 5.7   (a) Example 5.1 Bode Diagram for

$$G(s) = \frac{25}{(s + 1)(0.1s + 1)(0.05s + 1)};$$

(b) Example 5.2 Bode Diagram for

$$G(s) = \frac{50(0.2s + 1)}{s(s + 1)(0.02s + 1)};$$

(c) Example 5.3 Bode Diagram for

$$G(s) = \frac{500(0.2s + 1)(0.1s + 1)}{s^2(s + 1)(0.02s + 1)};$$

(d) Example 5.4 Bode Diagram for

$$G(s) = \frac{50(0.05s + 1)}{s(0.1s + 1)(0.02s + 1)[(s/200)^2 + (0.02s/200) + 1]}.$$

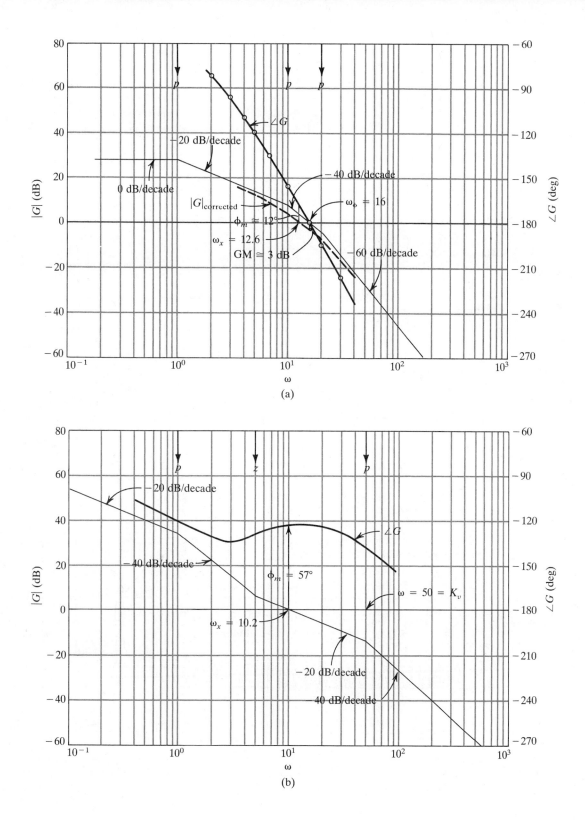

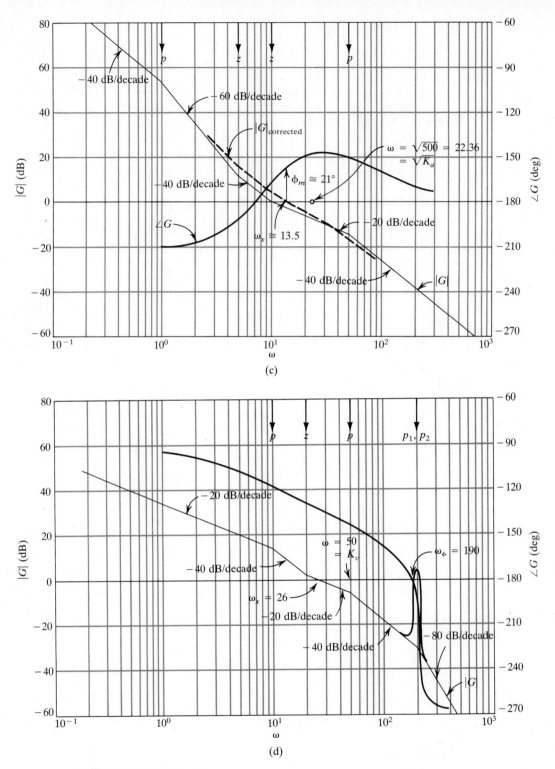

FIGURE 5.7 (*Continued*)

5. The $\angle G$ curve is constructed using the phase angle template and adding as previously explained.

From Figure 5.7(a), the phase margin is 12° and the gain margin is 3 dB. Thus, the system is stable and $\zeta \simeq 0.1$. Then

$$T_s = \frac{4}{(0.1)(12.6)} = 3.175 \text{ s.}$$

 EXAMPLE 5.2

Draw the Bode diagram for

$$G(s) = \frac{50(0.2s + 1)}{s(s + 1)(0.02s + 1)}.$$

The diagram is shown in Figure 5.7(b). The procedure follows:

1. The term $G(s)$ has three poles at 0, 1, and 50, and one zero at 5. Because there is one pole at the origin, this is a type-one system. The pole at the origin cannot be marked on the plot, but the other two poles and the zero are marked at the appropriate values of ω.

2. The lowest frequency asymptote has a slope of -20 dB/decade and is drawn so that its extension would intersect the 0-dB axis at $\omega = K_v = 50$.

3. The asymptote slope changes from -20 dB/decade to -40 dB/decade at the first corner frequency, $\omega = 1$; then to -20 dB/decade at $\omega = 5$; and back to -40 dB/decade at $\omega = 50$.

4. The gain crossover is marked at $\omega_x = 10.2$ at the *asymptote* crossing. For this problem, the magnitude curve is *not* corrected because the corrections would be very small in the region of the gain crossover.

5. The $\angle G$ curve is constructed using the template and adding.

From Figure 5.7(b), the phase margin is 57° and the gain margin is not defined. The system is stable and well damped with $\zeta \simeq 0.6$. Thus, the settling time is

$$T_s = \frac{4}{(0.6)(10.2)} = 0.654 \text{ s.}$$

 EXAMPLE 5.3

Draw the Bode diagram for

$$G(s) = \frac{500(0.2s + 1)(0.1s + 1)}{s^2(s + 1)(0.02s + 1)}.$$

The diagram is shown in Figure 5.7(c). The procedure follows:

1. The term $G(s)$ has four poles at 0, 0, 1, and 50 and two zeros at 5 and 10. The two poles at the origin classify the system as type two and cannot be marked on the

plot. The other two poles and the two zeros are marked at the appropriate frequencies.

2. The lowest frequency asymptote is drawn with a slope of -40 dB/decade, and is located such that its extension would intersect the 0-dB axis at $\omega = \sqrt{K_a} = \sqrt{500} = 22.36$.

3. The asymptote slope changes from -40 dB/decade to -60 dB/decade at the first corner frequency, $\omega = 1$; then to -40 dB/decade at $\omega = 5$; to -20 dB/decade at $\omega = 10$; and to -40 dB/decade at $\omega = 50$.

4. The $|G|$ curve is corrected over part of the frequency range in order to determine the gain crossover more accurately. The gain crossover is marked at $\omega = 13.5$.

5. The $\angle G$ curve is constructed using the template.

From Figure 5.7(c), the phase margin is $21°$, so $\zeta \simeq 0.2$. The system is stable and the settling time should be

$$T_s = \frac{4}{(0.2)(13.5)} = 1.48 \ s.$$

 EXAMPLE 5.4

Draw the Bode diagram for

$$G(s) = \frac{50(0.05s + 1)}{s(0.1s + 1)(0.02s + 1)\left[\left(\dfrac{s}{200}\right)^2 + \dfrac{0.02s}{200} + 1\right]}.$$

The diagram is shown in Figure 5.7(d). The procedure follows:

1. The term $G(s)$ has three real poles at 0, 10, and 50 and a pair of lightly damped complex poles with $\zeta = 0.01$ and $\omega_n = 200$. There is one zero at 20. The pole at the origin designates the system as type one and cannot be marked on the diagram. The other poles and the zero are marked at the appropriate frequencies. Note that the corner frequency for the complex poles is $\omega = \omega_n = 200$ and is marked as *two* poles.

2. The lowest frequency asymptote has a slope of -20 dB/decade and is drawn so that its extension would cross the 0-dB axis at $\omega = K_v = 50$.

3. The asymptote slope changes from -20 dB/decade to -40 dB/decade at $\omega = 10$; then to -20 dB/decade at $\omega = 20$; to -40 dB/decade at $\omega = 50$; and to -80 dB/decade at $\omega = 200$.

4. The gain crossover frequency is marked at $\omega_x = 26$. This is the asymptote crossover; the magnitude curve has not been corrected because it was estimated that the correction would be negligible at the gain crossover frequency.

5. Whenever the transfer function has lightly damped complex poles, the resulting resonance peak may cross the 0-dB axis. If this happens, the closed loop may be unstable. Therefore, the curve should be corrected at the corner frequency of the

complex poles and perhaps at a few adjacent frequencies. For any complex pole pair;

$$|G| = \frac{1}{\sqrt{\left(1 - \frac{\omega}{\omega_n}\right)^2 + \left(\frac{2\zeta\omega}{\omega_n}\right)^2}}$$

at the corner frequency ω_n, $\omega/\omega_n = 1$, and $|G|$ reduces to

$$|G|_{\omega=\omega_n} = \frac{1}{2\zeta}.$$

For this problem,

$$20\log_{10}|G| = 20\log_{10}\left|\frac{1}{0.02}\right| = 34 \text{ dB}.$$

The resonance is marked on Figure 5.7(d) and it does rise above the 0-dB line. For poles with $\zeta = 0.01$, very careful calculation for frequencies near resonance is needed to obtain plots for stability interpretation. Figure 5.7(d) does not provide such a plot, only the point at $\omega = \omega_n$ is accurate. However, this problem was studied in Example 3.11 where the resonance has been calculated in detail and is shown on the polar plot of Figure 3.13.

5.6   EVALUATING THE CLOSED-LOOP FREQUENCY RESPONSE: THE NICHOLS CHART

Preceding sections have treated the Bode diagram of the open-loop transfer function. When the closed-loop frequency response (often called the *closed-loop Bode diagram*) is needed, it is calculated readily with the computer, which is the preferred method. If the computer is not available, graphical computation using the Nichols chart is convenient, as will be discussed. When the data are experimental, i.e., frequency response has been measured (as with a spectrum analyzer) but equations are not available, the Nichols chart is very useful in manipulating the data and helps to derive a transfer function equation.

The Nichols chart is a nomograph consisting of curves that are maps of the M and N circles of Figure 3.19 on a new coordinate system. The Nichols coordinate system plots $|G|$ in decibels as ordinate versus $\angle G$ as abscissa, with origin at $|G| = 0$ dB and $\angle G = -180°$. This origin is a map of the $-1 + j0$ point of the polar coordinate (Nyquist) plot. Figure 5.8 illustrates the coordinate system (not the nomogram) with a plot of the Bode diagram data from Figure 5.4(a). Note that the phase and gain margins are readily identified.

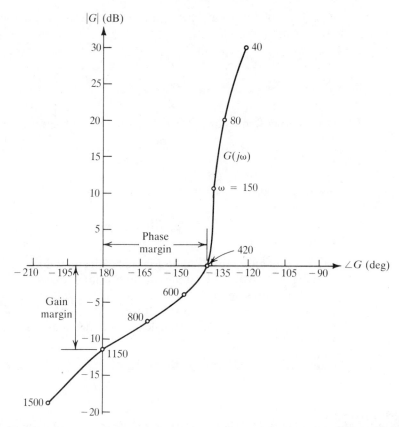

FIGURE 5.8   Nichols Coordinate System with Sample $G(j\omega)$. Plot is from Bode Diagram of Figure 5.4(a).

Consider the circles of Figures 3.19(a) and 3.19(b). To map a point, $A + jB$, from this coordinate system to the Nichols system, we note that

$$|G| = \sqrt{A^2 + B^2},$$

$$\angle G = \tan^{-1} B/A. \tag{5.10}$$

Using these relationships and points on the circles as defined by Eqs. 3.16 through 3.19, each circle can be mapped onto Nichols coordinates with the result shown in Figure 5.9. When the data from an open-loop Bode diagram is plotted on the Nichols coordinates, as shown in Figure 5.8 and again in Figure 5.10(a), the magnitude of G and angle of G are determined on the Bode plot for a selected value of ω, the Nichols plot is entered using the rectangular coordinate scales, and the value of ω is recorded next to the plotted point. This procedure is repeated until a sufficient number of points has been plotted, and a smooth curve is drawn through the points. This curve [the $G(j\omega)$ curve] crosses both M and N contours. At any intersection between the $G(j\omega)$ curve

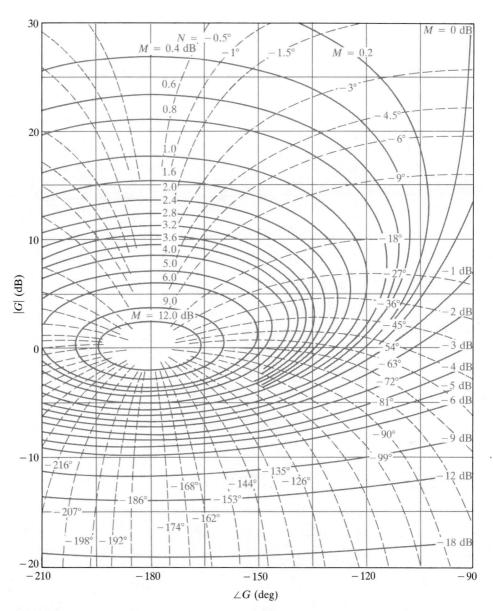

FIGURE 5.9   Nichols Chart.

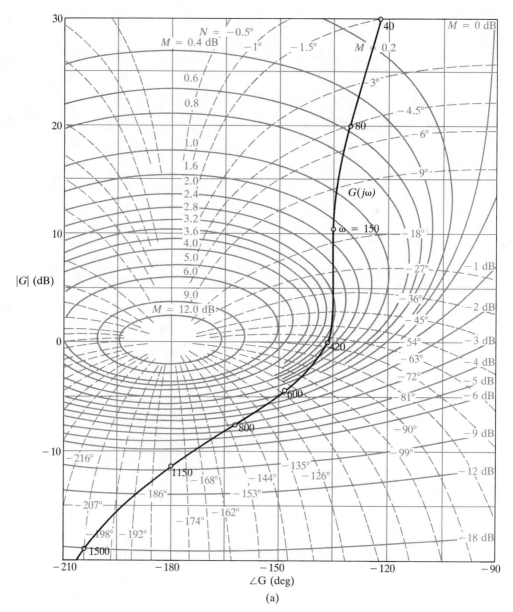

(a)

FIGURE 5.10 (a) Nichols Chart with Plot of $G(j\omega)$ from Figure 5.8 and (b) Closed-Loop Frequency Response from Nichols Chart of (a).

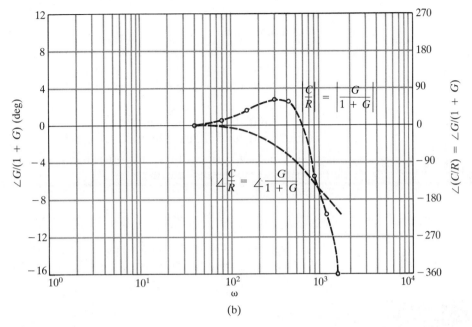

FIGURE 5.10 (*Continued*)

and an M curve we read off[d]

$$M = \left| \frac{G}{1 + G} \right| \tag{5.11a}$$

and also read off the value of ω at that point. In like manner, at any intersection between the $G(j\omega)$ curve and an N curve, we read off

$$N = \angle \frac{G}{1 + G} \tag{5.11b}$$

and the corresponding value of ω. This is repeated to get enough points for the closed-loop Bode diagram as shown in Figure 5.10(b).

When the data are empirical (e.g., the closed-loop frequency response is measured), the Nichols chart is used to obtain the open-loop Bode diagram, to which asymptotes are fitted to obtain the transfer function equation. The procedure follows:

1. Plot the test data on a Bode diagram and smooth it. (Replotting on the Nichols chart tends to magnify scatter.)

---

[d] The nomogram constructs $G/(1 + G)$ from G. For systems that do not have unity feedback, the Bode diagram is usually a plot of the loop transfer function GH. When this is used, the nomogram constructs $GH/(1 + GH)$. If one knows H, the closed-loop frequency response is calculated from $(1/H)[GH/(1 + GH)] = G/(1 + GH)$.

2. Replot the data on the Nichols chart, entering the points with the M and N contours as coordinates.

3. Read out $|G|$ and $\angle G$ for each point plotted in step 2 (and intermediate points if desired). Record the value of ω at each point.

4. Replot the data obtained in step 3 on the Bode diagram.

5. Fit asymptotes to the magnitude curve of step 4. This is a trial and error procedure, but is aided by the constraint that the slopes of all asymptotes must be integral multiples of 20 dB/decade.

6. Read off the error coefficient and all corner frequencies from the asymptotic plot. Write out the transfer function.

7. Check the result:

   a. Multiply $-90°$ times the difference between the number of poles and number of zeros. The result should equal $\angle G(\omega \to \infty)$.

   b. The term $G(j\omega)$ was obtained using $|G|$ only. Using $G(j\omega)$, calculate $\angle G(j\omega)$ and see if it fits the curve obtained from the Nichols chart.

5.7 SUMMARY

To use the Nyquist criterion for analysis and design, the system frequency response must be calculated. The Bode diagram is a convenient tool for these calculations, and procedures were given to construct the diagram quickly. Phase margin and gain margin are evaluated from the diagram providing stability analysis and means for predicting the time response. The Nichols chart is a nomograph for calculating the closed-loop frequency response from the open-loop response and vice versa. It can be used to analyze experimental closed-loop frequency response data from which the open-loop response is obtained, fitted with asymptotes, and interpreted to find the poles, zeros, and gain of the open-loop transfer function.

REFERENCES

1. Savant Jr., C. J. *Basic Feedback Control System Design.* p. 135. New York: McGraw Hill Book Co. (1958).

2. Bower, J. L.; and Schultheis, P. M. *Introduction to the Design of Servomechanisms.* p. 88. New York: John Wiley and Sons (1958).

BIBLIOGRAPHY

Bode, H. W. "Relations Between Attenuation and Phase in Feedback Amplifier Design." *Bell System Tech. J.* 19 July 1940.

Bode, H. W. *Network Analysis and Feedback Amplifier Design.* New York: Van Nostrand (1948).

Eveleigh, V. W. *Introduction to Control System Design.* New York: McGraw-Hill Book Co (1972).

Franklin, G. F.; Powell, J. D.; and Emani-Naeini, A. *Feedback Control of Dynamic Systems.* Reading, Mass.: Addison Welsey (1986).

PROBLEMS

5.1 For each of the following:

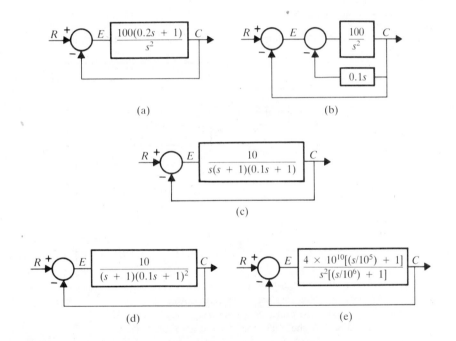

(a)

(b)

(c)

(d)

(e)

a. Draw the Bode diagram.
b. Mark the phase margin and gain margin.
c. Record the gain crossover frequency and the phase crossover frequency.
d. Record *your estimate* of the closed-loop bandwidth ω_b.
e. Calculate the settling time.

5.2 For the system of Problem Figure 5.1(c):
a. Adjust the gain (raise or lower) to provide a phase margin of 30°.
b. Obtain the Nyquist polar plot.
c. Obtain the closed-loop frequency response.
d. What is the resonant frequency?

5.3   For the system of Problem Figure 5.1(e):
  a.   Use the Nichols chart (and/or the computer) and determine the closed-loop frequency response.
  b.   What is the resonant frequency?
  c.   What is the bandwidth?

5.4   For the following system, draw the Bode diagram, check stability, and evaluate the settling time.

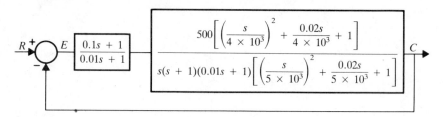

5.5   For the following system, draw the Bode diagram, check stability, and evaluate the settling time.

5.6   Given:

$$G(s) = \frac{170\left(\dfrac{s}{10} + 1\right)}{s\left(\dfrac{s}{1.75} + 1\right)\left(\dfrac{s}{60} + 1\right)^2}.$$

  a.   Draw the open-loop Bode diagram.
  b.   What is the gain crossover frequency?
  c.   What is the phase crossover frequency?
  d.   What are the phase margin and gain margin?
  e.   Is the closed-loop system stable?
  f.   Determine values for each of the following:
    1.   maximum overshoot for a step input
    2.   bandwidth
    3.   settling time.

5.7   Given:

$$G(s) = \frac{20}{s(0.1s + 1)(0.01s + 1)}$$

  a.   What are the phase margin and gain margin?
  b.   For a unit step input, what are the peak overshoot and the settling time?

5.8   For the system defined on the Nichols plot below, what is the expected settling time?

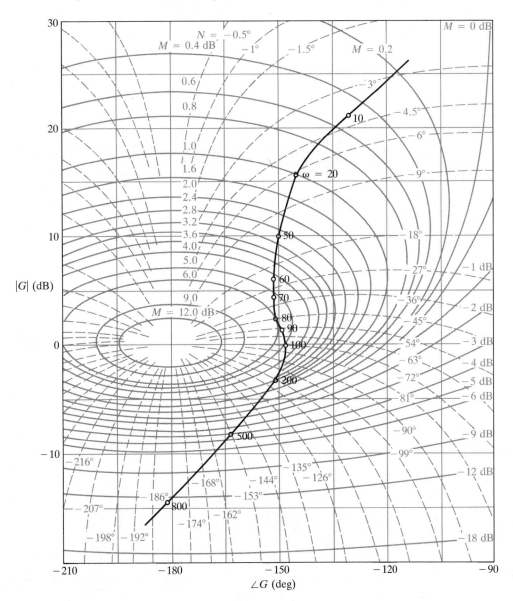

5.9   Draw the Bode diagram for each of the following loop transfer functions:

a.
$$G(j\omega) = \frac{100}{j\omega(0.1j\omega + 1)}.$$

b.
$$G(j\omega) = \frac{100}{j\omega(0.01j\omega + 1)^2}.$$

c.
$$G(j\omega) = \frac{225(0.1j\omega + 1)}{(j\omega)^2(0.01j\omega + 1)}.$$

d.
$$G(j\omega) = \frac{1000}{(j\omega + 1)(0.01j\omega + 1)(0.0005j\omega + 1)}.$$

For each plot, record gain crossover frequency, phase crossover frequency, gain margin, and phase margin.

5.10 a. Draw the Bode diagram for

$$G(s) = \frac{100(0.02s + 1)}{(s + 1)(0.1s + 1)(0.01s + 1)^2}.$$

b. Mark the following on the Bode diagram, recording the numerical values:
1. gain crossover frequency
2. phase margin
3. phase crossover frequency
4. gain margin.

5.11 Given:

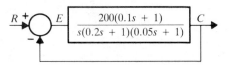

a. Draw the Bode diagram.
b. Is the system stable?

5.12 From the asymptotic diagram below, determine $G(s)$.

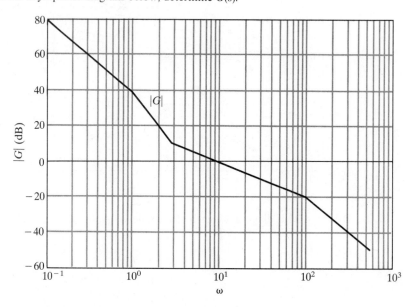

5.13 a. From the following Bode diagram for G, determine:

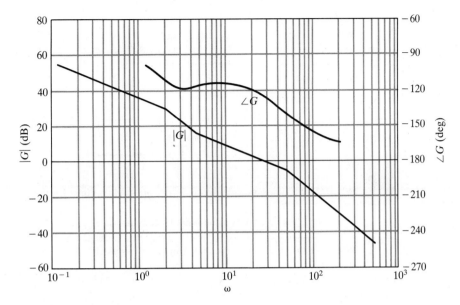

1. system type number
2. error coefficient
3. phase margin (mark this on the diagram).

b. What is the settling time of the closed loop? Show work.

5.14 Given:

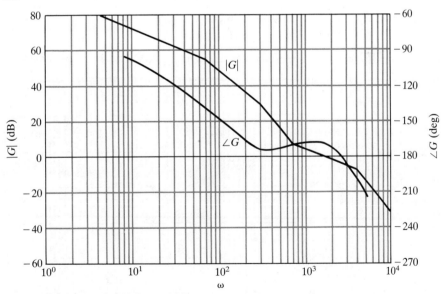

a. What is the gain crossover frequency?
b. What is the phase crossover frequency?
c. What is the phase margin?
d. What is the gain margin?

e.   Is the system stable?
f.   What is the type number of the system?
g.   What is the numerical value of the error coefficient?
h.   By inspection of the diagram, write down the transfer function $G(s)$.
i.   What is your estimate of the bandwidth of the closed-loop system?

5.15   a.   Draw the Bode diagram.

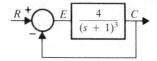

b.   What are the phase margin and gain margin? (Mark them on the Bode diagram.)
c.   Is the closed-loop system stable?
d.   Draw the Nichols curve on the Nichols chart.
e.   Obtain the closed-loop frequency response and plot on a Bode diagram.
f.   What are the resonant frequency and the bandwidth? (Mark these on *BOTH the Nichols chart AND the closed-loop frequency response.*)
g.   What are ζ and ω_n?
h.   What is the settling time?

5.16   a.   Draw the asymptotic plot for $|G(j\omega)|$ on the Bode diagram.

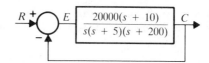

b.   Draw $\angle G(j\omega)$ on the Bode diagram.
c.   Is the system stable? How do you know?

5.17   Determine $G(s)$ from the Bode diagram below. Plot the phase curve.

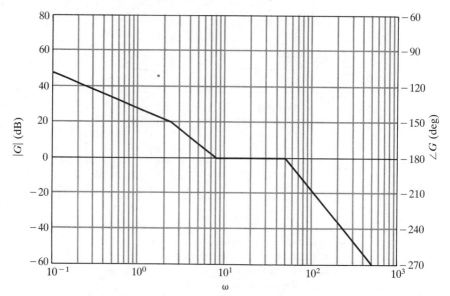

a. Determine the equation for $G(s)$.
b. Construct the phase curve.
c. Mark on the diagram (give values):
   1. phase crossover frequency
   2. gain crossover frequency
   3. phase margin and gain margin.

5.18 Determine $G(s)$. Plot the phase curve.

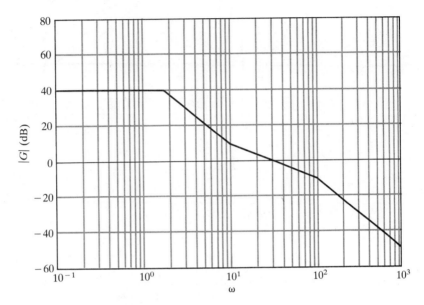

5.19 Determine $G(s)$. Plot the phase curve.

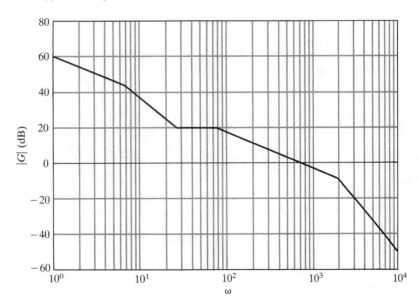

5.20   a.   Determine $G(s)$. Plot the phase curve.

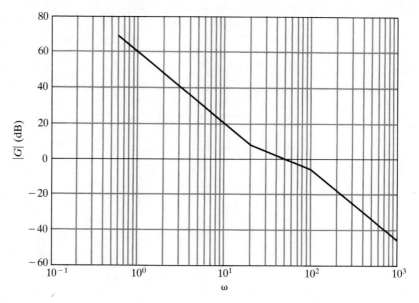

b.   Determine $G(s)$. Plot the phase curve.

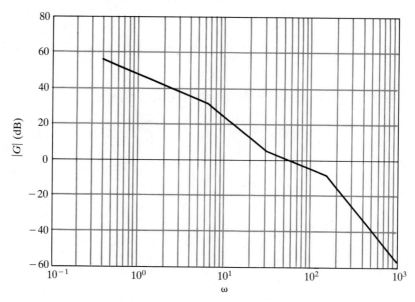

5.21   A feedback control system has unity feedback and forward transfer function as follows:

$$G(s) = \frac{10^7(s + 5)^2}{s^2(s + 1)(s + 500)^2}.$$

Is the system stable?

5.22 a. Draw the Bode diagram.
  b. On the diagram, mark the gain crossover, phase crossover, phase margin, and gain margin. Also list their numerical values.
  c. Is the system stable?

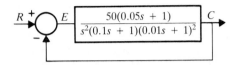

5.23 a. Determine the closed-loop frequency response.
  b. What is ω_r, the resonant frequency?
  c. What is the settling time of the system?

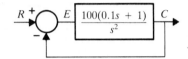

5.24 a. In the following diagram, assume all poles and zeros are in the left half plane. What is $G(s)$?
  b. Add the phase angle curve and determine the phase and gain margins.
  c. Predict the settling time.

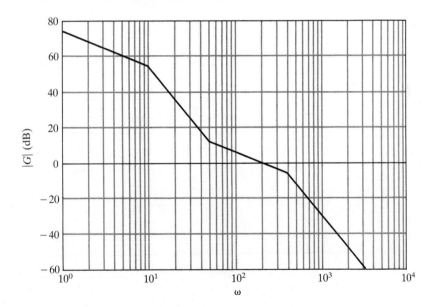

5.25 a. Draw the Bode diagram [magnitude versus ω; phase versus ω].
  b. Mark on the diagram and and record the numerical values for:
    1. gain crossover frequency
    2. phase crossover frequency

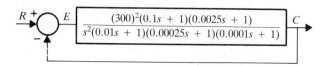

$$\frac{(300)^2(0.1s + 1)(0.0025s + 1)}{s^2(0.01s + 1)(0.00025s + 1)(0.0001s + 1)}$$

3. phase margin
4. gain margin.

c. Is the closed-loop system stable?
d. What are the bandwidth and resonant frequency of the closed-loop system?
e. What is the settling time of the closed-loop system?

5.26 Given:

$$G(s) = \frac{100\left(\dfrac{s}{7} + 1\right)}{(s + 1)^2\left(\dfrac{s}{70} + 1\right)\left(\dfrac{s}{200} + 1\right)}$$

What is the resonant frequency of the closed-loop system? Show all work.

5.27 Two Bode diagrams are given below. Each contains the asymptotic magnitude plot of an open-loop transfer function.

a. By inspection of each plot, write down the respective transfer functions, specifying numerical values as obtained from the plot.
b. There are no open-loop poles in the right half s plane. Complete the Bode diagram by adding the phase curve.
c. From your plot determine values for the following:
   1. phase margin
   2. gain margin
   3. settling time.

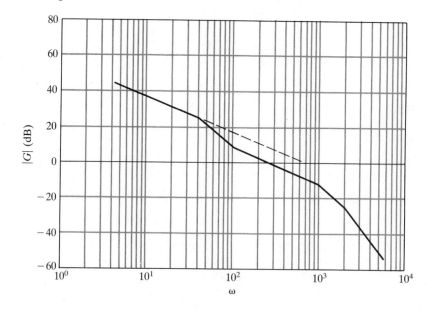

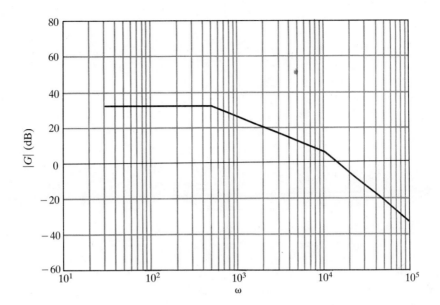

6 Root Locus Methods for Analysis

6.1 INTRODUCTION

When all of the coefficients of a polynomial are known numerically, the polynomial may be factored (by a computer) for its roots, which are fixed numbers. If any one parameter of the system is not defined numerically, then an algebraic symbol representing the parameter would appear in one or more of the coefficients of the polynomial, and the polynomial cannot be factored until a numerical value is assigned to that parameter. If we assign more than one numerical value to the undefined parameter, a set of roots can be evaluated for each such value chosen. Root points determined for several values of a parameter may be plotted on the s plane and if there are N roots then each of the N roots traces out a continuous curve as the parameter is varied; for an N'th-order polynomial there are then N planar curves on the s plane. Many computer centers have subroutines for calculating and plotting *root loci* and, in general, such subroutines use a root solver and iterate the calculation for specified values of the parameter.

Such root loci can be very valuable in the analysis of dynamic systems, and are also used for design. By inspection of the curves, we determine:

1. whether any loci cross the $j\omega$ axis into the right half s plane (If such a crossing exists, it defines a stability limit for the system.)
2. the frequency (value of ω) at such a stability limit
3. whether the loci go through any area on the s plane where we want dominant complex roots for our system (i.e., can we get what we want!).

We can also determine the value of the parameter at the stability limit, the value of the roots for any specified value of the parameter, and the value of the parameter required to place roots at any selected point on the loci. These latter items, however, are not done by inspection of the curves, but by inspection of the computer printout or by separate calculations.

When two parameters of the system are adjustable, it is possible to calculate and plot a *family* of root loci for the pair of parameters. If there are three parameters, then a three-dimensional space is required for proper representation, but the result is not suitable for engineering analysis and design.

Root locus curves were first obtained by Walter Evans,[1] who used graphical methods. Although use of the computer to obtain the root loci is highly recommended, these graphical methods are still valuable, partly because they permit us to *sketch* the loci with sufficient accuracy for many analysis and design problems, but more importantly because the graphical rules provide valuable insight when choosing design methods.

6.2   MATHEMATICAL BASIS FOR THE CONSTRUCTION OF ROOT LOCI

We have seen that the characteristic equation of a feedback control system is

$$1 + GH(s) = 0. \tag{6.1a}$$

This may be rewritten

$$GH(s) = -1 \tag{6.1b}$$

and since s is a complex variable we may rewrite Eq. (6.1b) in parametric form

$$|GH(s)| = 1.00 \tag{6.2a}$$

$$\angle GH(s) = (2N - 1)\pi, \tag{6.2b}$$

where N is an integer.

For a specific value of s to be a root of the characteristic equation, it must satisfy *both* Eqs. (6.2a) and (6.2b). Clearly, we would expect each of these equations to be satisfied by an infinite number of values of s, i.e., each equation must define one or more curves on the s plane. Since the roots are those values of s that satisfy *both* equations, then the root points are points where these curves intersect.

If we can find an easy way to construct the curves of *either* Eq. (6.2a) or (6.2b), the problem of finding the roots of the polynomial is simplified; we simply test points *on* the curve we know until we find a point that satisfies the other equation. Such a point would be the point of intersection had we calculated both curves. Evans found rather simple methods for finding the angle curves of Eq. (6.2b), and it is these curves that are called the *root loci*. They are exactly the same curves calculated by a computer as described in Sec. 6.1.

6.3   DEVELOPMENT OF RULES FOR CONSTRUCTING THE ROOT LOCI DRAWING THE LOCI

Following Eq. (6.2b), we start with the loop transfer function $GH(s)$. We must have this transfer function, with both numerator and denominator in factored[a] form. We then plot the poles and zeros on the s plane. Figure 6.1 shows an s plane plot with a number of poles and zeros, illustrative of a $GH(s)$ function. All poles and zeros are shown in the left half plane; this is not required but is convenient for the illustration.

The first step in developing construction rules is to select an arbitrary test point and use it to define the angles of the individual vectors that contribute to $\angle GH(s)$. We could, of course, read off the coordinates of the test point, substitute them for s in the

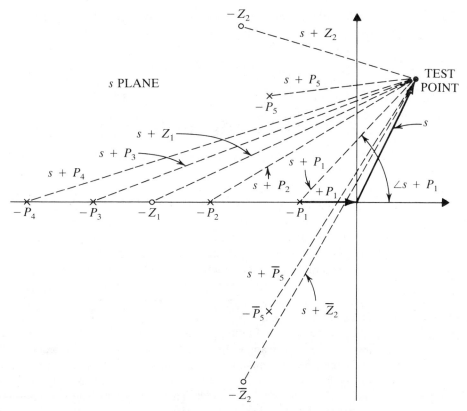

FIGURE 6.1   An s Plane Plot to Demonstrate Use of a Test Point with Eq. (6.2b) to Develop Root Locus Construction Rules.

---

[a] In general, computer programs do not require factored form; polynomial form is acceptable, even required.

$GH(s)$ equation, and evaluate both $|GH(s)|$ and $\angle GH(s)$, but this would not help us to develop construction rules.

By definition, a vector drawn from the origin to the test point defines the test value of s. Using this we can construct vectors for each factor of the $\angle GH(s)$ equation. For example, the plotted pole $-P_1$ was obviously obtained from a factor $(s + P_1)$. In Figure 6.1, a line from $-P_1$ to the origin is a vector $+P_1$, and the apparent vector addition gives us the vector $s + P_1$ marked on the figure and, of course, the length of the $s + P_1$ vector, to scale, is the magnitude of the $s + P_1$ factor. At this point, we are concerned only with the angles of the factors. It is clear that a vector from each pole to the origin defines a vector "$+P$," and in like manner from the zeros we get vectors "$+Z$." When added to the s vector, each pole and zero generates a vector as shown by the dotted lines in Figure 6.1.

Each of the vectors defines a factor and the angle of that factor. Equation (6.2b) states that the test point is a point on the root locus if the algebraic summation of all the angles thus defined evaluates to an odd multiple of π. We can (and do) use this basic rule to locate points on the root locus, but more importantly we have a graphical way of representing the summation that we can use to develop more general rules so that sketches of the loci can be made without using test points. Let us use the concept to develop a rule for the location of root locus segments on the real axis of the s plane.

6.3A   Root Loci on the Real Axis of the s Plane

Any polynomial of odd order must have at least one real root; often it has more than one. Even-order polynomials are not required to have real roots, but if they are of order 4 or higher they frequently do have real roots. The real roots can only exist on clearly defined subsections of the real axis, and the angle rule permits us to locate these sections easily. Consider the s plane plot of Figure 6.2, which displays the pole-zero array of Figure 6.1. Clearly, any point on the real axis that might be a root *must* satisfy the angle equation (6.2b). Therefore, we use the concept of a *test point* to determine the angle summation at every point on the real axis and thus develop a rule for locating line segments that do indeed satisfy Eq. (6.2b). We choose the test point initially far to the right, beyond any poles or zeros of $GH(s)$, and we draw in vectors from all zeros and poles to the test point.

By inspection of Figure 6.2, we see that, since complex poles and zeros are conjugates, the vector angles (for the test point on the real axis) are conjugates and so add to *zero* for any and all points on the real axis. Considering the real zeros and poles, for the test point location of Figure 6.2, each vector is at an angle of zero degrees, so their sum is zero and $\angle GH(s) = 0$. This is obviously true for all points on the real axis to the right of P_1, so none of these points can be roots. Before proceeding, and in order to formulate an easily remembered rule, let us number all poles and zeros on the real axis, calling the right-most one "1" and numbering in sequence as we proceed to the left without regard to the nature (zero or pole) of the singular point. This is shown in Figure 6.2. Next, let the test point sweep from right to left. Observe that for the test point to the right of 1 all vectors are at zero degrees so none of the points tested satisfies the angle rule; they are *not* points on the root locus. When the test point passes 1,

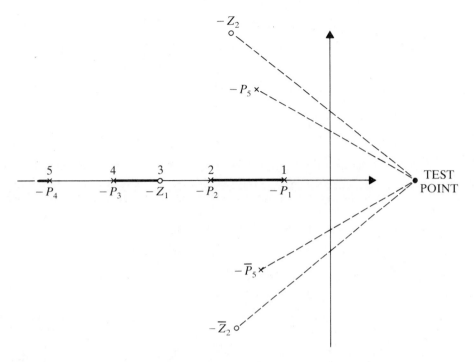

FIGURE 6.2   Rule for Root Locus Segments on the Real Axis of the s Plane.

the vector from the pole P_1 changes its angle to $180°$; for all points between 1 and 2, the angle summation is $\angle GH(s) = -180°$ and all points between 1 and 2 form a segment of root locus. As the test point sweeps past 2, a second vector reverses, and the angle summation becomes $\angle GH(s) = -0°$; points between 2 and 3 are *not* part of a root locus.

Passing point 3, another vector reverses and points between 3 and 4 form another root locus segment. In like manner, points between 4 and 5 do not satisfy the angle requirement and that segment is not a root locus, but all points from 5 to $-\infty$ are on a root locus.

As a result of the above we can formulate a simple rule.

Rule 1   To obtain real axis root loci, we number the poles and zeros on the real axis consecutively starting with the number 1 for the right-most singular point and proceed to the left. Those segments of the real axis to the left of odd-numbered singular points are root locus segments.

6.3B   Terminal Points on the Root Loci

To develop other construction rules, it is helpful to return to the transfer function and the characteristic equation and study some of the consequences of a variable parameter. The basic form of the transfer function is

$$GH(s) = \frac{K(s + Z_1)(s + Z_2)\cdots}{s^N(s + P_1)(s + P_2)\cdots}. \tag{6.3}$$

Observe that the angle of $GH(s)$ is determined only by the zeros and poles; the parameter K (the gain in this case) does not contribute to the angle summation. Next, consider the characteristic equation, which is

$$s^N(s + P_1)(s + P_2) + \cdots K(s + Z_1)(s + Z_2)\cdots = 0. \tag{6.4}$$

Clearly, for different values of K, this equation has different roots. Since the points on the root locus (whatever the curves are) are not changed by K, it is clear that the effect of changing K is to move the roots along the loci to new locations. We can use Eq. (6.4) to study the root locations for extreme values[b] of the parameter K, i.e., $K = 0$ and $K = +\infty$. Consider Eq. (6.4) when $K = 0$:

$$s^N(s + P_1)(s + P_2) + \cdots = 0. \tag{6.5}$$

Note that Eq. (6.5) is already in factored form, the roots are seen to be the *poles* of the transfer function. Therefore, we can state the following rules.

Rule 2   Root loci start at the poles of the open-loop transfer function at which points the value of the parameter is zero ($K = 0$).

Rule 3   The number of branches of the root loci is equal to the order of the polynomial and is equal to the number of the open-loop poles (providing $\#P > \#Z$).

   When K is increased ($K > 0$), the roots are no longer at the poles but must move to new locations. The only permissible locations are points that satisfy the angle summation requirement, so the roots must trace out the root loci as they move. To determine where the loci end, consider Eq. (6.4) as $K \to \infty$. One portion of the polynomial,

$$K(s + Z_1)(s + Z_2)\cdots, \tag{6.6}$$

will be infinite for all values of s *except* those values that are the negatives of the zeros, i.e., $s = -Z_1, -Z_2$, etc. For example, if $s = -Z_1$, then

$$K(-Z_1 + Z_1)(-Z_1 + Z_2) = \infty(0)(-Z_1 + Z_2) \tag{6.7}$$

and Eq. (6.7) is indeterminate; $\infty(0)$ may be finite: The other portion of Eq. (6.4),

$$s(s + P_1)(s + P_2)\cdots, \tag{6.8}$$

clearly evaluated to a finite number for $s = -Z_1$. Thus, we expect some branches of the root loci to terminate on finite zeros. For all other values of s, Eq. (6.6) evaluates to

---

[b] In this text, we consider only positive values of the parameter K and only positive values of any other parameter we wish to vary. Root loci for negative values of a parameter are readily obtained and, in general, they correspond to the use of positive feedback.

infinity as $K \to \infty$, and Eq. (6.8) must also evaluate to ∞ if Eq. (6.4) is to be satisfied. We thus conclude that:

Rule 4   As $K \to \infty$, all finite zeros will be terminating points for branches of the root locus. All excess branches terminate at ∞.

From Rule 4, it is clear that the number of branches of root locus that go to infinity is:

Rule 5   The number of branches of root locus terminating at infinity is equal to $\#P - \#Z$.

6.3C   Asymptotes to the Loci

To determine where these loci go to infinity, we apply the test point procedure. Let the test point be placed at a great distance from the nearest finite pole or zero. Then the vectors from all poles and zeros to the test point are essentially parallel as shown in Figure 6.3. Clearly, the angles of all vectors are equal in value at any location of the test point so the sum of all angles is just the angle of any vector multiplied by the number of excess poles. For all points on the root locus, it is required that $\angle GH(s) = (2N - 1)\pi$; this condition applies as the loci approach infinity. Since the $\angle GH(s \to \infty)$ is a function of the number of excess poles, we may calculate the angles for each case:

EXAMPLE

$$2 \quad \text{excess poles}$$
$$2.\alpha = +(2N - 1)180°$$
$$\alpha = +90°, -90°.$$

EXAMPLE

$$3 \quad \text{excess poles}$$
$$3.\alpha = +(2N - 1)180°$$
$$\alpha = +60°, -60°, 180°.$$

Table of Results

Excess Poles	Asymptote Angles						
1	180°						
2	+90°	−90°					
3	+60°	−60°	180°				
4	+45°	−45°	+135°	−135°			
5	+36°	−36°	108°	−108°	108°		
6	+30°	−30°	+90°	−90°	+150°	−150°	
7	+25.7°	−25.7°	+77.1°	−77.1°	+128.6°	−128.6°	180°

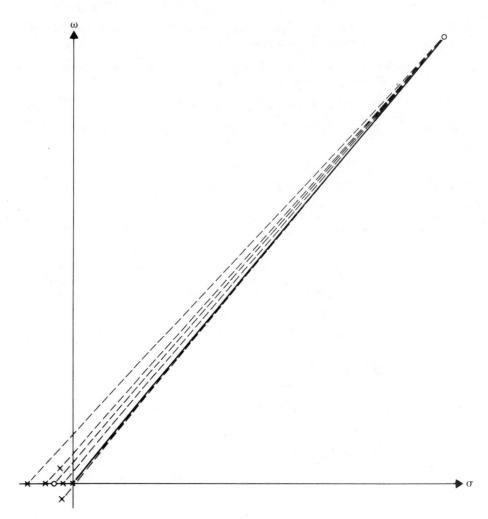

FIGURE 6.3   Application of the Test Point Concept to the Evaluation of Asymptote Angles. For the case shown (test point at a remote but finite distance), the total angle spread is about 6°. Clearly, if the test point is at an infinite distance, the angle spread → 0° and the vectors are parallel.

These are the limiting angles of the vectors from the poles and zeros of the loop transfer function. They are straight lines, and clearly they are the asymptotes to the loci at infinity.

Rule 6   Those branches of the root locus that proceed to infinity arrive there asymptotic to straight lines.

The angles of these asymptotes are found from

$$\text{angles of asymptotes} = \frac{(2N-1)\pi}{\#P - \#Z}.$$

We note that Rule 6 does not give enough information to permit drawing the asymptotes. We need to know the location of one point on each straight line. It is clear that asymptotes are symmetrical with respect to the real axis, and this is a natural result because complex roots must occur in conjugate pairs, so root loci must be symmetrical about the real axis. Therefore, each *pair* of asymptotes intersects the real axis at a common point. For systems with many asymptotes, it is intuitively clear that all of them intersect the real axis at a single point. This may be seen by considering a test point at a very great distance from the cluster of poles and zeros, all of which are a relatively small distance from the origin. If we could look from the test point toward the poles and zeros, the cluster would be so small that it would appear as a single point, and all vectors to the test point, including the asymptote, would appear as a single line. It is the single point from which the asymptote emanates that we wish to evaluate quantitatively so we can locate it and from it draw the asymptotes. There are at least two ways to derive the equation, both of them starting from the concept that there is a single point from which all asymptotes start and that this point, called a *centroid*, is inside of the cluster of poles and zeros at some kind of an average location.

An algebraic derivation may be based on the test point concept—note that, if all of the asymptotes start at one point on the real axis, the asymptotes obtained are exactly the same as would be obtained with a multiple real pole at this point where the multiplicity, of course, is equal to the number of excess poles. Thus, in addition to the actual loop transfer function $GH(s)$, we can write an *equivalent* transfer function, i.e., a transfer function that is equivalent to the actual $GH(s)$ when s is a very large value:

$$G_{eq}(s) = \frac{K}{(s + \sigma_x)^{n-m}}$$

$$= \frac{K}{s^{n-m} + (n-m)\sigma_x s^{n-m-1} + \cdots}, \tag{6.9}$$

where $n = $ number of poles and $m = $ number of zeros. The actual $GH(s)$ would be of the form

$$GH(s) = \frac{K\left(s^m + \sum\limits_{k=1}^{m} Z_k s^{m-1} + \cdots\right)}{s^n + \sum\limits_{j=1}^{n} P_j s^{n-1} + \cdots}. \tag{6.10}$$

In Eq. (6.10), we now divide the numerator polynomial into the denominator to put $GH(s)$ into an algebraic form similar to that of Eq. (6.9):

$$GH(s) = \frac{K}{s^{n-m} + \left(\sum\limits_{j=1}^{n} P_j - \sum\limits_{k=1}^{m} Z_k\right)s^{n-m-1} + \cdots}. \tag{6.11}$$

Since the two transfer functions are identical for large values of s, we may equate coefficients; all we need is

$$(n - m)\sigma_x = \sum_{j=1}^{n} P_j - \sum_{k=1}^{m} Z_k \tag{6.12}$$

from which the value of the asymptote centroid is:

Rule 7   The asymptote centroid is located at

$$\sigma_x = \frac{\sum_{j=1}^{n} P_j - \sum_{k=1}^{m} Z_k}{n - m}$$

$$= \frac{\text{summation of poles} - \text{summation of zeros}}{\text{number of poles} - \text{number of zeros}}.$$

6.3D   Intersections with the Real Axis

We have seen that root loci start on poles, and for many closed-loop systems the poles are on the real axis, but the closed-loop roots are complex. Clearly, some of the loci that start on the real axis must somewhere leave this axis. There are several ways to evaluate these points. We consider one simple procedure, which depends on the rule of Eq. (6.2a):

$$|GH(s)| = 1.00. \tag{6.2a}$$

Before proceeding, consider Figure 6.2 and observe that the two right-most singular points are the poles P_1 and P_2, and that the segment of the real axis between them is a root locus segment. The root loci that leave P_1 and P_2 cannot terminate on this segment and so must leave the real axis at some point. To apply Eq. (6.2a), qualitatively, we simply observe that the *transfer function gain* is zero *at* the poles, but increases as the roots move away from the poles. It is clear that only one value of gain is defined at each point on the locus. Again observing Figure 6.2, as the gain is raised, the roots move from P_1 and P_2 toward each other and at some point they will meet, i.e., both roots will occupy the same point for one specific value of gain. This gain value is the largest associated with any point on the real axis segment. Therefore, any increase in gain causes the roots to leave the real axis and enter the complex plane.

Rule 8   Root loci that emerge from the real axis do so at points such that the root locus gain is maximum for that segment of real axis.

This rule defines the location of the emergence point. There are several ways to calculate the location when solving a specific problem. We start by rewriting the magnitude requirement:

$$\left| \frac{K \Pi(s + Z)}{s^n \Pi(s + P)} \right| = 1.0. \tag{6.13a}$$

From which

$$K = \frac{|s^n||\Pi|s + P|}{\Pi|s + Z|}.$$ **(6.13b)**

We are interested only in real roots, for which $s = \sigma$. So we may write

$$\frac{dK}{d\sigma} = 0 = \frac{d}{d\sigma} \frac{|s^n||\Pi|s + P|}{\Pi|s + Z|}.$$ **(6.14)**

This equation is the direct theoretical approach, but not the simplest computationally. It is clear that upon differentiation Eq. (6.14) provides a polynomial that must be factored to find the desired value of σ; this polynomial is, at best, only one order lower than that of the transfer function—except for simple cases, we need a computer to factor it. It is also clear that the solutions to Eq. (6.14) may give *minima* as well as maxima. When the minima are found, they represent points at which the root locus *enters* the real axis, as will be demonstrated.

A simple numerical calculation procedure enables us to find the desired emergence point and/or the immergence point. It is based on Eq. (6.13b) and the *a priori* knowledge that we are looking for the largest value of K in a designated section of the real axis. We simply choose points on the real axis, substitute the numerical value of σ in Eq. (6.13b), and evaluate K. The calculations involve real numbers only. Repeating several times for selected points is usually sufficient to find the maximum value of K, which is the desired point.

 EXAMPLE 6.1

$$G(s) = \frac{K(s + 5)(s + 10)}{s(s + 1)(s + 20)^2}.$$

The poles and zeros are plotted in Figure 6.4(a) and the root loci on the real axis are shown. It is clear that the loci must emerge from the segment between 0 and 1. It is also clear that root loci must enter the axis between -5 and -10. We expect to find one point of maximum gain between 0 and -1 and one point of *minimum* gain between -5 and -10.

At any point on the root locus,

$$|G(s)| = \left| \frac{K(s + 5)(s + 10)}{s(s + 1)(s + 20)^2} \right| = 1.0,$$

$$K = \frac{|s||s + 1||s + 20|^2}{|s + 5||s + 10|}.$$

Evaluating K for selected points on the real axis, we obtain Table 6.1. From this table, the maximum and minimum values of K, and the value of s at which they occur, are

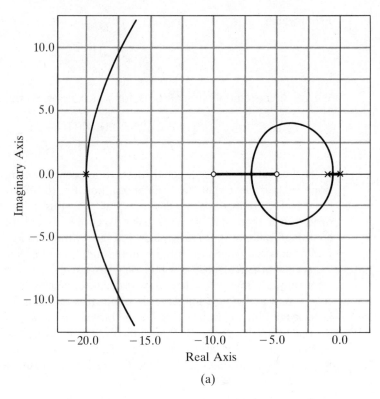

(a)

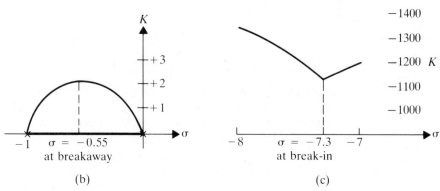

(b)   (c)

FIGURE 6.4   (a) Root Locus for:

$$G(s) = \frac{K(s + 5)(s + 10)}{s(s + 1)(s + 20)^2}$$

and Use of the K Plot: (b) Breakaway Near Origin (See Table 6.1) and (c) Break-in Near $\sigma = -7.3$.

TABLE 6.1   Calculated K

s	$\|s\|$	$\|s + 1\|$	$\|s + 20\|$	$\|s = 20\|^2$	$\|s + 5\|$	$\|s = 10\|$	K
-0.4	0.4	0.6	19.6	384.16	4.6	9.6	2.0878
-0.45	0.45	0.55	19.55	382.2	4.55	9.55	2.177
-0.5	0.5	0.5	19.5	382.25	4.5	9.50	2.224
-0.55	0.55	0.45	19.45	378.3	4.45	9.34	$\rightarrow$2.2265
-0.6	0.6	0.4	19.4	376.36	4.4	9.4	2.184
-7.0	7.0	6.0	13.0	169.0	2.0	3.0	1183.0
-7.3	7.3	6.3	12.7	161.29	2.3	2.7	$\rightarrow$1137.6
-7.5	7.5	6.5	12.5	156.25	2.5	2.5	1218.75
-7.7	7.7	6.7	12.3	151.29	2.7	2.3	1256.85
-8.0	8.0	7.0	12.0	144.0	3.0	2.0	1344.0

readily seen by inspection. These values of K are also plotted as ordinates versus σ as abscissa in Figures 6.4(b) and (c). The maximum and minimum points on the K curves are evident.

6.3E   Intersections with the Imaginary Axis

For most feedback control systems, stability is an important consideration. The points at which the root loci cross the imaginary axis define the stability limits. Several methods are available for locating them:

1. A graphical approach may be used. Select test points on the imaginary axis and evaluate $\angle G(s)$ by summing the angles measured at the poles and zeros. Points at which $\angle G(s) = (2N - 1)\pi$ are, of course, points on the root locus. Care is required for accuracy.

2. The Routh criterion may be used. Basic application of the Routh array determines the gain at the stability limit using the s^1 row of the array. By substituting this value of gain in the auxiliary equation (formed from the s^2 row), the value of ω at the stability limit is evaluated.

3. An alternative to the Routh array is to ask whether any point on the imaginary axis can satisfy (be a root of) the characteristic equation? To answer this question, substitute $j\omega$ for s in the characteristic equation, require that reals and imaginaries go to zero independently, and solve the resulting two equations simultaneously.

■ EXAMPLE 6.2

$$G(s) = \frac{K}{s(s + 1)(s + 2)}\bigg|_{H(s) = 1}.$$

The characteristic equation is

$$s^3 + 3s^2 + 2s + K = 0$$

and the Routh array is

s^3	1	2
s^2	3	K
s^1	$(6 - K)/3$	0
s^0	K	.

From the s^1 row, $6 - K = 0$ and $K = 6$ at the stability limit. From the s^2 row, the auxiliary equation is $3s^2 + K = 0$. Substituting $K = 6$,

$$3s^2 + 6 = 0,$$
$$s = \sqrt{-2} = \mp j\sqrt{2},$$
$$\omega = \sqrt{2}.$$

 EXAMPLE 6.3

A characteristic equation is $s^3 + 3s^2 + 2s + K$. Let $s = j\omega$:

$$(j\omega)^3 + 3(j\omega)^2 + 2j\omega + K = 0$$
$$-j\omega^3 - 3\omega^2 + 2j\omega + K = 0.$$

1. Summation of reals $-3\omega^2 + K = 0$.
2. Summation of imaginaries $-\omega^3 + 2\omega = 0$.

From 2, $\omega = 0$; $\omega = \mp\sqrt{2}$ and from 1, $-3(\sqrt{2})^2 + K = 0$ and $K = 6$.

6.3F   Direction of the Locus at Complex Poles and Zeros

We can determine the direction of the locus at a complex pole or zero by application of the angle summation rule. At any point on the locus,

$$\angle GH(s = s_{\text{test}}) = (2N - 1)\pi - \sum_{i=1}^{m} \angle(s_{\text{test}} + Z_1) + \sum_{j=1}^{n} (s_{\text{test}} + P_j). \qquad (6.15)$$

We can, in general, use $180°$ instead of $(2N - 1)\pi$ and, if the particular point at which we wish to evaluate the angle is, for example, a complex pole P_x, then the angle summations becomes

$$-\angle(s_{\text{test}} + P_x) = 180° - \sum_{i=1}^{m} \angle(s_{\text{test}} + Z_1) + \sum_{\substack{j=1 \\ j \neq x}}^{n} \angle(s_{\text{test}} + P_j). \qquad (6.16)$$

6.4   DRAWING THE LOCI

As discussed in Sec. 6.1, digital computer programs for calculating root loci are readily available and provide more accurate curves than are possible with the graphical procedures. The advantage of the graphical construction rules is that they help us to *sketch* the loci rapidly and with reasonable accuracy, which is sufficient for the analysis of many problems and also is sufficient to guide initial design efforts. The choice between sketches and computer-drawn loci thus depends on the intended use, and perhaps on time, since the sketches can be made rapidly. The only tools needed for the sketches are rectangular coordinate paper, a straight edge, and a protractor or (preferably) a spirule.[c]

To sketch the loci, the following procedure is suggested:

1. Choose your coordinate scale, which must be the same[d] for both abscissa and ordinate so that angles may be measured.

2. Plot the open-loop poles and zeros.

3. Mark the root locus segments on the real axis.

4. Locate the asymptote centroid and draw in the asymptotes.

5. Sketch the complex plane segments of the loci. Note that the real axis emergence points, imaginary axis crossing points, and angles at complex poles and zeros are *not* determined. For most problems they are not needed in the analysis and design. Of course, they can always be evaluated if the analyst wishes to do so.

6. If accurate placement of the locus is needed in selected areas, or if there is doubt about the correctness of the sketch, apply the angle summation rule. This is done by selecting one or more points on the sketched locus and using the spirule to measure and sum the angles at that point. If the sum is not 180°, the test point is moved off of the sketch and the angle summation repeated to find correct points. The sketch is then redrawn through these points.

Consider the examples shown in Figures 6.5(a), (b), and (c). For the two-pole case of Figure 6.5(a), the asymptotes are identical to the loci and the solution is, therefore, exact. For Figure 6.5(b), a pole is added, increasing the asymptotes to 3 and changing their angles. Also, as seen from the plot, the addition of the pole "pushes" the complex part of the locus away from the introduced pole. In sketching the curve, the point of emergence is moved a little to the right [estimated in Figure 6.5(b), not calculated]. The sketched locus was checked at the ω axis crossover by applying the Routh criterion. It is seen that the error is not very large. In Figure 6.5(c), a zero is added to the second-order system of 6.5(a). This decreases the number of asymptotes from 2 to 1. It also "pulls" the

---

[c] Spirules are available from The Spirule Company, 9728 El Venado, Whittier, California.

[d] Computer plots need not observe this rule since angle measurement is not involved.

complex part of the locus toward the introduced zero and also pulls the point of emergence [estimated on Figure 6.5(c)] toward the zero. In general, "like" singular points tend to push or repel, unlike ones tend to pull or attract. Computer plotted root loci for the systems of Figures 6.5(b) and (c) are given on Figures 6.5(d) and (e).

Figure 6.5(f) illustrates the procedure for checking the accuracy of the sketched root locus. For the given transfer function, the poles and zero were plotted, the real axis loci marked, the asymptote centroid located, and the asymptotes drawn. The locus was then sketched as shown. A check point was selected and the vectors from poles and zero drawn in for illustration. The angle of each vector is marked, and when these values are added the total angle at the check point is found to be $\angle G(\text{check point}) = -203.5°$,

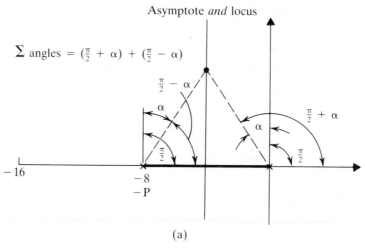

(a)

FIGURE 6.5   (a) Root Locus for $G(s) = K/s(s + 8)$.
There are two asymptotes at $\mp 90°$:

$$\sigma_x = -\frac{0 + 8 - 0}{2} = \frac{-8}{2} = -4.$$

(b) Root Locus for $G(s) = K/s(s + 8)(s + 16)$.
There are three asymptotes at $\mp 60°$, $180°$:

$$\sigma_x = -\frac{0 + 8 + 16 - 0}{3} = -\frac{24}{3} = -8.$$

(c) Root Locus for $G(s) = K(s + 12)/s(s + 8)$. (d) Computer-Drawn Root Locus for (b).
(e) Computer-Drawn Root Locus for (c). (f) Checking by Angle Summation.
Root locus for

$$G(s) = \frac{K(s + 10)}{s(s + 8)(s + 16)(s + 80)}.$$

There are three asymptotes at $\mp 60°$, $180°$:

$$\sigma_x = -\frac{0 + 8 + 16 + 80 - 10}{4 - 1} = \frac{-94}{3} = -31.333.$$

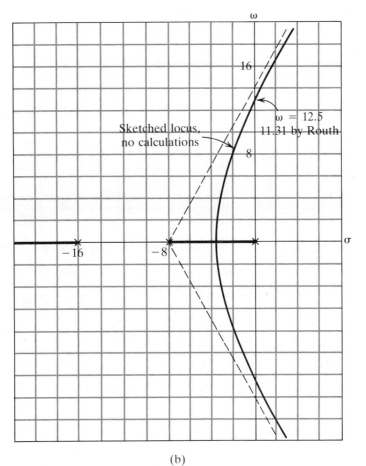

(b)

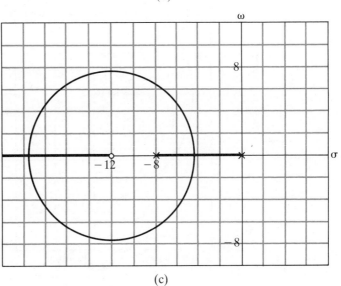

(c)

FIGURE 6.5 (*Continued*)

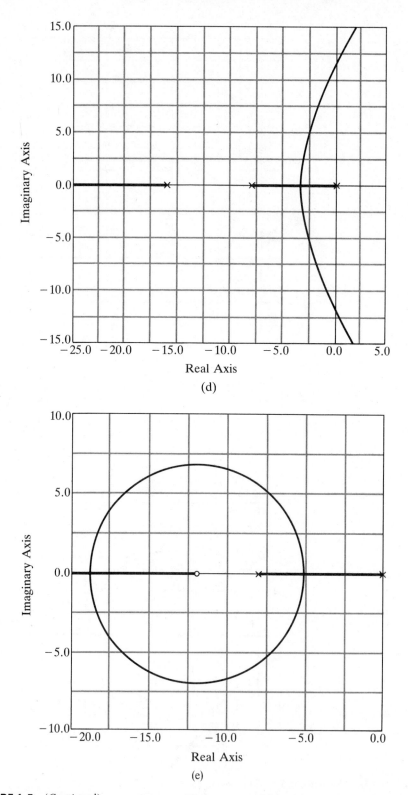

(d)

(e)

FIGURE 6.5 (*Continued*)

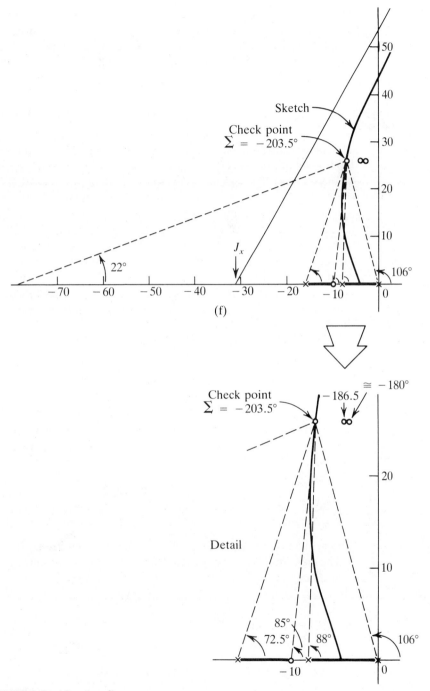

FIGURE 6.5   (*Continued*)

which is too large. To locate the correct point, we search along a *horizontal*[e] line and select a new test point to the right of the locus (since there are more poles than zeros, the net angle *decreases* as the test point moves to the right). From the figure the angle at the second point is $-186.5°$, and at the third point it is nearly $-180°$.

6.5   LOCATING THE ROOTS ON THE LOCI

The curves called the *root loci* are completely defined by the angle rule of Eq. (6.2b), and thus the set of all points which *might* be roots of the characteristic equation are defined by these curves. The specific points in the set that are roots are those which also satisfy the magnitude requirement of Eq. (6.2a). The number of roots is equal to the order of the characteristic equation, which is normally equal to the number of poles of $G(s)$; note that the number of branches of the root locus is equal to the number of poles and there will be one root on each branch of the locus. One way to find the roots is to find the set of points that satisfies the magnitude requirement of Eq. (6.2a) and plot the corresponding constant magnitude contours. The intersections of the root loci with the magnitude contours are points which satisfy both Eqs. (6.2a) and (6.2b), and thus are roots of the characteristic equation. While this can be done, it is quite laborious and simpler methods are available.

In general, the loop transfer function used to plot the root loci is of the form

$$G(s) = \frac{K(s + Z_1)(s + Z_2)\cdots}{s^N(s + P_1)(s + P_z)\cdots}. \qquad \textbf{(6.17a)}$$

The zeros and poles of Eq. (6.17a) determine the root loci. The value of K is not needed to draw the curves. Clearly, however, the value of K determines where the root points are located on the loci. Two types of problems may be encountered in control system analysis:

1. For a specified point on the root locus to become a root what value of K is needed?

2. A value of K is specified for the system. Where are the roots?

Both problems are solved by application of the magnitude rule of Eq. (6.2a).

The magnitude rule requires that in order for a point on the root locus to be a root it must also satisfy

$$|GH(s)| = 1.00. \qquad \textbf{(6.2a)}$$

Write Eq. (6.17a) in this form:

$$\left| K \frac{(s + Z_1)(s + Z_2)\cdots}{s^N(s + P_1)(s + P_2)\cdots} \right| = 1, \qquad \textbf{(6.17b)}$$

---

[e] A horizontal line is usually convenient, but not necessary.

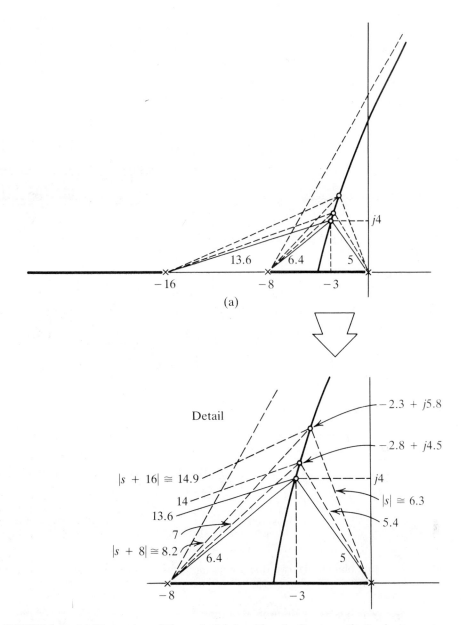

FIGURE 6.6   (a) Illustration of Example 6.4, Locating the Roots on the Loci:

$$G(s) = \frac{K}{s(s + 8)(s + 16)}$$

1.   What is K for a real root at $s = -20$?

$$K = |s||s + 8||s + 16| = |-20||-20 + 8||-20 + 16| = 960.$$

2.   What is K for roots at $s = -2.3 \pm j5.8$? From plot:

$$K \simeq (6.3)(8.2)(14.9) \simeq 770.$$

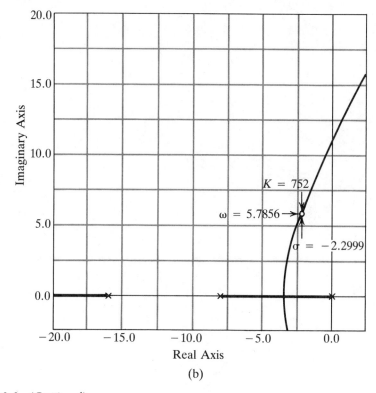

(b)

FIGURE 6.6   (*Continued*)

3. Find root locations if $K = 500$:

    a. Since $500 < 770$, roots are closer to the poles than for 2. Try $s = -3 + j4$.

    b. At $s = -3 + j4$, $K = (5)(6.4)(13.6) = 435$. Therefore, the roots are between $-2.3 + j5.8$ and $-3 + j4$.

    c. At $s = -2.8 \mp j4.5$, $K = (5.4)(7)(14) = 529.2$. Conclusion: Roots are at *approximately* $s = -2.9 \mp j4.2$. If a more accurate answer is desired (without computer assistance), substitute the coordinates of the trial point into the equation for K and carry out the complex arithmetic.

(b) Computer-Drawn Root Locus for (a), where K, σ and ω are Indicated at One Point.

which is easily manipulated to:

$$K = \frac{|s^N||s + P_1||s + P_2|\cdots}{|s + Z_1||s + Z_2|\cdots}. \qquad (6.17c)$$

For the first problem mentioned above, knowing the desired value of s for the root we use it to evaluate each of the terms on the right side of Eq. (6.17c) and thus obtain the value of K. For the second problem, the procedure is exactly the same but several trials are involved. We do not know the value of s, so we guess, picking a point on the curve

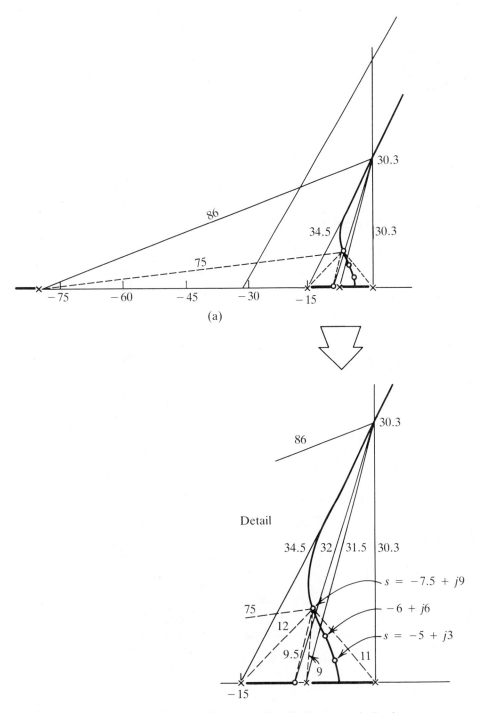

FIGURE 6.7 (a) Illustration for Example 6.5, Locating the Roots on the Loci:

$$G(s) = \frac{K(s + 10)}{s(s + 8)(s + 16)(s + 80)}.$$

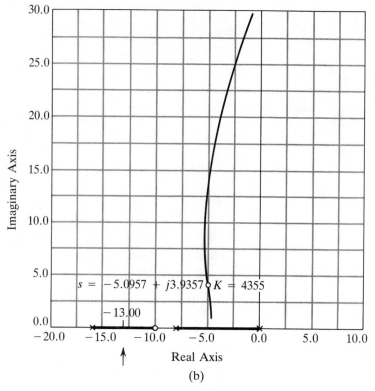

$s = -5.0957 + j3.9357$ $K = 4355$

-13.00

FIGURE 6.7 *(Continued)*

The characteristic equation of the closed loop is:

$$s^4 + 10^4 s^3 + 2048 s^2 + (10240 + K)s + 10K = 0.$$

At the stability limit, $\omega = 30.3$.

1. What is K for a real root at $s = -13$?

$$K = \frac{(13)(5)(3)(67)}{3} = 4355.$$

2. If $K = 4355$, where are the complex roots?

    a. Try $s = 0 + j30.3$:

$$K = \frac{(30.3)(31.5)(34.5)(86)}{32} = 88495.$$

    b. Try $s = -7.5 + j9$:

$$K = \frac{(11)(9)(12)(75)}{9.5} = 9379.$$

    c. Try $s = -4.5 + j0$:

$$K = \frac{(4.5)(3.5)(11.5)(75.5)}{5.5} = 2486.4.$$

FIGURE 6.7   *(Continued)*

___

d.   Try $s = -5.0 \mp j3$:

$$K = \frac{(5.5)(4)(11.5)(75)}{5.83} = 3255.$$

e.   Try $s = -6 + j6$:

$$K = \frac{(8.49)(5.29)(11.66)(74.24)}{7.21} = 5392.$$

Note that convergence is slow. Next trial might be at $s = -5.5 + j4.5$.

(b) Computer-Drawn Root Locus for (b), where Root Locations for $K = 4355$ are Shown.

___

and evaluating[f] K for that point. If the value of K thus found is smaller than the specified value, we select a second point on the locus that is farther from the pole on which that locus segment started. If the second point chosen does not give the required K, we proceed by choosing a third point, interpolating or extrapolating as needed. The process converges rapidly. When the root loci are calculated by computer, then, of course, we need only inspect the tabulated values for K and for s. Examples 6.4 and 6.5 are shown in Figures 6.6(a) and 6.7(a), with computer solutions in Figures 6.6(b) and 6.7(b).

6.6   ROOT LOCI FOR A PARAMETER OTHER THAN K: PARTITIONING

Consider Eqs. (6.3) and (6.4). Equation (6.3) gives the basic form of the loop transfer function and Eq. (6.4) shows the characteristic polynomial obtained from Eq. (6.3). Either equation permits calculation of the root loci—if Eq. (6.3) is used, the graphical constructions apply; if Eq. (6.4) is used, the digital computer is more convenient. In either case, all parameter values are known and fixed except one, K. If the digital computer is used, a starting value of K is chosen and then incremented as desired. For each value of K, the computer finds the roots. When these root points are plotted on the s plane, the curves obtained are the root loci *for the parameter K*, and they are precisely the same as the ones obtained graphically from Eq. (6.3). It is clear, however, that the digital computer requires that *all* parameters be specified numerically in order for it to find the roots. It neither knows nor cares which parameter has been changed in giving it data. Thus, we can increment any parameter we choose and the computer will find the roots and provide root loci for the chosen parameter.

___

[f] Evaluation may be numerical or graphical. We may substitute the coordinates of the s point in the right side of Eq. (6.17c) and carry out the complex arithmetic, or we may note that each term represents the length of a vector which may be measured on the plot, evaluated to scale, and the resulting real numbers substituted in Eq. (6.17c).

The same result can be obtained graphically. We need only choose the parameter for which we want root loci, then manipulate the algebra so that the value of that parameter is not needed in the graphical procedures, i.e., we must rearrange the transfer functions so that the chosen parameter becomes the multiplying factor. A simple algebraic procedure to do this is called *partitioning*. The steps for partitioning for a chosen parameter are as follows:

1. Obtain the characteristic polynomial.

2. Partition (separate) the terms of the polynomial into two groups:

    a. Group 1 is to consist of all terms that do *not* contain the designated parameter.

    b. Group 2 is to consist of all terms that do contain the parameter as a common factor.

    c. Divide both sides of the equation by group 1. The equation is now in the form

    $$1 + \frac{\text{group 2}}{\text{group 1}} = 0$$

    and this defines a new transfer function form $G(s)$ for which the desired parameter is (algebraically speaking) the gain.

    d. Factor the polynomials of groups 2 and 1 to find the zeros and poles needed for the graphical procedures.

    e. Construct the root loci.

For example, if a single-loop servo has $G(s) = 100/s(s + P)$ and we want to find the root loci for P, the characteristic equation is

$$s^2 + Ps + 100 = 0.$$

Partitioning,

$$(s^2 + 100) + Ps = 0.$$

Dividing,

$$1 + \frac{Ps}{s^2 + 100} = 0.$$

Factoring the denominator,

$$1 + \frac{Ps}{(s + j10)(s - j10)} = 0.$$

Plot the poles and zero and construct the root locus as shown in Figure 6.8.

 **EXAMPLE 6.4**

A commonly encountered case where partitioning is useful is in the adjustment of velocity feedback. Given the system of Figure 6.9(a), find the root loci for k. The

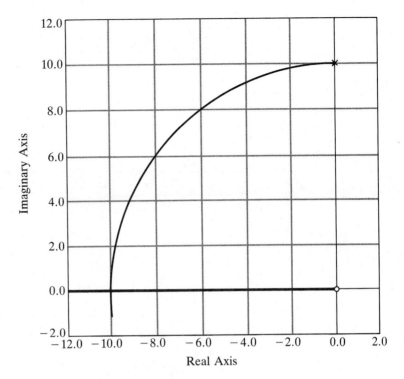

FIGURE 6.8   Computer-Drawn Root Locus for Partitioned Equation:

$$G(s) = \frac{Ps}{(s + j10)(s - j10)}.$$

characteristic equation is

$$s^3 + 15s^2 + (50 + 750k)s + 750 = 0.$$

Partitioning,

$$s^3 + 15s^2 + 50s + 750 + 750ks = 0.$$

Dividing,

$$1 + \frac{750ks}{s^3 + 15s^2 + 50s + 750} = 0.$$

Factoring the denominator,

$$1 + \frac{750ks}{(s + 15)(s + j\sqrt{50})(s - j\sqrt{50})} = 0$$

The root loci are shown in Figure 6.9(b).

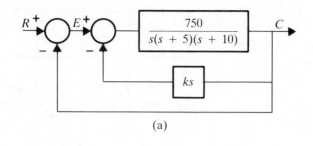

(a)

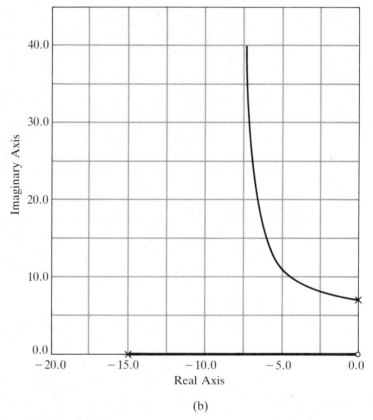

(b)

FIGURE 6.9 Example 6.4: (a) Block Diagram of a Control System and (b) Computer-Drawn Root Loci for the System of (a).

<hr />

6.7   TWO PARAMETER STUDIES: FAMILIES OF ROOT LOCI

The root locus method is basically a one parameter method that can be used to study the effect (on root locations) of *any* chosen parameter. It is then easy to see that we can use the method to study the effects of two parameters by simply doing the calculations

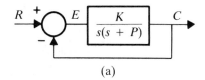

(a)

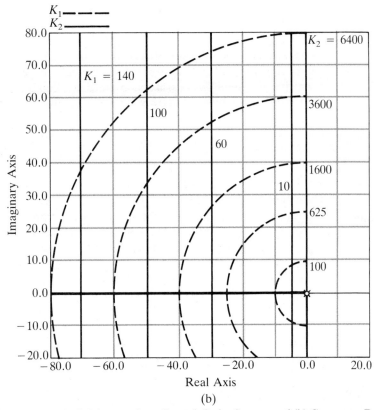

(b)

FIGURE 6.10   (a) Block Diagram for a Second-Order System and (b) Computer-Drawn, Two Parameter Root Loci for the System of (a) where $K1 = P$, and $K2 = K$.

---

sequentially. Consider the second-order system of Figure 6.8, but change the forward gain from 100 to a variable K. The block diagram is then as shown in Figure 6.10(a). There are two variable parameters K and P, and the characteristic equation is

$$s^2 + Ps + K = 0.$$

Choose one of the variables for the first study, set the other to zero, and partition. In this case, let K be the variable and $P = 0$:

$$s^2 + K = 0$$

$$1 + \frac{K}{s^2} = 0,$$

and the root locus for K is on the imaginary axis as shown, with values of K marked at selected root points. Next, let P be the variable and choose any desired value for K. Let $K = 25$ and partition for P:

$$s^2 + Ps + 25 = 0$$

$$s^2 + 25 + Ps = 0$$

$$1 + \frac{Ps}{(s + j5)(s - j5)} = 0.$$

Note that this is the solution already found in Figure 6.8. The root loci are shown in Figure 6.10(b) and form a family of curves that defines the values of the two parameters for any linear system. In general, a computer solution seems preferable because of the amount of labor required for graphical construction.

6.8   SUMMARY

Root loci are curves on the s plane along which the roots of a polynomial move as a parameter (such as a gain) is varied from 0 to ∞. There is one locus for each root. The loci may be calculated by computer or constructed graphically. When constructed for a control system, the open-loop transfer function is used to obtain the loci of the closed-loop roots. Each locus starts on a pole and terminates on either a finite zero or at infinity.

Rules for constructing the loci are derived and their use demonstrated. Loci can be determined for any parameter, and families of loci for two parameters can also be obtained.

The loci may be used to analyze stability and transient performance. They show the effects of parameter adjustment and thus aid design.

REFERENCES

1.   Evans, W. R. "Control System Synthesis by Root Locus Method." *AIEE Trans.* (1950).

BIBLIOGRAPHY

Evans, W. R. "Graphical Analysis of Control Systems." *AIEE Trans.*, 67 (1948).

———. *Control System Dynamics.* New York: McGraw-Hill Book Co. (1954).

Thaler, G. J.; Elliott, D. W.; and Heseltine, J. C. W. "Feedback Compensation using Derivative Signals." *AIEE Trans. Part I, Routh's Criteria; Root Loci, AIEE Trans. Appl. Ind.* 82. Part II (1963).

———; and Han, K. W. "High Order System Analysis and Design Using Root Locus Method." *J. Franklin Institute* 281 (1966).

———; Ross, E. R.; and Warren, T. C. "Design of Servo-Compensation Based on the Root Locus Method." *AIEE Trans. Appl. Ind.* 79. Part II (1960).

Wojcik, C. "Analytic Representation of Root Locus." *Trans. ASME, J. Basic Eng. Series D* 86 (March 1964).

PROBLEMS

6.1 For systems with $G(s)$ defined as shown in the block diagram, draw the root loci for each of the following $G(s)$ functions. Calculate and record values for the asymptote centroid, number of asymptotes, and value of ω at which the locus crosses the imaginary axis. Also draw the asymptotes on the plots.

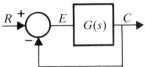

a. $\dfrac{K}{s(s+5)^2}$

b. $\dfrac{K(s+10)}{s(s+5)}$

c. $\dfrac{K}{s(s+1)(s+2)}$

d. $\dfrac{K(s+1)}{s(s+2)(s+3)(s+4)}$

e. $\dfrac{K(s+1)^2}{s^3}$

f. $\dfrac{K(s+3)}{(s+1)(s+2)}$

g. $\dfrac{K(s+5)}{s^2(s+3)}$

h. $\dfrac{K(s+2)}{s(s+1)(s+7)}$

i. $\dfrac{K(s+7)}{s(s+1)(s+2)}$

j. $\dfrac{K}{s(s+3+j2)(s+3-j2)}$

k. $\dfrac{K(s+2)}{s(s+1)(s+3+j10)(s+3-j10)}$

l. $\dfrac{K}{s(s^2+2s+25)}$

m. $\dfrac{K(s^2+2s+25)}{s(s+1)(s^2+4s+25)}$

n. $\dfrac{K(s+5)}{s(s+10)(s^2+2s+25)}$

o. $\dfrac{K(s+15)}{(s+5)(s+2+j6)(s+2-j6)}$

p. $\dfrac{K(s+9)(s+11)}{s(s+1)(s+2)(s+25)}$

6.2 Draw the root loci and locate one complex root using the magnitude rule.

$$R \xrightarrow{+} \overset{E}{\bigcirc} \boxed{\dfrac{500(s+10)}{(s+1)(s+5)(s+100)}} \xrightarrow{C}$$

6.3 It is required that roots be located such that $\zeta = 0.707$.
  a. What value of K is needed?
  b. What are the coordinates of the root point?

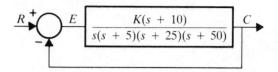

6.4 For the following system, *sketch the root loci* as a function of K.

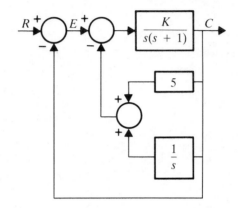

6.5 Using root locus, show how to estimate the locations of the roots of $s^4 + 100s^2 + 10\,000 = 0$.

6.6 For

$$G(s) = \frac{K}{(s + 1)(s + 1 - j1)(s + 1 + j1)},$$

where does the root locus cross the imaginary axis? Give the value of ω.

6.7 For

$$G(s) = \frac{K(s + 1)(s + 5)}{s^2(s + 2)},$$

at what points does the root locus leave the real axis? Where does it reenter the real axis? Give values of σ.

6.8 For

$$G(s) = \frac{K(s + 2)}{s(s + 1)(s + 5)(s + 8)},$$

draw the root loci. At some point the imaginary coordinate is $\omega = 1.5$; what is the σ coordinate?

6.9 Given:

$$G(s) = \frac{K}{s(s+1)(s+2)},$$

find the value of K required to place a real root at $\sigma = -5$. For this value of K, where are the complex roots?

6.10 Given:

$$G(s) = \frac{K(s+1)}{s(s+2)(s+3)(s+4)}.$$

If $K = 10$, where are the roots?

6.11 If

$$G(s) = \frac{9(s+5)}{s(s+1)},$$

find the locations of the roots on the root locus.

6.12 Given:

$$G(s) = \frac{K(s^2 + 0.5s + 100)}{s(s+1)(s+2)}.$$

a. Sketch the root loci.
b. If the loci cross the imaginary axis, determine values of ω at the axis intersections.
c. If we want complex roots with $\zeta = 0.707$, what must be the value of K?

6.13 If

$$G(s) = \frac{K(s+2)}{(s+1)(s+3)^3},$$

at what value of ω does the root locus cross the imaginary axis?

6.14 If

$$G(s) = \frac{K(s+1.2)}{s(s+1)(s+2)},$$

sketch the root loci. Also draw the K versus σ curves and determine the points at which the loci leave the real axis and reenter it.

6.15 For the system with velocity feedback below, determine the value of k required to give complex roots with $\zeta = 0.707$. Give the coordinates (σ, ω) of the root point.

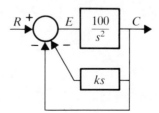

6.16   For each of the following, draw root loci as functions of k.

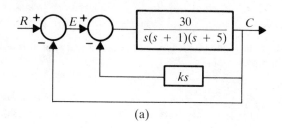

(a)

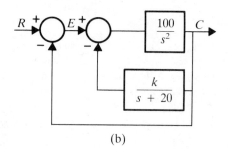

(b)

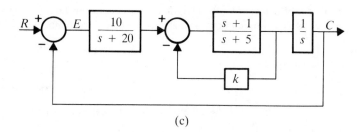

(c)

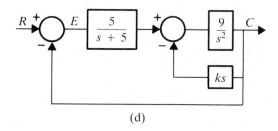

(d)

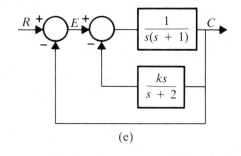

(e)

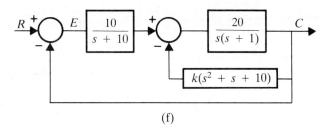

(f)

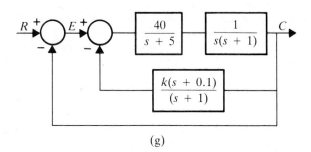

(g)

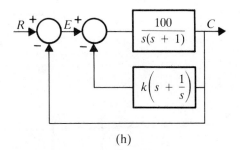

(h)

6.17   Sketch the root loci as functions of the time constant.

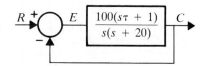

6.18   Analyze the effect of varying the parameter P on the stability and transient response of the system.

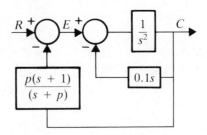

6.19   Given:

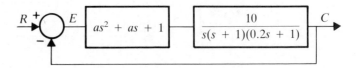

a.   Is it possible to adjust a to obtain complex roots with $\zeta = 0.5$?
b.   If so, what is the value of ω for these roots?
c.   What is the effect of adjusting a on the steady-state accuracy of the system?

6.20   For the given system, draw root loci as a function of the parameter Z.

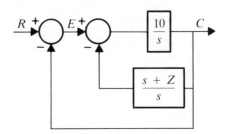

6.21   Again, draw root loci as a function of Z.

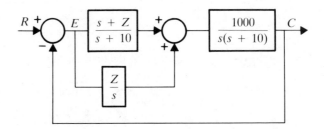

6.22 Determine the root locus as a function of A.

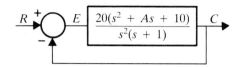

6.23 Determine the effect of varying k on the stability and transient response of the system.

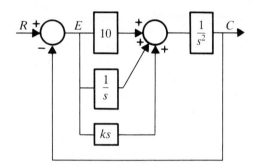

6.24 Determine the effect of varying k on the transient response and steady-state accuracy of the system.

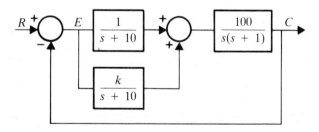

6.25 Show the effect of varying the parameter P on the locations of the roots of the system.

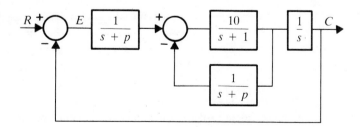

6.26   Given:

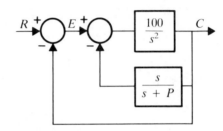

$$\frac{K(s + 100)}{(s + 5)^2(s + 20)^2}$$

a. Locate the asymptote centroid. Draw the asymptotes and sketch the root loci.
b. At what value of ω does the root locus cross the imaginary axis?
c. At what value of σ does the root locus *reenter* the real axis?

6.27   Given:

a. Draw the root loci as a function of P.
b. What is the maximum value of ζ obtainable by adjusting P?
c. What value of P provides the maximum ζ?

6.28   Given:

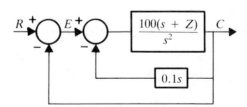

a. Draw the root loci.
b. It is desired to have complex roots with $\zeta = 0.3$. What value of k is needed?
c. What are the coordinates of the roots?

6.29   Given:
a. Partition the characteristic equation and sketch the root loci as a function of Z.
b. Evaluate the breakaway point on the real axis.
c. Complex roots are desired, and there is a minimum possible value of ζ. Find (evaluate) ζ_{min}.
d. What value of Z provides this ζ_{min}?

7

State Variable Analysis

INTRODUCTION

The *state* of a system may be defined as the minimum amount of information required such that (given the input to the system) the response of the system may be completely determined for all future time. For dynamic systems (including control systems), the response of the system is defined by the differential equations that model the system, the initial conditions, and the forcing function. Consider the following differential equation:

$$y^n + A_{n-1}y^{n-1} + A_{n-2}y^{n-2} + \cdots + A_2\ddot{y} + A_1\dot{y} + A_0y = R(t). \qquad \textbf{(7.1a)}$$

Let us rewrite this in the form

$$y^n = R(t) - A_0y - A_1\dot{y} - A_2\ddot{y} - \cdots - A_{n-2}y^{n-2} + A_{n-1}y^{n-1}. \qquad \textbf{(7.1b)}$$

It is clear that if we know the forcing function $R(t)$ and also know the value of y and $(n-1)$ derivatives at some instant $t = t_0$, we can calculate the value of the n'th derivative y^n at $t = t_0$, and can proceed to calculate values for all of the terms for all future time. However, if the initial value of y or any of the $(n-1)$ derivatives is not known, we do not have enough information. The differential equation is of n'th order; n derivatives are defined, and n initial conditions must be known. We may choose to call each of the variables, y and each of the first $(n-1)$ derivatives, a *state variable*. The number of state variables or *states* is then equal to the order of the differential equation, which is normally equal to the number of energy storage elements in the system. To provide a systematic mathematical approach to analysis of the characteristics of the system, it is convenient to describe the system by a set of simultaneous first-order differential equations with each equation defining one state. This set of equations is called the *state equations*. Equation (7.1a) can be put in this form (shown later) using y and the $(n-1)$ derivatives as state variables. These particular variables—y and the derivatives—are also called *phase variables*. Note that they are not the only state variables possible; any system may be described by many possible sets of state variables, which may be chosen as a matter of convenience or at the preference of the analyst.

7.2   DEFINING STATE VARIABLES

7.2A   Finding State Equations from n'th-Order Differential Equations

The choice of states for a given problem depends, in part, on the form in which the basic system information is given, in part on the type of data desired from the analysis, and in part on convenience. Let us consider several cases. If the information available is given in the form of Eq. (7.1a), i.e., an n'th-order differential equation, the choice of y and the derivatives as states is convenient. The $(n - 1)$ state equations are simply definitions. Using X_1, X_2, and X_3 as symbols for the state variables, we define

$$X_1 = y$$
$$\dot{X}_1 = X_2 = \dot{y}$$
$$\dot{X}_2 = X_3 = \ddot{y}$$
$$\vdots$$
$$\dot{X}_{n-2} = X_{n-1} = y^{n-2}$$
$$\dot{X}_{n-1} = X_n = y^{n-1}.$$

(7.2a)

The n'th state equation is obtained from Eq. (7.1b) by substituting the definitions into it:

$$\dot{X}_n = R(t) - A_0 X_1 - A_1 X_2 - A_2 X_3 - \cdots - A_{n-2} X_{n-1} - A_{n-1} X_n.$$

(7.2b)

For convenience in notation, as well as for computational purposes, the state equations are usually summarized in vector matrix form:

$$\frac{d}{dt}\begin{bmatrix} X_1 \\ X_2 \\ X_3 \\ \vdots \\ X_{n-1} \\ X_n \end{bmatrix} = \begin{bmatrix} 0 & 1 & 0 & 0 & 0 & \cdots & 0 \\ 0 & 0 & 1 & 0 & 0 & \cdots & 0 \\ 0 & 0 & 0 & 1 & 0 & \cdots & 0 \\ \vdots & & & & & & \\ 0 & 0 & \cdots & & & & 1 \\ -A_0 & -A_1 & -A_2 & \cdots & & & -A_{n-1} \end{bmatrix}$$
$$\times \begin{bmatrix} X_1 \\ X_2 \\ X_3 \\ \vdots \\ X_{n-1} \\ X_n \end{bmatrix} + \begin{bmatrix} 0 \\ 0 \\ 0 \\ \vdots \\ 0 \\ 1 \end{bmatrix} R(t).$$

(7.3)

To avoid writing out the matrices at every step in the analysis, shorthand notation for Eq. (7.3) is

$$\dot{X} = \mathbf{A}X + Bu, \tag{7.4}$$

where

X = state vector
$\mathbf{A}$ = matrix of coefficients
Bu = forcing function vector.

In analyzing physical systems, variables other than derivatives of a chosen state may be of interest and importance. If the state variable method is to be used, it is often important that the physical variables of interest be chosen as states.

 EXAMPLE 7.1

Consider the electric motor with inertia and friction load, as shown in Figure 7.1(a). The electric circuit equation, obtained by application of Kirchoff's voltage law, is

$$V - IR - L\dot{I} - K_B\dot{\theta} = 0. \tag{7.5}$$

The mechanical equilibrium equation, by Newton's law is

$$J\ddot{\theta} + f\dot{\theta} = K_T I, \tag{7.6}$$

where $J = J_M + J_L$
$f = f_M + f_L$
J_M, J_L = motor and load inertias
f_M, f_L = motor and load viscous friction coefficients
R, L = motor armature resistance and inductance
K_T = motor torque constant
K_B = motor back emf (generator) constant
θ = shaft angular position
I = armature current
V = armature applied voltage.

Three state variables are required because the equations are third order. Readily measured physical quantities are:

$$\text{shaft position}—\theta = X_1$$
$$\text{shaft velocity } \dot{\theta} = \dot{X}_1 = X_2 \tag{7.7}$$
$$\text{armature current}—I = X_3.$$

Substituting in Eq. (7.5), for I, $\dot{I}$, and $\dot{\theta}$,

$$V - RX_3 - L\dot{X}_3 - K_B X_2 = 0. \tag{7.8a}$$

Substituting in Eq. (7.6) for $\ddot{\theta}$, $\dot{\theta}$, and I,

$$J\dot{X}_2 + fX_2 = K_T X_3. \tag{7.8b}$$

These two equations are rearranged, and the definition of Eq. (7.7) i.e., $\dot{X}_1 \triangleq X_2$, is used

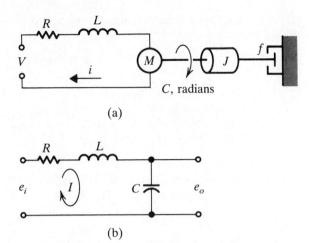

(a)

(b)

FIGURE 7.1 (a) An Armature-Controlled dc Motor and (b) an R-L-C Filter.

as the third state equation giving

$$\dot{X}_1 = X_2$$

$$\dot{X}_2 = -\frac{f}{J}X_2 + \frac{K_T}{J}X_3 \tag{7.8c}$$

$$\dot{X}_3 = -\frac{K_B}{L}X_2 - \frac{R}{L}X_3 + \frac{V}{L}$$

$$\begin{bmatrix} \dot{X}_1 \\ \dot{X}_2 \\ \dot{X}_3 \end{bmatrix} = \begin{bmatrix} 0 & 1 & 0 \\ 0 & -f/J & K_T/J \\ 0 & -K_B/L & -R/L \end{bmatrix} \begin{bmatrix} X_1 \\ X_2 \\ X_3 \end{bmatrix} + \begin{bmatrix} 0 \\ 0 \\ 1/L \end{bmatrix} V. \tag{7.9}$$

■ EXAMPLE 7.2

To obtain the state equations from differential equations, first consider the circuit of Figure 7.1(b). From Kirchoff's laws we obtain

$$L\frac{di_L}{dt} + i_L R + v_c = 0$$

$$v_c = \frac{1}{C}\int i_L \, dt.$$

Differentiating the second equation and rearranging

$$\frac{di_L}{dt} = -i_L\frac{R}{L} - \frac{v_c}{L}$$

$$\frac{dv_c}{dt} = i_L\frac{1}{C}.$$

Choosing i_L and v_c as the state variables X_1 and X_2, we obtain

$$\dot{X}_1 = -X_1 \frac{R}{L} - \frac{X_2}{L}$$

$$\dot{X}_2 = X_1 \frac{1}{C}.$$

In matrix form,

$$\begin{bmatrix} \dot{X}_1 \\ \dot{X}_2 \end{bmatrix} = \begin{bmatrix} -(R/L) & -(1/L) \\ 1/C & 0 \end{bmatrix} \begin{bmatrix} X_1 \\ X_2 \end{bmatrix}.$$

7.2B Finding State Equations from Transfer Functions

Frequently the mathematical description of a device or a system is known in transfer function form. Customarily the numerator and denominator of the transfer function are known in factored form. If not, they must be factored when state equations using physical variables are to be used. Assume that the transfer function of the device is

$$\frac{C}{E} = \frac{K(s + Z)}{s(s + P_1)(s + P_2)(s^2 + \alpha s + \beta)}. \tag{7.10}$$

This may be rearranged as a product of ratios:

$$\frac{C}{E} = \frac{1}{s} \cdot \frac{K}{s + P_1} \cdot \frac{1}{s + P_2} \cdot \frac{s + Z}{s^2 + \alpha s + \beta}. \tag{7.11}$$

In Eq. (7.11) the choice of the ratios was arbitrary. As far as the mathematical manipulations are concerned, this arbitrary selection is acceptable; but if it is desirable to choose physical variables as state variables, then the choice of the ratios should be guided (if possible) by grouping the poles and zeros so that those associated with a specific physical device are allocated to the same ratio. It is also desirable to arrange the ratios in the sequence in which the signal progresses through the system. A convenient way to do this is to use block diagrams as shown in Figure 7.2. Only two blocks are shown in Figure 7.2(a), designating the components involved. Each block contains the transfer function of the component and it is seen that the motor block transfer function has three poles while the amplifier function has two. Since state equations are first-order equations, there must be a state variable for each pole and it is convenient (where possible) to subdivide the diagram into a set of cascaded blocks, each block containing a transfer function factor as shown in Figure 7.2(b). When it is not convenient to factor a transfer function into factors containing a single pole—as in the case of the amplifier in Figure 7.2(b) where the denominator quadratic is assumed to have complex roots—state equations are readily defined as will be shown.

A basic technique that is helpful in choosing state variables from transfer functions is to develop block diagrams (or signal flow graphs) in terms of *integrators*. The

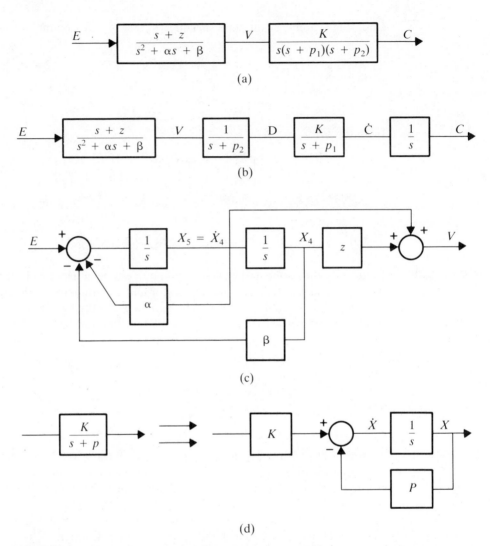

FIGURE 7.2   Block Diagram Formulation for Deriving State Equations: (a) Block Diagram Designating Components (b) Rearrangement of Block Diagram to Simplify Choice of State Variables (c) Expanded Amplifier Block and (d) Diagramming a Single Pole Transfer Function in Integrator Form.

states are then the *outputs* of the integrators. For example, the single-pole transfer function block of Figure 7.2(b) may be diagrammed as shown in Figure 7.2(d). A quadratic factor for complex poles may be diagrammed as shown in Figure 7.2(c).

The important point here is that in the process of setting up the block diagram, as in Figure 7.2(b), the sequence of blocks can be chosen so that many of the symbols at the terminals of the blocks represent physically measurable signals. Then, by choosing these quantities as state variables, the resulting mathematical model (and any simulation model) provides results that are checked easily by direct measurement of the physical variable in the system itself.

 EXAMPLE 7.3

For the system of Figure 7.2 (since the output device is a motor), it is apparent that

$$C = \text{shaft angular position} \triangleq X_1$$
$$\dot{C} = \text{shaft angular velocity} \triangleq X_2$$
$$D \triangleq X_3 \text{ is not measurable and has no simple physical significance}$$
$$V = \text{voltage applied to motor, but is not a state.}$$

Two more states may be defined by expanding the amplifier block as shown in Figure 7.2(c):

$$\frac{X_1(s)}{X_2(s)} = \frac{1}{s}; \qquad\qquad sX_1(s) = X_2(s), \qquad\qquad \dot{X}_1(t) = X_2(t)$$

$$\frac{X_2(s)}{X_3(s)} = \frac{K}{s + P_1}; \qquad (s + P_1)X_2(s) = KX_3(s), \qquad \dot{X}_2(t) = -P_1 X_2(t) + KX_3(t)$$

$$\frac{X_3(s)}{V(s)} = \frac{1}{s + P_2}; \qquad (s + P_2)X_3(s) = V(s), \qquad \dot{X}_3(t) = -P_2 X_3(t) + V(t)$$

$$V(s) = ZX_4 + X_5 \qquad\qquad\qquad\qquad\qquad\qquad\qquad\qquad \textbf{(7.12)}$$

Thus,

$$\dot{X}_3(t) = -P_2 X_3(t) + ZX_4(t) + X_5(t)$$
$$\dot{X}_4(t) = X_5(t)$$
$$\dot{X}_5(t) = E(t) - \beta X_4(t) - \alpha X_5(t).$$

In vector matrix form, Eq. (7.12) becomes:

$$\frac{d}{dt}\begin{bmatrix} X_1 \\ X_2 \\ X_3 \\ X_4 \\ X_5 \end{bmatrix} = \begin{bmatrix} 0 & 1 & 0 & 0 & 0 \\ 0 & -P_1 & K & 0 & 0 \\ 0 & 0 & -P_2 & Z & 1 \\ 0 & 0 & 0 & 0 & 1 \\ 0 & 0 & 0 & -B & -\alpha \end{bmatrix}\begin{bmatrix} X_1 \\ X_2 \\ X_3 \\ X_4 \\ X_5 \end{bmatrix} + \begin{bmatrix} 0 \\ 0 \\ 0 \\ 0 \\ 1 \end{bmatrix} E. \qquad \textbf{(7.13)}$$

In general, when the transfer function numerator is a polynomial, a suitable procedure for obtaining state equations is as follows:

1. Define a new transfer function by replacing the numerator with 1.0 and draw the block diagram (or signal flow diagram).

2. Construct the diagram for the original transfer function by introducing feedforward paths as defined by the original numerator.

3. Define states and write state equations from the diagram.

 EXAMPLE 7.4

Given

$$G = \frac{C}{E} = \frac{(As^2 + Bs + D)}{s^4 + Xs^3 + Ys^2 + Zs + W},$$

Define

$$G' = \frac{Q}{E} = \frac{1}{s^4 + Xs^3 + Ys^2 + Zs + W}.$$

Cross multiply and rearrange:

$$\ddddot{Q} = E - WQ - Z\dot{Q} - Y\ddot{Q} - X\dddot{Q}.$$

The signal flow diagram is given in Figure 7.3. Now note that

$$G = G'(As^2 + Bs + D)$$

$$\frac{C}{E} = \frac{Q}{E}(As^2 + Bs + D).$$

Thus,

$$C = DQ + B\dot{Q} + A\ddot{Q}$$

and feedforward paths are added to the basic diagram. The states are defined from the

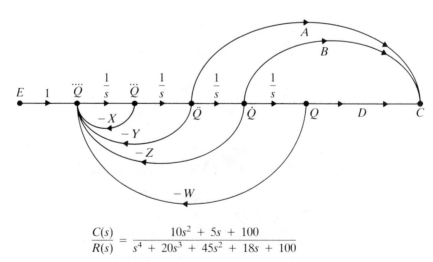

$$\frac{C(s)}{R(s)} = \frac{10s^2 + 5s + 100}{s^4 + 20s^3 + 45s^2 + 18s + 100}$$

FIGURE 7.3 Signal Flow Diagram.

signal flow graph as follows:

$$U_1 = Q$$
$$\dot{U}_1 = U_2 = \dot{Q}$$
$$\dot{U}_2 = U_3 = \ddot{Q}$$
$$\dot{U}_3 = U_4 = \dddot{Q}$$
$$\dot{U}_4 = \ddddot{Q} = E - sU_4 - yU_3 - ZU_2 - WU_1.$$

From which

$$\frac{d}{dt}\begin{bmatrix} U_1 \\ U_2 \\ U_3 \\ U_4 \end{bmatrix} = \begin{bmatrix} 0 & 1 & 0 & 0 \\ 0 & 0 & 1 & 0 \\ 0 & 0 & 0 & 1 \\ -W & -Z & -Y & -X \end{bmatrix} + \begin{bmatrix} 0 \\ 0 \\ 0 \\ 1 \end{bmatrix} E.$$

The output is

$$C = DU_1 + BU_2 + CU_3.$$

 EXAMPLE 7.5 STATE EQUATIONS OF A LIQUID LEVEL SYSTEM

A system of two tanks connected by a pipe and with another outlet pipe in tank 2 is shown in Figure 7.4, where

$Q_{\text{in}}, Q_{12}, Q_{\text{out}}$ = Flow rates (e.g., ft^3/s)
A_1, A_2 = cross-sectional areas of the tanks
H_1, H_2 = heads (heights of liquid level above a reference)
R_{12}, R_{out} = effective resistance of pipes in opposing flow.

The first step toward obtaining the state equations is to derive the differential equations from the basic laws of physics. In general, the basic relationship for the

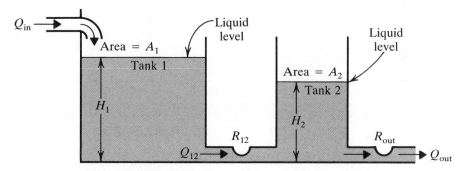

FIGURE 7.4 A Two-Tank System.

linearized system is $Q = H/R$. Then the flow between tanks is

$$Q_{12} = \frac{H_1}{R_{12}} - \frac{H_2}{R_{12}} = \frac{H_1 - H_2}{R_{12}}.$$

Since the input flow Q_{in} is assumed constant, the water level in tank 1 will change (unless $Q_{in} \equiv Q_{12}$), and the rate of this change is

$$Q_{in} - Q_{12} = Q_{in} - \frac{H_1 - H_2}{R_{12}} = A_1 \frac{dH_1}{dt}.$$

The output flow is $Q_{out} = H_2/R_{out}$. Thus, there will also be a change in water level in tank 2, which is defined by

$$\frac{H_1 - H_2}{R_{12}} - \frac{H_2}{R_{out}} = A_2 \frac{dH_2}{dt}.$$

We can now choose state variables. Note that there are two tanks, each storing energy, so two state variables, are needed. Logical choices for the states are H_1 and H_2 since each is a measure of the energy stored. Let $H_1 = X_1$ and $H_2 = X_2$. Substituting,

$$Q_{in} - \frac{X_1}{R_{12}} + \frac{X_2}{R_{12}} = A_1 \dot{X}_1$$

$$\frac{X_1}{R_{12}} - \frac{X_2}{R_{12}} - \frac{X_2}{R_{out}} = A_2 \dot{X}_2.$$

Rearranging,

$$\dot{X}_1 = - \frac{X_1}{A_1 R_{12}} + \frac{X_2}{A_2 R_{12}} + \frac{Q_{in}}{A_1}$$

$$\dot{X}_2 = \frac{X_1}{R_{12} A_2} - \frac{X_2}{R_{12} A_2} - \frac{X_2}{R_{out} A_2}.$$

In matrix form,

$$\begin{bmatrix} \dot{X}_1 \\ \dot{X}_2 \end{bmatrix} = \begin{bmatrix} \dfrac{1}{A_1 R_{12}} & \dfrac{1}{A_1 R_{12}} \\ \dfrac{1}{A_2 R_{12}} & -\dfrac{1}{A_2 R_{12}} - \dfrac{1}{A_2 R_{out}} \end{bmatrix} \begin{bmatrix} X_1 \\ X_2 \end{bmatrix} + \begin{bmatrix} \dfrac{1}{A_1} \\ 0 \end{bmatrix} Q_{in}.$$

7.2C   Another Manipulation of the Transfer Function

Still another canonical form for the state equation may be obtained from the transfer function by applying the partial fraction expansion.

The transfer function representation of any system is of the form

$$\frac{Y(s)}{X(s)} = \frac{K \sum_{j=0}^{m} a_j s^j}{\sum_{k=0}^{n} a_k s^k} \quad (n > m). \tag{7.14}$$

The denominator can always be factored, giving

$$\frac{Y(s)}{X(s)} = \frac{K \sum_{j=0}^{m} a_j s^j}{(s + \lambda_1)(s + \lambda_2)\cdots(s + \lambda_n)}. \tag{7.15}$$

Using the partial fraction expansion (assuming the numerator degree is less than the denominator degree),

$$\frac{Y(s)}{X(s)} = \frac{C_1}{(s + \lambda_1)} + \frac{C_2}{(s + \lambda_2)} + \cdots \frac{C_n}{(s + \lambda_n)}, \tag{7.16}$$

from which

$$Y(s) = \frac{C_1 X(s)}{(s + \lambda_1)} + \frac{C_2 X(s)}{(s + \lambda_2)} + \cdots \frac{C_n X(s)}{(s + \lambda_3)}. \tag{7.17}$$

Each term in Eq. (7.17) is then used to define a state equation, i.e.,

$$Z_1 = \frac{C_1 X(s)}{s + \lambda_1}; \quad (s + \lambda_1)Z_1 = C_1 X(s)d; \quad \dot{Z}_1(t) = -\lambda_1 Z_1(t) + C_1 X(t)$$

$$Z_2 = \frac{C_2 X(s)}{s + \lambda_2}; \quad (s + \lambda_2)Z_2 = C_2 X(s)d; \quad \dot{Z}_2(t) = -\lambda_2 Z_2(t) + C_2 X(t). \tag{7.18}$$

$$\vdots$$

These equations may be put in vector matrix form:

$$\frac{d}{dt}\begin{bmatrix} Z_1 \\ Z_2 \\ Z_3 \\ \vdots \end{bmatrix} = \begin{bmatrix} -\lambda_1 & 0 & 0 & - & - \\ 0 & -\lambda_2 & 0 & - & - \\ 0 & 0 & -\lambda_3 & - & - \end{bmatrix} \begin{bmatrix} Z_1 \\ Z_2 \\ Z_3 \end{bmatrix} + \begin{bmatrix} C_1 \\ C_2 \\ C_3 \end{bmatrix} X. \tag{7.19}$$

Note that the preceding methods of formulating the state equations may be applied when the original description of the system is in differential equation form or is an open-loop or closed-loop transfer function. When the state equations of the open-loop system have been obtained, they are converted readily to closed loop in several ways. Figure 7.5 defines the closed-loop system when the plant is the motor of Figure 7.1(a) and Equations (7.7), (7.8), and (7.9). Note that

$$V = K(IN - X_1). \tag{7.20}$$

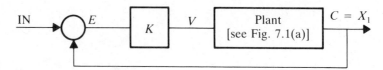

FIGURE 7.5 A Closed-Loop System.

Substituting into Eq. (7.8),

$$\dot{X}_3 = -\frac{K_B}{L} X_2 - \frac{R}{L} X_3 + \frac{K(IN - X_1)}{L}. \qquad (7.21)$$

Then Eq. (7.9) becomes

$$\frac{d}{dt}\begin{bmatrix} X_1 \\ X_2 \\ X_3 \end{bmatrix} = \begin{bmatrix} 0 & 1 & 0 \\ 0 & -f/J & K_T/J \\ -K/L & -K_B/L & -R/L \end{bmatrix}\begin{bmatrix} X_1 \\ X_2 \\ X_3 \end{bmatrix} + \begin{bmatrix} 0 \\ 0 \\ K/L \end{bmatrix} IN. \qquad (7.22)$$

Alternately substituting in Eq. (7.9),

$$\frac{d}{dt}\begin{bmatrix} X_1 \\ X_2 \\ X_3 \end{bmatrix} = \begin{bmatrix} 0 & 1 & 0 \\ 0 & -f/J & K_T/J \\ 0 & -K_B/L & -R/L \end{bmatrix}\begin{bmatrix} X_1 \\ X_2 \\ X_3 \end{bmatrix} + \begin{bmatrix} 0 \\ 0 \\ 1 \end{bmatrix}\frac{K(IN - X_1)}{L}. \qquad (7.23)$$

Rearranging

$$\frac{d}{dt}\begin{bmatrix} X_1 \\ X_2 \\ X_3 \end{bmatrix} = \begin{bmatrix} 0 & 1 & 0 \\ 0 & -f/J & K_T/J \\ 0 & -K_B/L & -R/L \end{bmatrix}\begin{bmatrix} X_1 \\ X_2 \\ X_3 \end{bmatrix} - \begin{bmatrix} 0 \\ 0 \\ 1 \end{bmatrix}\frac{KX_1}{L} + \begin{bmatrix} 0 \\ 0 \\ 1 \end{bmatrix}\frac{K \cdot IN}{L}. \qquad (7.24)$$

Combining

$$\frac{d}{dt}\begin{bmatrix} X_1 \\ X_2 \\ X_3 \end{bmatrix} = \begin{bmatrix} 0 & 1 & 0 \\ 0 & -f/J & K_T/J \\ -K/L & -K_B/L & -R/L \end{bmatrix}\begin{bmatrix} X_1 \\ X_2 \\ X_3 \end{bmatrix} + \begin{bmatrix} 0 \\ 0 \\ 1 \end{bmatrix}\frac{K \cdot IN}{L}. \qquad (7.25)$$

Thus far, the text has developed state equations for systems with one input and one output only. Many systems have a number of inputs and outputs, as indicated by Figure 7.6. The outputs are not necessarily states, but are functions of the states. In addition to the state matrix equation

$$\dot{X} = \mathbf{A}X + Bu, \qquad (7.26)$$

another equation is required to relate the states (and the inputs) to the outputs. This output equation is also a vector matrix equation:

$$y = \mathbf{C}^T X + Du. \qquad (7.27)$$

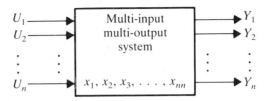

FIGURE 7.6 A Multivariable System.

Note that the system of Figure 7.5 is single input, single output. The state equations are (7.22) or (7.25), but a separate output equation has not been defined. This is accomplished easily by inspection; i.e., it is clear that the output is

$$y = \mathbf{C}^T X = \begin{bmatrix} 1 & 0 & 0 \end{bmatrix} \begin{bmatrix} X_1 \\ X_2 \\ X_3 \end{bmatrix} = X_1. \qquad (7.28)$$

EXAMPLE 7.6

The steering dynamics of an ocean-going ship are given by the transfer function block diagram[a] of Figure 7.7. The ship has the usual rudder (δ) and a set of anti-roll fins (Δ). Use of the rudder to change course results in both a heading (ψ) change and in a roll (θ) disturbance. If the fins are deflected, a rolling moment results, but there is no effect on the ship heading.

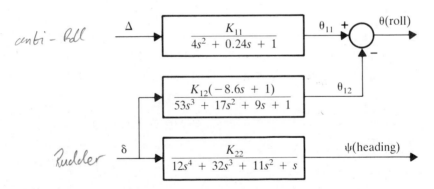

FIGURE 7.7 Example 7.6: Transfer Function Block Diagram for a Ship Moving at Constant Speed.

---

[a] In engineering practice, such a model *may* be determined by testing, and analysis of test data, without benefit of a differential equation.

To convert the transfer function model to state variables:

A. $\quad \dfrac{\theta_{11}}{\Delta} = \dfrac{K_{11}}{4s^2 + 0.24s + 1}$

$$4\ddot{\theta}_{11} + 0.24\dot{\theta}_{11} + \theta_{11} = K_{11}\Delta.$$

Define

$$\theta_{11} = X_1$$
$$\dot{\theta}_{11} = X_2;$$

then

$$\dot{X}_1 = X_2$$

$$\dot{X}_2 = \frac{K_{11}\Delta}{4} - \frac{X_1}{4} - \frac{0.24X_2}{4}.$$

B. $\quad \dfrac{\theta_{12}}{\delta} = \dfrac{K_{12}(-8.6s + 1)}{53s^3 + 17s^2 + 9s + 1}$

$$53\dddot{\theta}_{12} + 17\ddot{\theta}_{12} + 9\dot{\theta}_{12} + \theta_{12} = -8.6K_{12}\dot{\delta} + K_{12}\delta.$$

Define

$$\theta_{12} = X_3$$
$$\dot{\theta}_{12} = \dot{X}_3 = X_4$$
$$\ddot{\theta}_{12} = \dot{X}_4 = X_5$$
$$\dddot{\theta}_{12} = \dot{X}_5 = \frac{K_{12}\delta}{53} - \frac{8.6K_{12}\dot{\delta}}{53} - \frac{X_3}{53} - \frac{9X_4}{53} - \frac{17X_5}{53}.$$

C. $\quad \dfrac{\psi}{\delta} = \dfrac{K_{22}}{12s^4 + 32s^3 + 11a^2 + s}$

$$12\ddddot{\psi} + 32\dddot{\psi} + 11\ddot{\psi} + \dot{\psi} = K_{22}\delta.$$

Define

$$\psi = X_6$$
$$\dot{\psi} = \dot{X}_6 = X_7$$
$$\ddot{\psi} = \dot{X}_7 = X_8$$
$$\dddot{\psi} = \dot{X}_8 = X_9$$
$$\ddddot{\psi} = \dot{X}_9 = \frac{K_{22}\delta}{12} - \frac{X_7}{12} - \frac{11X_8}{12} - \frac{32X_9}{12}.$$

Combining the results of **A**, **B**, and **C** in matrix form:

$$
\begin{bmatrix} \dot{X}_1 \\ \dot{X}_2 \\ \dot{X}_3 \\ \dot{X}_4 \\ \dot{X}_5 \\ \dot{X}_6 \\ \dot{X}_7 \\ \dot{X}_8 \\ \dot{X}_9 \end{bmatrix} =
\begin{bmatrix}
0 & 1 & 0 & 0 & 0 & 0 & 0 & 0 & 0 \\
-2.5 & -0.06 & 0 & 0 & 0 & 0 & 0 & 0 & 0 \\
0 & 0 & 0 & 1 & 0 & 0 & 0 & 0 & 0 \\
0 & 0 & 0 & 0 & 1 & 0 & 0 & 0 & 0 \\
0 & 0 & -0.189 & -1.698 & -3.21 & 0 & 0 & 0 & 0 \\
0 & 0 & 0 & 0 & 0 & 0 & 1 & 0 & 0 \\
0 & 0 & 0 & 0 & 0 & 0 & 0 & 1 & 0 \\
0 & 0 & 0 & 0 & 0 & 0 & 0 & 0 & 1 \\
0 & 0 & 0 & 0 & 0 & 0 & -0.0833 & -0.917 & -2.67
\end{bmatrix}
$$

$$
\times
\begin{bmatrix} X_1 \\ X_2 \\ X_3 \\ X_4 \\ X_5 \\ X_6 \\ X_7 \\ X_8 \\ X_9 \end{bmatrix} +
\begin{bmatrix}
0 & 0 \\
0.25K_{11} & 0 \\
0 & 0 \\
0 & 0 \\
0.189K_{12} & -1.623K_{12} \\
0 & 0 \\
0 & 0 \\
0 & 0 \\
0.0833K_{22} & 0
\end{bmatrix}
\begin{bmatrix}
0 & \Delta & 0 & 0 & \delta & 0 & 0 & 0 & \delta \\
0 & 0 & 0 & 0 & \dot{\delta} & 0 & 0 & 0 & 0
\end{bmatrix}
$$

$$
y = \begin{bmatrix} \theta \\ \psi \end{bmatrix} = \begin{bmatrix} X_1 & X_3 \\ 0 & X_6 \end{bmatrix} \begin{bmatrix} 1 & 0 \\ -1 & 1 \end{bmatrix}.
$$

7.3   TRANSFER FUNCTIONS FROM STATE VARIABLES

Equation (7.9) gives the state variable description of a component, the dc motor of Figure 7.1(a). To obtain the transfer function of the motor from this vector matrix equation, we must first transform the equations (for all initial conditions zero), solve for the motor output, and form the ratio of output/input, which is the transfer function. Using the abbreviated matrix notation,

$$\dot{X} = \mathbf{A}X + Bu,$$

transforming,

$$sX(s) = \mathbf{A}X(s) + Bu(s). \tag{7.29}$$

Rearranging,

$$sX(s) - \mathbf{A}X(s) = Bu(s), \tag{7.30}$$

which gives

$$(s\mathbf{I} - \mathbf{A})X(s) = Bu(s) \tag{7.31}$$

$$X(s) = (s\mathbf{I} - \mathbf{A})^{-1}Bu(s). \tag{7.32}$$

But by Eq. (7.28)

$$y(s) = \mathbf{C}^T X(s).$$

Therefore,

$$y(s) = \mathbf{C}^T(s\mathbf{I} - \mathbf{A})^{-1}Bu(s) \tag{7.33a}$$

$$\frac{y(s)}{u(s)} = \mathbf{C}^T(s\mathbf{I} - \mathbf{A})^{-1}B. \tag{7.33b}$$

Expanding Eq. (7.33b),

$$G(s) = \frac{y(s)}{u(s)} = \begin{bmatrix} 1 & 0 & 0 \end{bmatrix}[s\mathbf{I} - \mathbf{A}]^{-1}\begin{bmatrix} 0 \\ 0 \\ 0 \\ 1/L \end{bmatrix}$$

$$[s\mathbf{I} - \mathbf{A}] = \begin{bmatrix} s & -1 & 0 \\ 0 & s + f/J & -K_t/J \\ 0 & +K_v/L & s + R/L \end{bmatrix}$$

$$[s\mathbf{I} - \mathbf{A}]^T = \begin{bmatrix} s & 0 & 0 \\ -1 & s + f/J & K_V/L \\ 0 & -K_T/J & s + R/L \end{bmatrix}$$

$$\det[s\mathbf{I} - \mathbf{A}] = s(s + f/J)(s + R/L) + \frac{sK_V K_T}{JL} = \Delta$$

$$\text{adj}[s\mathbf{I} - \mathbf{A}] = \text{cof}[s\mathbf{I} - \mathbf{A}]^T$$

$$\text{cof}[s\mathbf{I} - \mathbf{A}]^T = \begin{bmatrix} (s + f/J)(s + R/L) + \dfrac{K_V K_T}{JL} & (s + R/L) & K_T/J \\ 0 & s(s + R/L) & +sK_T/J \\ 0 & -K_V s & s(s + f/J) \end{bmatrix}$$

$$[s\mathbf{I} - \mathbf{A}]^{-1} = \frac{\text{cof}[s\mathbf{I} - \mathbf{A}]^T}{\det[s\mathbf{I} - \mathbf{A}]}$$

$$G(s) = \mathbf{C}^T[s\mathbf{I} - \mathbf{A}]^{-1}B$$

$$G(s) = \begin{bmatrix} 1 & 0 & 0 \end{bmatrix}\begin{bmatrix} \dfrac{(s + f/J)(s + R/L) + \dfrac{K_V K_T}{JL}}{\Delta} & \dfrac{(s + R/L)}{\Delta} & \dfrac{K_{T/J}}{\Delta} \\ \dfrac{0}{\Delta} & \dfrac{s(s + R/L)}{\Delta} & \dfrac{sK_T/J}{\Delta} \\ \dfrac{0}{\Delta} & \dfrac{-sK_V}{\Delta} & \dfrac{s(s + f/J)}{\Delta} \end{bmatrix}$$

$$\times \begin{bmatrix} 0 \\ 0 \\ 1/L \end{bmatrix}.$$

Expanding,

$$G(s) = \begin{bmatrix} \dfrac{(s + f/J)(s + R/L) + \dfrac{K_V K_T}{JL}}{\Delta} & \dfrac{(s + R/L)}{\Delta} & \dfrac{K_T/J}{\Delta} \end{bmatrix} \begin{bmatrix} 0 \\ 0 \\ 1/L \end{bmatrix}.$$

Therefore,

$$G(s) = \frac{K_T/JL}{s(s + f/J)(s + R/L) + sK_V K_T/JL}. \tag{7.34}$$

To verify Eq. (7.34), a derivation starting with Eqs. (7.5) and (7.6) gives

$$V(s) - (R + sL)I(s) - K_V s\theta(s) = 0$$

$$(Js^2 + fs) - K_T I(s) = 0.$$

Combining,

$$V(s) - (R + SL)\frac{1}{(K_T)}(J^2 s + fs)\theta(s) - K_V s\theta(s) = 0$$

$$G(s) = \frac{\theta}{V}(s) = \frac{1}{(1/KT)(R + sL)(Js^2 + fs) + K_V s}$$

$$= \frac{K_T/J}{s(R + sL)(s + f/J) + K_V K_T s/J}$$

$$= \frac{K_T/JL}{s(s + R/L)(s + f/J) + K_V K_T s/JL}.$$

7.4   ANALYSIS WITH STATE VARIABLES: STABILITY, CONTROLLABILITY, AND OBSERVABILITY

A primary requirement for a control system is that it must be stable. For linear systems, this means that all of the roots of the characteristic equation must be in the left half of the s plane. Matrix manipulations as in Eqs. (7.29) through (7.34) develop the characteristic polynomial, which is det $(s\mathbf{I} - \mathbf{A})$. Factoring the polynomial is a separate problem. However, computer programs that find the roots (eigenvalues) given the state formulation are readily available. Thus, stability is readily checked.

Two fundamental properties of a dynamic system that are important in all applications but not available from the transfer function approach are *controllability* and *observability*: These properties are easily defined:

A system $\dot{X} = f(X, u, t)$ is completely *controllable* if any initial state $X(t_0)$ can be transferred to any chosen final state $X(t_f)$ in finite time $t_0 < t < t_f$ by use of some forcing function $u(t)$.

In other words, if we can make the system change its state (output) from one chosen value to another by manipulating the available input, then we can control the system. Most engineering systems are controllable in this sense, probably because we use components that we know are controllable. However, there are many exceptions and often the lack of controllability is not intuitively obvious. We need a mathematical test.

Consider Eq. (7.32) and take the inverse transform:

$$X(t) = e^{At}X(0) + \int_0^t s^{A(t-\tau)}Bu(\tau)\,d\tau, \tag{7.35}$$

where $e^{At} = \mathcal{L}^{-1}(s\mathbf{I} - \mathbf{A})^{-1}.$
            $X(0) =$ the inital state.

To obtain a check on controllability, choose the desired final state as

$$X(t_f) = 0.$$

Then

$$0 = e^{At}X(0) + \int_0^{t_f} e^{A(t-\tau)}Bu(\tau)\,d\tau, \tag{7.36a}$$

from which

$$X(0) = \int_0^{t_f} e^{-A\tau}Bu(\tau)\,d\tau, \tag{7.36b}$$

but $e^{-A\tau}$ can be written as a power series:

$$e^{-A\tau} = \sum_{k=0}^{n-1} \alpha_k(\tau)A^k; \tag{7.37}$$

thus,

$$X(0) = -\sum_{k=0}^{n-1} A^k B \int_0^{t_f} \alpha_k(\tau)u(\tau)\,d\tau. \tag{7.38}$$

For each k, the integral will have a value.

$$-\int_0^{t_f} \alpha_k(\tau)\omega(\tau)\,d\tau = Q_k. \tag{7.39}$$

We may then write

$$X(0) = \sum_{k=0}^{n-1} A^k B Q_k \tag{7.40a}$$

or

$$X(0) = [B \quad AB \quad A^2B \quad \cdots \quad A^{n-1}B]\begin{bmatrix} Q_0 \\ Q_1 \\ \vdots \\ Q_{n-1} \end{bmatrix}. \tag{7.40b}$$

This equation will be satisfied only if the vectors B, AB, $\cdots$ are linearly independent, which will be true if the matrix

$$[B \quad AB \quad A^2B \quad \cdots \quad A^{n-1}B] \tag{7.41}$$

is of rank n.

 EXAMPLE 7.7

A feedback system has the state equations

$$\dot{X}_1 = -3X_1 - X_2 + u$$

$$\dot{X}_2 = 2X_1$$

$$Y = X_1 - X_2 + u.$$

In matrix form,

$$\begin{bmatrix} \dot{X}_1 \\ \dot{X}_2 \end{bmatrix} = \begin{bmatrix} -3 & -1 \\ 2 & 0 \end{bmatrix} \begin{bmatrix} X_1 \\ X_2 \end{bmatrix} + \begin{bmatrix} 1 \\ 0 \end{bmatrix} u$$

$$y_0 = [1 \quad -1] \begin{bmatrix} X_1 \\ X_2 \end{bmatrix} + u.$$

The order of the system is 2, so $n = 2$. Checking for controllability,

$$\begin{bmatrix} 1 & \vdots & \begin{bmatrix} -3 & 1 \\ 2 & 0 \end{bmatrix} \begin{bmatrix} 1 \\ 0 \end{bmatrix} \end{bmatrix} = \begin{bmatrix} 1 & \vdots & -3 \\ 0 & \vdots & 2 \end{bmatrix},$$

which has rank 2, so the system is controllable.

 EXAMPLE 7.8

Another feedback system has state equations

$$\dot{X}_1 = -2X_1 + 3X_2 + u$$

$$\dot{X}_2 = X_1 + u$$

$$y = X_1.$$

In matrix form,

$$\begin{bmatrix} \dot{X}_1 \\ \dot{X}_2 \end{bmatrix} = \begin{bmatrix} -2 & 3 \\ 1 & 0 \end{bmatrix} \begin{bmatrix} X_1 \\ X_2 \end{bmatrix} + \begin{bmatrix} 1 \\ 1 \end{bmatrix} u$$

$$y = [1 \quad 0] \begin{bmatrix} X_1 \\ X_2 \end{bmatrix}$$

The order of this system is 2, so $n = 2$. Checking for controllability,

$$\begin{bmatrix} 1 & \vdots & \begin{bmatrix} -2 & 3 \\ 1 & 0 \end{bmatrix} \begin{bmatrix} 1 \\ 1 \end{bmatrix} \end{bmatrix} = \begin{bmatrix} 1 & \vdots & 1 \\ 1 & \vdots & 1 \end{bmatrix}$$

and the system is uncontrollable.

The property of *observability* has comparable importance and may be defined as follows. A system $\dot{X} = \mathbf{A}X + \mathbf{B}u$, $y = \mathbf{C}X + \mathbf{D}u$ is completely observable if all of its states can be determined over a finite time interval by observing the time variation of the output.

A test for observability is that the $n \times nm$ matrix,

$$[\mathbf{KK}] = [\mathbf{C}^T | \mathbf{A}^T\mathbf{C}^T | \mathbf{A}^{T^2}\mathbf{C}^T | \cdots | \mathbf{A}^{T^{n-1}}\mathbf{C}^T], \tag{7.42}$$

must have rank n.

The transfer function of a system describes those portions of the system that are both controllable and observable.

 EXAMPLE 7.9

The system of Example 7.8 has the matrices

$$\mathbf{A} = \begin{bmatrix} -3 & -1 \\ 2 & 0 \end{bmatrix} \quad \mathbf{A}^T = \begin{bmatrix} -3 & 2 \\ -1 & 0 \end{bmatrix} \quad \mathbf{C} = [1 \quad -1] \quad \mathbf{C}^T = \begin{bmatrix} 1 \\ -1 \end{bmatrix}.$$

Then

$$[\mathbf{C}^T \;\vdots\; \mathbf{A}^T \quad \mathbf{C}^T] = \begin{bmatrix} 1 & \vdots & \begin{bmatrix} -3 & 2 \\ -1 & 0 \end{bmatrix} \begin{bmatrix} 1 \\ -1 \end{bmatrix} \\ -1 & \vdots & \end{bmatrix}$$

$$= \begin{bmatrix} 1 & \vdots & -5 \\ -1 & \vdots & -1 \end{bmatrix},$$

which has rank 2, so the system is observable.

 EXAMPLE 7.10

The system of Example 7.9 has the matrices

$$\mathbf{A} = \begin{bmatrix} -2 & 3 \\ 1 & 0 \end{bmatrix} \quad \mathbf{A}^T = \begin{bmatrix} -2 & 1 \\ 3 & 0 \end{bmatrix} \quad \mathbf{C} = [1 \quad 0] \quad \mathbf{C}^T = \begin{bmatrix} 1 \\ 0 \end{bmatrix}.$$

Then

$$[\mathbf{C}^T \;\vdots\; \mathbf{A}^T \quad \mathbf{C}^T] = \begin{bmatrix} 1 & \vdots & \begin{bmatrix} -2 & 1 \\ 3 & 0 \end{bmatrix} \begin{bmatrix} 1 \\ 0 \end{bmatrix} \\ 0 & \vdots & \end{bmatrix}$$

$$= \begin{bmatrix} 1 & \vdots & -2 \\ 0 & \vdots & 3 \end{bmatrix},$$

which has rank 2, so the system is observable.

7.5 OBSERVERS

When all of the state variables are measurable, they may be fed back through appropriate gains thus obtaining desired values for the roots of the characteristic equation. This design procedure is called the *pole placement* problem and is treated in

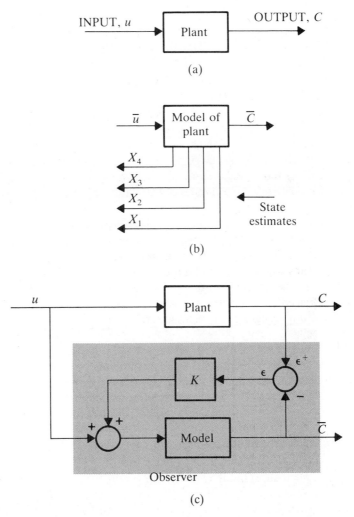

FIGURE 7.8 Models and Observers.

Chapter 8. A preliminary problem is that of measuring all of the states. In many systems, direct measurement of the states is not possible, but proper use of a special filter permits us to obtain signals that are good estimates of the states. This filter is called an *observer*.

Consider Figure 7.8(a). The "plant" is a single input, single output system[b] for which we know the system equations and parameter values, but only the output can be measured. If we build a filter that is an exact model of the plant, we can design this filter so that all of its states can be measured, as indicated in Figure 7.8(b). If plant and model

---

[b] Multi-input multi-output observes are beyond the scope of this text.

were to be operated in an identical fashion, these state signals from the model might be used as measures of the states in the plant. However, one cannot guarantee that the initial conditions and disturbances are the same for both plant and model, so use of the simple model to "observe" the states of the plant is not a reliable procedure. But if we drive both plant and model with the same input u, we can force the model output to approach the system output very rapidly by using a feedback loop. Consider Figure 7.8(c), which shows the model with feedback loop, i.e., the *observer*. If the initial conditions in the plant are different from those in the model ($C \neq \hat{C}$), comparison provides an error signal ϵ. The feedback loop forces ϵ to zero, which can be done quickly by using a large value of K. When $\epsilon = 0$, there is no signal in the feedback loop, plant and model operate identically, and state signals obtained from (observed by) the model are valid estimates of the plant states and may be used in state variable feedback designs.

The feedback loop, which is called the *observer*, must be designed as any other feedback loop. Its primary purpose is to eliminate the differences due to initial conditions and to do it quickly. Thus, a wide bandwidth loop is wanted for fast transient response, but good damping is also desired. Therefore, a relatively large value of K is needed. If stability of the loop becomes a problem, a compensating filter can be placed in the feedback link, since this link is inoperative in steady state when the state variable signals are being generated.

■ EXAMPLE 7.11

Let the transfer function of plant (and model) be $G(s) = 1/s(s + 10)$. If the loop is closed as shown in Figure 7.8(c), with $K = 1.0$, the characteristic equation of the observer loop is $s^2 + 10s + 1 = 0$ with roots at $s = -9.898$ and -0.202. The settling time is dominated by the root at -0.202 and would be nearly 20 s.

If the observer gain is raised to 10, the characteristic equation becomes $s^2 + 10s + 10$ with roots at -8.875 and -1.125. For this pair, the settling time would be about $4/1.125 = 3.55$ s.

If a substantial reduction in observer settling time is required, the characteristic roots of the closed observer loop can be moved to arbitrarily chosen positions by using error feedforward as illustrated in Figure 7.9. From the diagram, the state equations for the observer are

$$\dot{\hat{X}}_1 = \hat{X}_2 + K_3\epsilon$$
$$\dot{\hat{X}}_2 = \hat{X}_3 + K_2\epsilon$$
$$\dot{\hat{X}}_3 = 100K_1\epsilon - 11\hat{X}_3 - 10\hat{X}_2 + 100u$$
$$\epsilon = X_1 - \hat{X}_1; \qquad u = 0.$$

Then

$$\dot{\hat{X}}_1 = -K_3\hat{X}_1 + \hat{X}_2 + K_3 X_1$$
$$\dot{\hat{X}}_2 = -K_2\hat{X}_1 - \hat{X}_3 + K_2 X_1$$
$$\dot{\hat{X}}_3 = -100K_1\hat{X}_1 - 10\hat{X}_2 - 11\hat{X}_3 + 100K_1 X - 1.$$

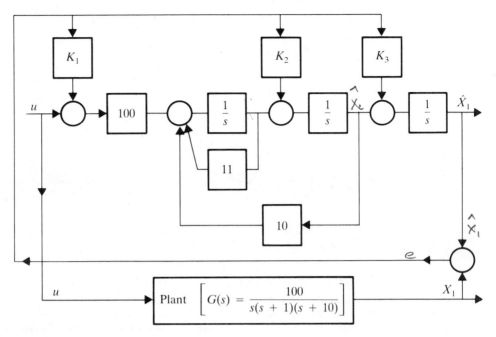

FIGURE 7.9   Observer Compensation.

In matrix form,

$$\begin{bmatrix} \dot{\hat{X}}_1 \\ \dot{\hat{X}}_2 \\ \dot{\hat{X}}_3 \end{bmatrix} = \begin{bmatrix} -K_3 & 1 & 0 \\ -K_2 & 0 & 1 \\ -100K_1 & -10 & -11 \end{bmatrix} \begin{bmatrix} \hat{X}_1 \\ \hat{X}_2 \\ \hat{X}_3 \end{bmatrix} + \begin{bmatrix} K_3 \\ K_2 \\ K_1 \end{bmatrix} X_1$$

$$[s\mathbf{I} - \mathbf{A}] = \begin{bmatrix} s + K_3 & -1 & 0 \\ K_2 & s & -1 \\ 100K_1 & +10 & s + 11 \end{bmatrix}$$

$$\det(s\mathbf{I} - \mathbf{A}] \Rightarrow s^3 + (11 + K_3)s^2 + (11K_3 + K_2 + 10)s + 100K_1 + 11K_2 + 10K_3 = 0.$$

By proper choice of K_1, K_2, and K_3, we can place roots of the observer wherever desired.

 EXAMPLE 7.12

For the plant shown in Figure 7.9, choose observer roots at $s = -1, -15$, and -20. Then the characteristic equation of the observer is

$$s^3 + 36s^2 + 335s + 300. = 0.$$

Equating coefficients,

$$11 + K_3 = 36 \rightarrow K_3 = 25$$
$$11K_3 + K_2 + 10 = 335 \rightarrow K_2 = 50$$
$$100K_1 + 11K_2 + 10K_3 = 300$$
$$100K_1 = 300 - 250 - 550$$
$$K_1 = \frac{-500}{100} = -5.$$

7.6 SUMMARY

The concept of *state* is introduced. The states of any system may be chosen in various ways and the result is always a set of first-order differential equations, which are conveniently summarized in matrix form. For engineering purposes, it is often desirable to choose as states those quantities that can be measured experimentally. State equations can be derived from transfer functions and vice versa. Controllability and observability are defined and illustrated. For some purposes, the states must be measured; when this is not possible, an observer may be designed and used to estimate the states.

BIBLIOGRAPHY

DeRusso, P. M. *State Variables for Engineers.* New York: John Wiley and Sons (1965).

Kalman, R. E. "On the General Theory of Control Systems. *Proc. First Int. Congress on Automatic Control*, Moscow, Vol. 1, London: Butterworth and Co., Ltd. (1960).

Kuo, B. C. *Automatic Control Systems.* 5th ed., Englewood Cliffs, N.J.: Prentice Hall (1987).

Luenberger, D. G. "Observing the State of a Linear System. Vol. MIL-8, *IEEE Trans. Military Electronics* (1964).

Schultz, D. G.; and Melsa, J. L. *State Functions and Linear Control Systems.* New York: McGraw-Hill Book Co. (1967).

Wiberg, D. M. *State Space and Linear Systems* in Schaum's Outline Series. New York: McGraw-Hill Book Co. (1971).

PROBLEMS

7.1 For each of the circuits given below, write state equations and define the **A** and **B** matrices.

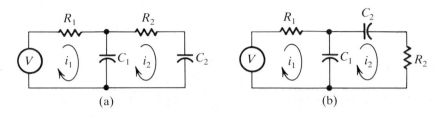

(a)   (b)

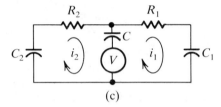

(c)

7.2 A dc shunt motor drives an inertia load. The shaft connecting motor to load is not rigid and must be considered a spring. The applied voltage is V (constant). Write the state equation.

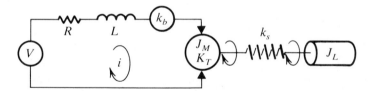

7.3 A hoist may be thought of as a motor lifting a mass in the presence of gravity. Referring to the figure below, write the appropriate equations and develop a state variable model.

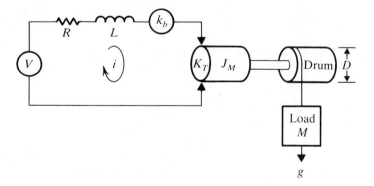

7.4 A vehicle suspension system may be modeled at one wheel as shown below. Write the state equations.

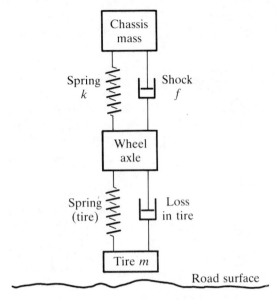

7.5 The D'Arsonval movement used in indicating instruments is a permanent magnet motor with shaft spring-loaded to the frame. From the diagram below, write the state equations.

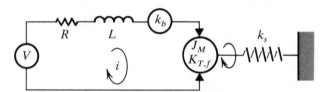

7.6 A second-order system has velocity feedback for damping. Derive state equations.

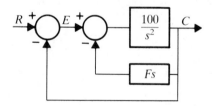

7.7 A third-order system includes a zero. What are the state equations?

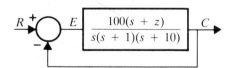

7.8 Given the system shown in block diagram form below, obtain the state equations.

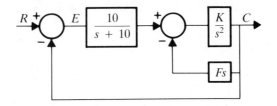

7.9 Obtain state equations for the closed-loop, unity feedback system if

$$G(s) = \frac{100}{s(s^2 + 0.1s + 10)}.$$

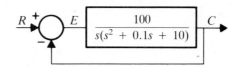

7.10 For the system of Problem 7.7, design an observer.

7.11 For the system of Problem 7.9, design an observer.

8 Design

INTRODUCTION

The design process includes selection of components, evaluation of component equations and parameters, definition of a basic or *minimum* system, setting the gain to meet specifications, stability analysis of the basic system, and design of compensation, which adjusts the dynamic performance of the system to meet all specifications. Selection of components and evaluation of their equations and parameter values is a very important and often difficult task, but it is also very dependent on the nature of the application; in many cases, the components are chosen to satisfy specifications not related to the control problem. It is not possible to discuss component selection without considering the application, and when the application is considered the problem becomes very specific; hence, one cannot develop general principles. In this text, selection of components is not considered. It is assumed that such selection has been made, that the equations and parameter values have been determined, and that these unchangeable parts of the system are called the *plant* and are described by a transfer function or by state equations.

The basic or *minimum* system is determined by closing a loop around the plant using unity feedback. Normally, sensors are assumed ideal (unit gain, no dynamics) and only an amplifier is added between error detector and plant. The gain is then set to meet steady-state accuracy requirements and some consideration may be given to the bandwidth requirements. Then stability analysis is carried out. Note that the accuracy demands of modern applications are so severe that we expect the basic system to be unstable—the results of the stability analysis indicate *how unstable* and *how much* must be done to stabilize the system and meet specifications.

Design of compensation involves two quite different steps. The first is to choose a way to modify the basic system. There are a variety of possible ways, and choosing a particular way is a matter of engineering judgment. There is no mathematical approach to this choice, nor is there necessarily a *best* way. Usually, any one of several possible choices permits satisfaction of the specifications, and the engineering choice is based on such features as cost, weight, space, heat generation, availability of suitable signals, etc.

The second step in the design process is the mathematical determination of parameter values for the chosen compensation scheme. The calculations required depend on the scheme chosen and are discussed later in this chapter.

Note that stability analysis is a very necessary preliminary step. It clearly shows how unstable the basic system is and thus indicates *how much* compensation must be done. It also permits extension of the analysis beyond testing stability. In particular, if one uses graphical methods (Bode diagram or root locus) and understands the basic effects of normal compensation schemes on the shape of these curves, the choice of a suitable scheme for stabilization is greatly simplified. Extension of analysis as a preparation for compensation design is therefore undertaken in the following sections.

8.2   ANALYSIS OF THE BASIC APPROACHES TO COMPENSATION

Unstable systems have roots in the right half plane. To stabilize an unstable system, the compensator (or controller) must move these roots into the left half plane. Furthermore, to provide acceptable transient response, the dominant roots (which are normally those moved from the right half to the left half plane) must be relocated in a suitable area of the plane. The *real* part of the complex roots determines the duration of the transient, i.e., the settling time, and the ζ line (radial line from origin) on which the roots lie determines the oscillating characteristics such as the amount of overshoot and number of oscillations.

In general terms, we can move the roots by:

1. Changing the gain.
2. Changing the plant or some parts of the plant.
3. Placing a dynamic element (usually a filter, but generally called a *compensator* or *controller*) in the forward transmission path.
4. Placing a dynamic element in a feedback path.
5. Feeding back all or some states.

Items (1) and (2) are seldom permissible and often not practical. They are not considered here. Items 3, 4, and 5 are all available and in common use. Selection of one or more of these methods is a matter of engineering judgment and depends on the nature of the problem, the specifications, the hardware available (including the plant), and the personal preference of the designer. The design techniques that may be used with each of the above approaches are, of course,

1. frequency response (Nyquist, Bode, Nichols)
2. root locus methods
3. state variables.

Conceptually, any one of the design methods may be used to design any type of compensator, but certain methods seem to work better than others on specific compensation schemes. Consider first the use of state feedback; solution of the state equations to place roots at specified locations (commonly called the *pole placement problem*) is accomplished readily by the computer, providing *all* states are fed back. When only some of the states are available to be fed back, the problem becomes more difficult. Whenever the compensation method introduces dynamics (such as a filter) and thus increases the number of states, the order of all matrices increases and programs for designing the compensator are not readily available. The frequency response and root locus may be used as tools for designing any of the compensation schemes, except that the root locus is not convenient when more than one or possibly two variable parameters are introduced.

8.3   CASCADE COMPENSATION: GENERAL CONCEPTS

A system is cascade compensated when the controller (compensator) is placed in the main forward transmission path, as shown in Figure 8.1. The cascade compensator, $G_c(s)$, may be thought of as a filter and, in fact, it is commonly realized as a filter. Mathematically then, $G_c(s)$ is a ratio of two polynomials. When the polynomials are factored the form of the filter transfer function is[a]

$$G_c(s) = K \frac{(s + Z_1)(s + Z_2) \cdots}{(s + P_1)(s + P_2) \cdots}.$$ **(8.1)**

When graphical design methods are used, the problem of selecting suitable locations for the poles and zeros is solved by trial and error. This approach works successfully because it converges quite rapidly, which is convenient for an engineer that has little experience with it. The Bode diagram approach is applied readily when a multisection filter (several zeros and poles) is required. The root locus method, while providing

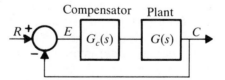

FIGURE 8.1   Block Diagram for Cascade Compensation.

---

[a] Equation (8.1) does not include any poles or zeros at the origin. Poles at the origin are often realized as part of the filter, but for purposes of analysis and design it is usually more convenient to include them in the plant transfer function. Zeros at the origin are not permitted in a cascade compensator because they would block the zero-frequency (dc) component of the error signal and the system would then be incapable of reaching the desired steady state.

excellent insight into complex system problems, is a good computational tool only when the compensator requirements are very modest.

To understand the use of a filter for compensation, it is both necessary and desirable for the effects of the filter on the Bode curves and on the root loci to be understood clearly. To demonstrate this, consider the form[b] of Eq. (8.1), i.e.,

$$G_c(s) = K \frac{(s + Z)}{(s + P)}, \tag{8.2}$$

where both Z and P are real. The filter of Eq. (8.2) may have:

$Z < P$ = high-pass filter, also called *phase lead* compensator
$P < Z$ = low-pass filter, also called *phase lag* compensator.

In the frequency domain, the gain and phase characteristics of these filters are shown in Figures 8.2(a) and (b). Observe that a

1. *high-pass filter* has a *positive phase*, which is maximum midway between the zero and pole and attenuates low frequencies, but the gain increases at 20 dB/decade between zero and pole.

2. *low-pass filter* has a *negative phase*, which is maximum midway between the zero and pole and does not attenuate low frequencies, but gain decreases at 20 dB/decade between pole and zero.

Clearly, when either filter is placed in cascade with the main transmission path, its characteristics will reshape both the gain and phase curves. Thus, all of the criteria we look for on the Bode diagram will be changed. In general, the introduction of the compensator changes the gain and phase crossover points, the gain margin and phase margin, and the bandwidth. Normally, the low-frequency gain (error coefficient) is not allowed to decrease; this means that when a lead compensator is used additional gain is also introduced so that the low-frequency attenuation of the filter is canceled.

The primary purpose of the compensator is to stabilize the system and provide damping for the transient. The graphical interpretation of these criteria is the requirement for positive phase and gain margins on the Bode diagram and, on the s plane, the roots must be moved into the left half plane.

On the Bode diagram, if the uncompensated system does not have adequate phase margin, it is clear that the introduction of positive phase shift over a selected range of frequencies could improve the phase margin. Clearly, this is the kind of adjustment that might be achieved with the lead filter of Figure 8.2(a). On the other hand, the negative phase available from the lag filter is not useful. When the lag filter is used, apparently the attenuation characteristic is the valuable characteristic and, since a good phase margin is mandatory, it appears that the attenuation characteristic must be used to achieve the desired phase margin. The graphical results are obvious; the lead

---

[b] It is not possible to build $(s + Z)$ only; there must always be an accompanying pole, however large. It *is* possible and easy to build $k/(s + P)$, but this is not used as a compensator in the usual sense, only in special cases.

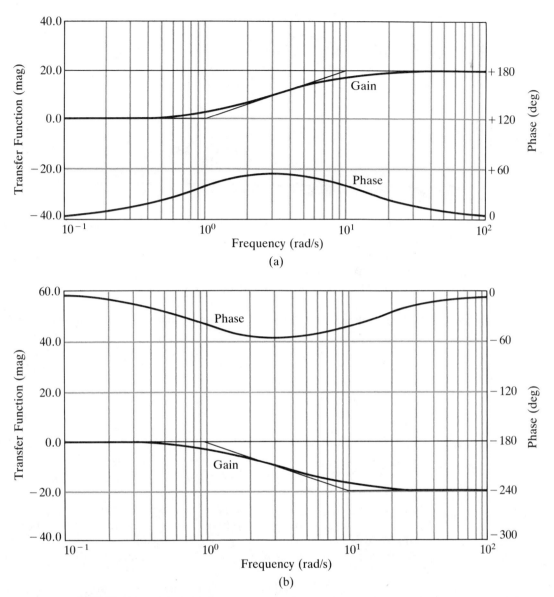

FIGURE 8.2 Bode Diagram: (a) High-Pass Filter and (b) Low-Pass Filter.

compensator is used to reshape the phase curve, while the lag compensator is used to reshape the gain curve.

On the root locus, when the lead compensator is used, since the zero is smaller than the pole, the asymptote centroid is moved to the left and the root loci are reshaped, being bent into the left half plane. When the lag compensator is used, the pole is smaller than the zero and this moves the asymptote centroid to the right and also would bend the loci to the right, which is not desirable. It should be observed, however, that the attenuation characteristic of the lag compensator reduces the root locus gain and this moves the roots along the loci. Since the lag compensator clearly does not reshape the root loci as desired, it must be this movement of the roots that permits stabilization.

8.4   CASCADE LEAD COMPENSATION

8.4A   Design on the Bode Diagram

Consider the system of Figure 8.3(a). The Bode diagram for the uncompensated system is sketched in Figure 8.3(b). Note that the phase margin is zero. Note also that $\angle G = -180°$ at all frequencies; positive phase shift must be introduced for stabilization and a phase lead compensator must be used.

From Figure 8.2(a), the lead filter attenuates low frequencies. To maintain the specified accuracy, the gain is usually increased by an amount equal to the lead compensator attenuation and this procedure is followed here. In terms of the Bode diagram gain curve, this means that the low-frequency asymptote of the uncompensated system is unchanged, but corner frequencies are introduced by the zero and pole of the filter with the asymptotes changing slopes at these corner frequencies.

The design problem is now clear. The locations of the zero and pole must be chosen so that the Bode diagram of the compensated system shows adequate phase margin. There is no *unique* or *best* solution. There are many[c] possible solutions, each of which may be acceptable. To illustrate some of the possibilities and to establish a general principle, consider the sketches of Figures 8.3(c), (d), and (e). In Figure 8.3(c), both the zero and pole are chosen at frequencies lower than the original gain crossover frequency; this moves the gain crossover to a much higher frequency but introduces maximum leading phase *between* zero and pole. The resulting phase margin is very small; this choice does not use the compensator phase effectively.

In Figure 8.3(d), both zero and pole are chosen at frequencies higher than the original gain crossover. This does not work either; the gain crossover is not moved, but the positive phase is introduced at too high a frequency and again the phase margin is small.

In Figure 8.3(e), the zero and pole are chosen such that the original gain crossover is between them. The new gain crossover, while displaced from the original, is in the same frequency range as the introduced positive phase and a substantial phase margin is obtained.

In general, the conditions of Figure 8.3(e) work best when using phase lead compensation; that is, with the gain crossover between compensator zero and pole.

---

[c] What is acceptable depends, of course, on the particular specifications. If the specifications are many and varied, the range of acceptable solutions may be very narrow, but this is not usual.

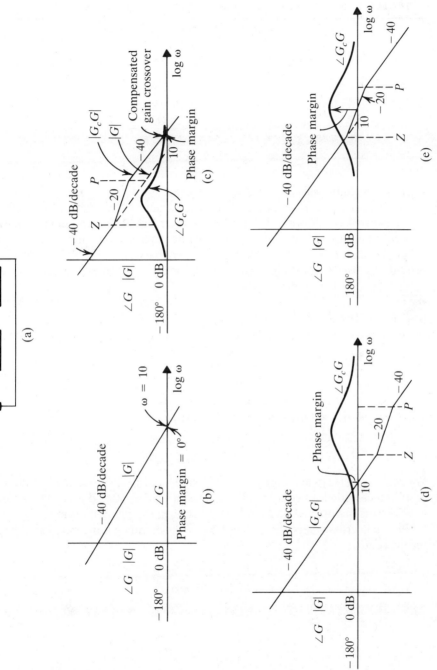

FIGURE 8.3 Cascade Compensation of a Simple Servo where the Transfer Function K/s^2 is Only Slightly Idealized. An Instrument Servo Motor Driven by a Current Source Amplifier has almost Exactly this Transfer Function: (a) Block Diagram; (b) Bode Diagram Sketch of the Uncompensated System; (c) Zero and Pole Both at Frequencies Less than the Original Gain Crossover Frequency; (d) Zero and Pole Both at Frequencies Greater than the Original Gain Crossover Frequency; and (e) Original Gain Crossover between Zero and Pole.

TABLE 8.1 Maximum Phase Lead Available Lead Compensator

$\alpha = Z/P$	$\angle\tan^{-1}\dfrac{\omega}{Z} - \tan^{-1}\dfrac{\omega}{P}$
0.6	15°
0.4	25°
0.2	42°
0.1	55°

Where $\omega_{\text{max phase}} = \sqrt{ZP}$.

There is no simple rule[d] for choosing the exact locations for the zero and pole; the separation between zero and pole determines the amount of positive phase contributed by the network (as shown in Table 8.1), but the choice of the zero and pole locations is also influenced by the phase margin of the uncompensated system, the rate of change of phase, and other factors peculiar to the particular problem. Trial-and-error procedures usually converge rapidly and thus have proven satisfactory. They will be applied later in this chapter.

 EXAMPLE 8.1

A single-loop, unity feedback control system has a forward transfer function:

$$G(s) = \frac{500}{s(0.1s + 1)}.$$

As shown in Figure 8.4, the phase margin is small, about 9°, and $\omega_x \simeq 70$ (uncompensated) so that the settling time is approximately

$$T_{su} \simeq \frac{4}{(0.1)(70)} = \frac{4}{7}\text{s}.$$

We wish to increase the damping substantially and to decrease the settling time. To do this, a cascade lead compensator is used, with $Z = 50$ and $p = 200$, and the error coefficient is maintained at $K_v = 500$. As shown, the phase margin is $\phi_{mc} \simeq 45°$ and the new gain crossover is at $\omega_{xc} \simeq 100$. Thus, the damping is increased and the settling time becomes

$$T_{sc} \simeq \frac{4}{(0.4)(100)} \simeq 0.1 \text{ s}.$$

---

[d] Some textbooks give a formula that is derived for the case of a second-order uncompensated system. When applied to higher order systems, results can be unsatisfactory.

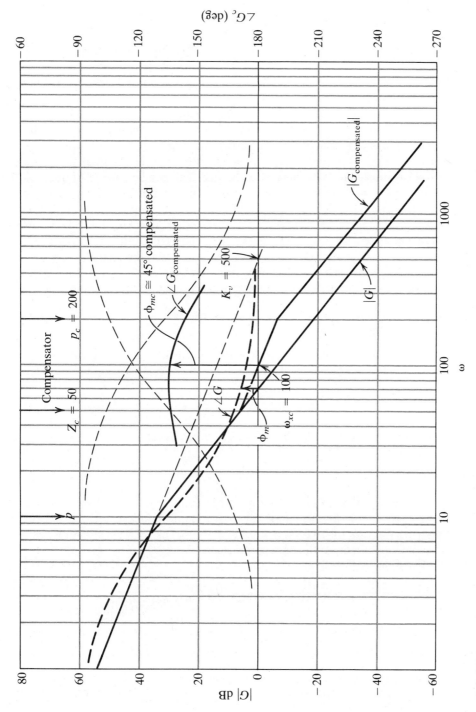

FIGURE 8.4   Example 8.1 Bode Diagram: Design of Compensation for a Second-Order Servo: $G(s) = 500/s\,(0.1s + 1)$.

8.4B   Design with Root Locus Methods

The root locus for the uncompensated system is shown in Figure 8.5(a). The loci and roots are on the imaginary axis and a lead compensator must be used to bend the loci into the left half plane. Where should the zero and pole be placed? Consider the test point in Figure 8.5(a) and assume that it is desired to move the *root* from $s = +j10$ to this location. To accomplish this, two conditions must be satisfied, $\angle GG_c(s) = (2N - 1)\pi$ (the angle condition) and $|GG_c(s)| = 1.0$ (the magnitude condition). The angle condition is usually easy to satisfy, many pole-zero pairs provide the required positive angle. However, to satisfy the magnitude requirement at the point, the gain must be adjusted to place the root as desired. This can be done, of course, but the gain required to place the root at the desired point may not be high enough to satisfy the accuracy specification (i.e., to keep the error coefficient constant). Trial and error is a possible approach [i.e., try selected zero-pole pairs checking the gain required at the designated point as shown in Figures 8.5(b), (c), and (d)], but convergence to the desired solution is not rapid; in fact, one is not guaranteed that the desired solution exists.[e] It is advisable, therefore, to investigate another technique.

For the system of Figure 8.3(a) the transfer function of the compensated system is

$$G_c G(s) = \frac{P}{Z} \frac{s + Z}{s + P} \frac{100}{s^2}, \tag{8.3}$$

where the multiplier p/z imbeds the gain adjustment, which keeps the error coefficient constant. The characteristic equation of the closed-loop system is then

$$1 + G_c G(s) = s^2(s + P) + 100\frac{P}{Z}(s + Z) = 0. \tag{8.4}$$

Now let us define $\alpha = Z/P$, from which $Z = \alpha P$. Substituting,

$$s^2(s + P) + \frac{100}{\alpha}(s + \alpha P) = 0 \tag{8.5}$$

Use of the definition $\alpha = Z/P$ is a matter of convenience—we have no guidance as to what values to choose for Z and P, but we know values of α that work (from Bode designs). For example, $0.05 < \alpha < 0.3$ is a usable range of values. Thus we can consider α a known number, so the only parameter in Eq. (8.4) is P.

Partitioning Eq. (8.5),

$$s^3 + Ps^2 + \frac{100s}{\alpha} + 100P = 0 \tag{8.6}$$

$$\frac{P(s^2 + 100)}{s(s^2 + 100/\alpha)} = 1 \tag{8.7}$$

---

[e] The designation of a specific *point* as a desired root point is a very precise requirement, especially when the error coefficient is also specified. It is a much more restrictive specification than most frequency domain specs.

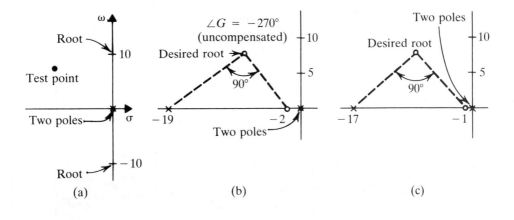

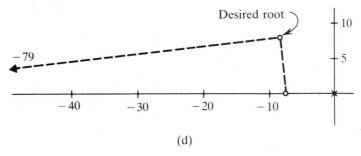

FIGURE 8.5 (a) Root Locus Study of Lead Compensation: Root Locus of Uncompensated System, Showing Roots when $G(s) = 100/s^2$; Root Locus Sketches Showing Design Trials for a Desired Root at $s = -8 + j8$: (b) Trial 1:

$$Z = 2 \qquad P = 1 \qquad K = \frac{|8 + j8|^2|8 + j8 - 19|}{|8 + j8 - 2|}$$

$$K = 174.08, \qquad K_a = (174.08)(2)/19 = 18.32;$$

(c) Trial 2:

$$Z = 1, \qquad P = 17 \qquad K = \frac{|8 + j8|^2|8 + j8 - 17|}{|8 + j8 - 1|}$$

$$K = 136.745, \qquad K_a = (136.745)(1)/17 = 8.04;$$

(d) Trial 3:

$$Z = 7, \qquad P = 79 \qquad K = \frac{|8 + j8|^2|8 + j8 - 79|}{|8 + j8 - 7|}$$

$$K = 1134.69 \qquad K_a = (1134.69)(7)/79 = 100.542.$$

Let us sketch the root loci, using $\alpha < 1.0$ because the compensator must introduce leading phase. The root loci as functions of α and P are shown in Figure 8.6(a). This family of root loci clearly indicates that the choice of a root point is *not arbitrary* if only one section of filter is to be used and if the error coefficient is not to be changed.

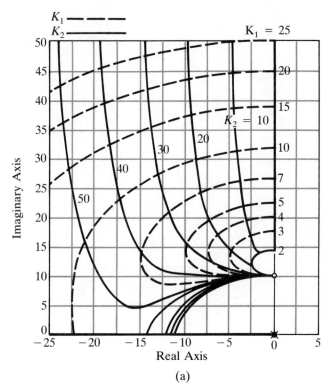

FIGURE 8.6 Two Parameter Root Loci, where $K_1 = 1/\alpha$: Usable Area Defines A (a) Root Relocation Lead Zone and (b) Root Relocation Lag Zone.

However, if we obtain a family of root loci for the two available parameters (using the computer), we need only inspect the plot to determine if an acceptable root location can be obtained; if so, we interpolate to find values for P and α, which completes the design; if not, then a more complex compensator is required.

The area within which the dominant roots can be placed using a single section of lead compensator and keeping the error coefficient unchanged is called a *root relocation lead zone*. In like manner, a root relocation lag zone can be defined for a lag compensator. Consider Eq. (8.7). For a lag compensator, $\alpha > 1.0$, and the family of root loci are given in Figure 8.6(b). For the given system, lag compensation does not work— all of the root loci are in the right half plane.

■ EXAMPLE 8.2

For the system of Example 8.1, use the two parameter root locus to design the lead compensator. Obtain complex roots with $\zeta \simeq 0.4$ and a settling time of 0.1 s. The characteristic equation of the compensated system is

$$s^3 + (10 + P)s^2 + (10P + 5000/\alpha)s + 5000P = 0,$$

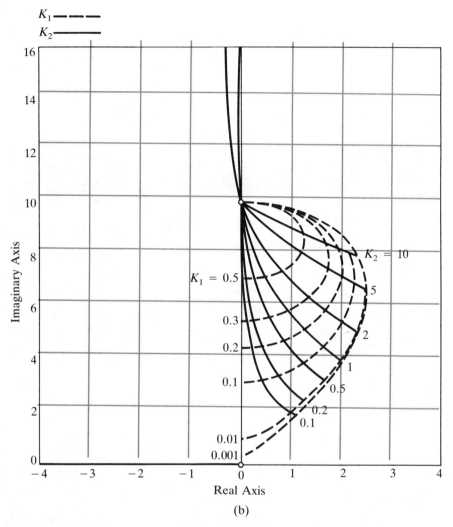

FIGURE 8.6 (*Continued*)

where $\alpha = Z/P$. The two parameter root loci are given in Figure 8.7. The $\zeta = 0.4$ line has been added and, for a settling time of 0.1 s, we need $\sigma = 40$, so that the complex roots would be at

$$s = -40 \mp j91.6.$$

Note that the Bode design of Example 8.1 used $p = 200$ with $\alpha = z/p = 50/200 = 1/4$ and $K_2 = 1/\alpha = 4$. From the curves of Figure 8.7, this would place the complex roots at $s = -61 \mp j87$.

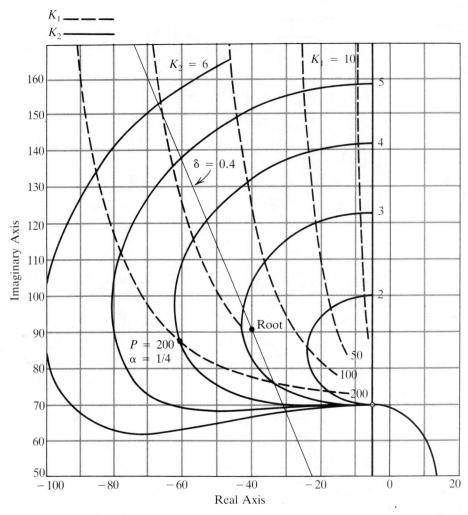

FIGURE 8.7 Example 8.2: Two Parameter Root Loci, where $K_1 = P$, $K_2 + 1/\alpha$, for the Lead Compensator Design for the System of Example 8.1.

8.5 CASCADE LAG COMPENSATOR DESIGN

For many systems, it is possible and even desirable to stabilize the system and obtain suitable dynamic response using a low-pass filter or lag compensator as shown in Figure 8.2(b). When such a filter is used, the phase curve is not helpful (lagging phase is destabilizing) and it is the attenuation (magnitude) characteristic that is used to obtain the desired effect. The philosophy is easily explained with the help of Figure 8.8. If the system is very unstable, as shown in Figure 8.8(a), it can be stabilized and given good phase margin if the gain crossover can be moved to a new location such as ω_L without changing the phase curve at that frequency. Of course, this can be done simply by reducing the forward gain, but this is usually undesirable—we want to keep the forward gain unchanged in order to preserve accuracy. We observe that a lag network (low-pass filter), as shown in Figure 8.2(b), reduces the high-frequency gain without changing the low-frequency gain. Thus, if we design a low-pass filter with enough attenuation to reduce the gain at ω_L by the required amount, the compensated system will have its gain crossover at ω_L. This is not sufficient, however. To obtain a suitable phase margin, we wish to keep the phase at ω_L essentially unchanged. Thus, in designing the lag compensator, the zero and pole must both be placed at frequencies much lower than the desired gain crossover frequency ω_L so that the negative phase introduced by the filter at ω_L is small. A reasonable rule is to place the zero about one decade below the desired gain crossover, i.e., $\omega_z \cong \omega_L/10$. The separation between zero and pole is determined by the amount of attenuation required.

 EXAMPLE 8.3

Let

$$G(s) = \frac{100}{s(s+1)(0.1s+1)}.$$

The Bode diagram is shown in Figure 8.8(b) and the system is seen to be very unstable. If the gain crossover were moved to $\omega_L = 0.7$, the phase of the uncompensated system

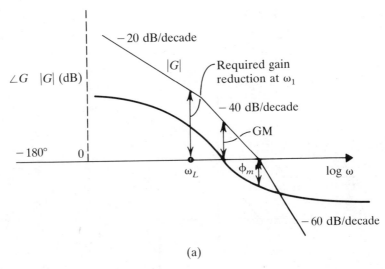

(a)

FIGURE 8.8 (a) Bode Diagram Sketch to Explain Lag Compensation and (b) Example 8.3 Bode Diagram Lag Compensation Design for

$$G(s) = \frac{100}{s(s+1)(0.1s+1)}.$$

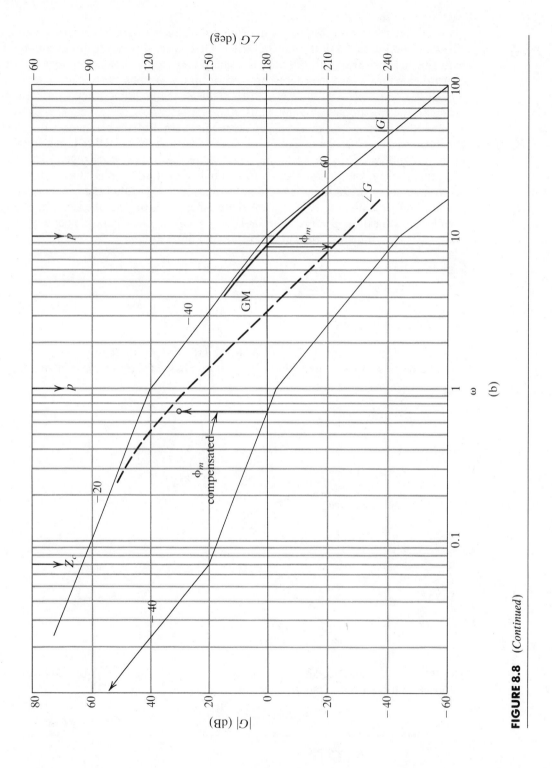

FIGURE 8.8   (*Continued*)

would provide a phase margin of about 50°. The attenuation required to put the gain crossover at ω_L is read from the diagram and is $\cong$ 43 dB. This can be provided by a lag compensator with pole and zero separated by slightly more than 2 decades. The compensator zero is introduced at $\omega_L/10 = 0.07$ and the phase curve is corrected. From the diagram, it can be seen that the net phase margin is about 45°. The equation of the compensator is

$$G_c = \frac{(s/0.07) + 1}{(s/0.00055) + 1}.$$

EXAMPLE 8.4

The root locus method provides considerable insight about the effects of lag compensation, as shown in Figure 8.9(a) and (b). Note that the pole and zero of the lag compensator are very close to the origin. For points in the s plane remote from the origin, each contributes essentially the same angle. Since these angles cancel in the angle summation, the root locus remains essentially unchanged except near the origin,

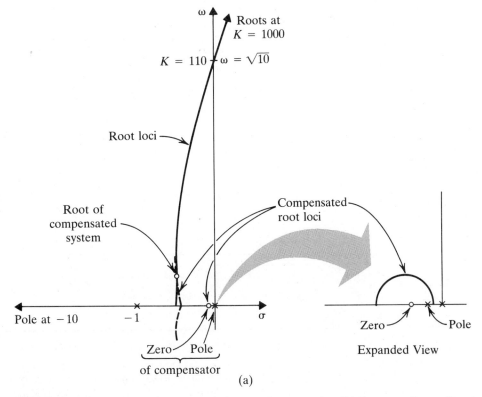

(a)

FIGURE 8.9 (a) Sketch of the Root Loci for Lag Compensation, (b) Computer Drawn Root Locus for the System of Figure 8.8(b) (Compensated), and (c) Step Response, Showing Long "Tail."

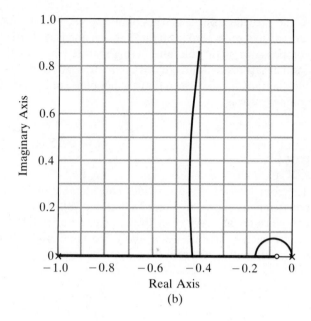

Real Axis

(b)

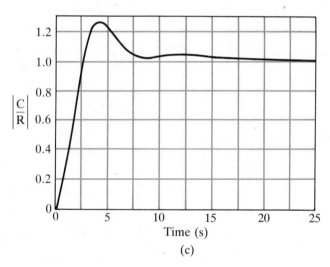

Time (s)

(c)

FIGURE 8.9 (*Continued*)

but the roots on these loci are moved to lower gain locations, i.e., closer to the poles and thus into the left half plane. The reason for this is not in the graphics, but in the algebra—insertion of the lag compensator reduces the root locus gain K (not the error coefficient). It should also be noted that the additional root generated by insertion of the compensator will be near the origin and very near to the zero. It is important that

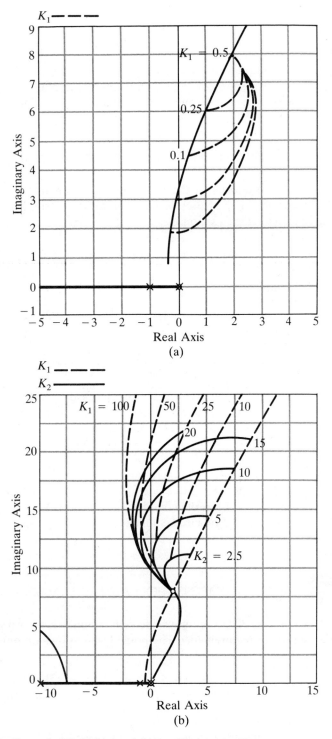

FIGURE 8.10 Example 8.3: (a) Lag and (b) Lead Relocation Zone.

this root be close to the zero since this reduces the residue at the root and makes its contribution to the system response negligibly small. If this is not the case, then the transient response may be dominated by this real root rather than the complex roots. The step response would then have a long "tail" due to the exponential contribution and this could make the settling time unacceptably long. A long "tail" is observed on the step response of Figure 8.9(c).

The root locus for the lag compensator can also be studied by partitioning, as was shown in Figure 8.3(c) for a second-order system. For the system of Figure 8.3, the characteristic equation is

$$s^4 + (11 + P)s^3 + (10 + 11P)s^2 + \left(100 + \frac{1000}{\alpha}\right)s + 1000P = 0.$$

Partitioning,

$$s^4 + 11s^3 + 10s^3 + \frac{1000}{\alpha}s + P(s^3 + 11s^2 + 10s + 1000) = 0$$

$$1 + \frac{P(s^3 + 11s^2 + 10s + 1000)}{s(s^3 + 11s^2 + 10s + 1000/\alpha)} = 0.$$

Note that the numerator is the characteristic equation of the uncompensated system. The denominator cubic is the same polynomial except for the different value of the gain term. Both cubics must be factored to apply the graphical construction[f] method, and it is seen that the zeros and poles will all lie on the root locus of the uncompensated system, as shown in Figure 8.10(a) and (b). Note that when P is small the root location will be close to the original root locus and, thus, for the lag compensator the roots will be as far as possible into the left half plane. Clearly then, one reason for wanting P to be small is to keep the complex roots as far to the left as we can with this kind of compensation. For a lead compensator with root loci as in Figure 8.10(b), a large value of P is used.

8.6   CASCADE COMPENSATION WITH SEVERAL SECTIONS OF FILTER

Design specifications may be so demanding that several sections of filter are needed to satisfy all requirements. The possible situations are numerous and the following list indicates a few possibilities:

1. The uncompensated system is very unstable. Lead compensation is required to get a fast response, but one zero-pole pair cannot provide enough phase margin.

---

[f] A computer program to plot the loci is preferable.

2.   The uncompensated system is very unstable. Lag compensation will be used, but cannot supply enough bandwidth so some lead compensation is also required.

3.   System is slightly unstable, but faster response is needed with nearly critical damping.

For such cases, if cascade compensation is used, several poles and zeros are required. This eliminates the root locus as a tool because too many variables are involved. The Bode diagram remains very practical, however, though the amount of trial and error required increases somewhat. Note that several basic questions do not have general answers. For instance, how many zeros and poles should be used? There is no precise answer; there may be a *minimum* number required to meet specifications, but a better overall design may result if a larger number is used. If the question is, how many lag sections and how many lead sections would be best?, again there is no precise answer. Yet experience and trial and error will usually result in a good system; one simply modifies the Bode diagram graphically to obtain the desired phase and gain margin and an acceptable gain crossover frequency.

The next step is to obtain the closed-loop frequency response (via the Nichol's chart or with the computer). If this appears acceptable, some simulation is advisable. If at any point in this sequence some result is not satisfactory, we return to the Bode diagram, modify it to correct (hopefully) the deficiency, and continues the sequence. The number of trials required depends on the complexity and difficulty of the problem and on the experience of the control engineer. Example 8.5 demonstrates the procedure on a rather simple problem.

   EXAMPLE 8.5

For this example, we use the system of Example 8.3, for which

$$G(s) = \frac{100}{s(s + 1)(0.1s + 1)},$$

but with more restrictive specifications: i.e., phase margin $\cong 45°$ and settling time $\cong 2.5$ s.

Using the second-order correlations of Chapter 4, a phase margin of $45°$ corresponds to $\zeta \cong 0.4$. The natural frequency, ω_n, is obtained from

$$T_s = \frac{4}{\zeta \omega_n} \qquad \omega_n = \frac{4}{\zeta T_s} = \frac{4}{(0.4)(2.5)} = 4 \text{ rad/s}.$$

Therefore, in designing the compensator on the Bode diagram, we should place the gain crossover in the vicinity of $\omega = 4$, and at that frequency more than $50°$ of positive phase must be introduced to achieve $45°$ phase margin.

As a beginning, design the lag compensator to first move the gain crossover to the same frequency as the phase crossover of the uncompensated system, which is $\omega = 3.2$. This is shown on the Bode diagram of Figure 8.11(a), and is achieved with a zero at $\omega = 0.32$ and a pole at $\omega = 0.032$. A single section of lead filter is added, placing the

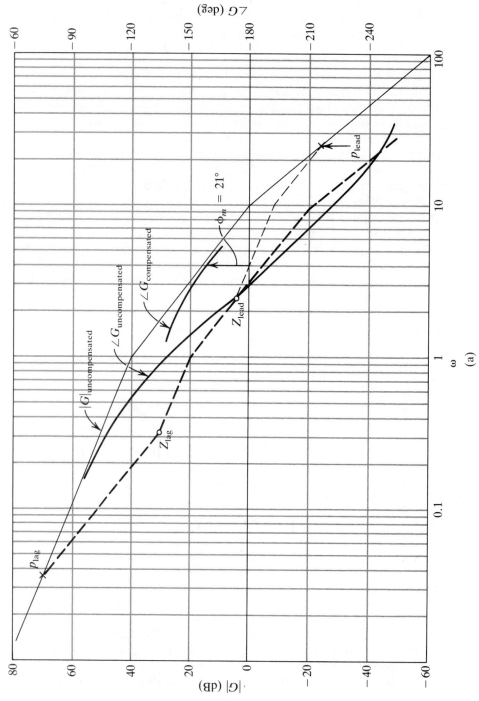

FIGURE 8.11 Example 8.5, Bode Diagram Design, Using Both Lag and Lead: (a) First Trial, (b) Second Trial, (c) Nichols Chart for Compensated System of (b), and (d) Step Response of Compensated System.

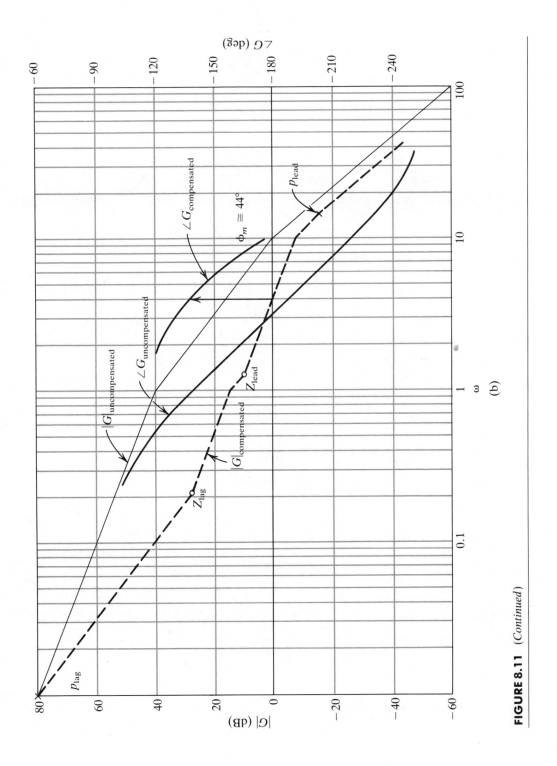

FIGURE 8.11   (*Continued*)

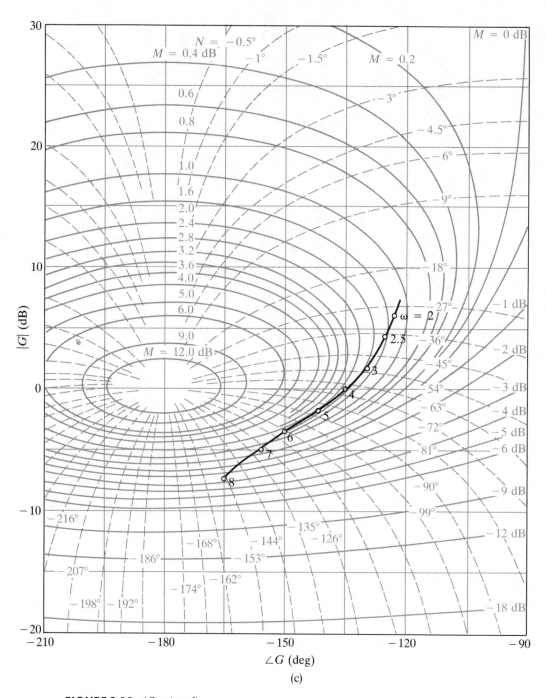

FIGURE 8.11 (*Continued*)

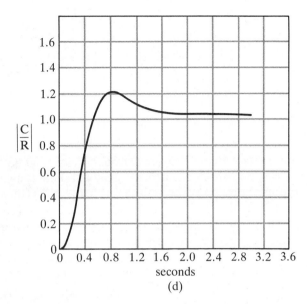

FIGURE 8.11 (*Continued*)

gain crossover at $\omega = 4$. To do this, the filter has a zero at $\omega = 2.5$ and a pole at $\omega = 25$. The phase curve is calculated for the compensated system as shown, and the phase margin is $\phi_m = 21°$, which is clearly inadequate and a second trial is needed.

What are the requirements and alternatives for a second trial? If we keep the gain crossover at $\omega = 4$ (note that this, too, was an assumption and may have to be modified later), we *must* introduce more positive phase at $\omega = 4$. This can be done if the zero of the lead compensator is moved to lower frequency and the lag compensator is also redesigned to maintain the gain crossover at $\omega = 4$. An alternative method would be to introduce an additional lead section, placing its zero so that the gain crossover is not changed appreciably. Of course, when a second lead section is added, it may be desirable to modify either the lag section or the first lead section or both.

A second trial is made as shown on the Bode diagram of Figure 8.11(b), moving the zero of the lead compensator to lower frequency and modifying the lag section. The resulting compensator is

$$G_c = \frac{[(s/0.22) + 1][s/1.3) + 1}{[(s/0.01) + 1][s/15) + 1]}.$$

This provides a phase margin of approximately 44° which is close to the desired value, so a check on the Nichol's chart is desired. This is shown in Figure 8.11(c), and we determine:

$$M_{p\omega} = 2.6 \text{ dB}, \qquad \omega_r = 4.5; \qquad \omega_b = 8.2.$$

To obtain a value for ω_n, the second-order conditions of Chapter 4 give:

$$\omega_n = \frac{\omega_r}{\sqrt{1 + 2\zeta^2}} = \frac{4.5}{\sqrt{1 - 2(0.4)^2}} = \frac{4.5}{\sqrt{1 - 0.32}} = 5.46.$$

Also from Figure 4.5,

$$\frac{\omega_b}{\omega_n}(\zeta = 0.4) = 1.37; \qquad \omega_n = \frac{\omega_b}{1.37} = \frac{8.2}{1.37} = 5.99.$$

Since these values for ω_n are reasonably close and both exceed the estimated value of $\omega_n = 4$, we conclude that the system will settle faster than required. Thus, the given specifications have been satisfied, although the system is clearly faster than required.

If desired, another trial might be made to reduce ω_n. A simple approach is to move the gain crossover to a lower frequency, probably by increasing the attenuation of the lag section. This is left to the student as an exercise. Assuming that the design is considered acceptable, a simulation check would be made. The step response of the compensated system is shown in Figure 8.11(d).

8.6A   PD, PI, and PID Controllers

When the controls engineer is designing a system, there are practical constraints on his design efforts. The development engineer may have considerable freedom of choice in his design efforts. The applications engineer, particularly those engaged in the process control industry, are usually expected to choose an available controller from one of the companies manufacturing a line of process controllers. He is expected to match the controller to the particular process by adjusting the parameters of the controller. The system characteristics that need to be adjusted are, of course, accuracy, response time, and damping.

As shown in previous chapters, increasing the gain also increases bandwidth and makes the response faster and more accurate, but usually decreases damping. Damping is improved by introducing a derivative signal and, if a substantial increase in accuracy is required, an integrator is used. Commercial controllers combine several of these concepts:

$$PD = \text{Proportional} + \text{Derivative} \rightarrow G(s) = K_p + K_D s$$

$$PI = \text{Proportional} + \text{Integral} \rightarrow G(s) = K_p + \frac{K_I}{s}$$

$$PID = \text{Proportional} + \text{Integral} + \text{Derivative} \rightarrow G(s)$$

$$= K_p + \frac{K_I}{s} + K_D s.$$

The adjustable parameters are the gains K_p, K_I, and K_D. It should also be noted that in the process control industry *derivative* is also called *rate* and *integral* is also called *reset*.

These controllers have been remarkably successful, especially the PID. This success has been due in no small measure to the adjustment procedures developed by Ziegler and Nichols.[1,2] Because of the demonstrated capability of the PID, it has been used in many applications, such as autopilots (for both ships and aircraft). This is not hard to understand if one considers the transfer functions of these controllers. For the PD controller,

$$G(s) = K_p + K_D s = K_D\left(s + \frac{K_p}{K_D}\right).$$

This controller simply introduces a free zero and the design problem is to place the zero where it will do the most good and adjust the gain accordingly. This is accomplished readily with the Bode or root locus methods.

For the PI controller,

$$G(s) = K_p + \frac{K_I}{s} = \frac{K_p(s + K_I/K_p)}{s}$$

and again the design problem is to place the zero at the most suitable location while adjusting the gain. Bode and root locus methods both work.

For the PID controller,

$$G(s) = K_p + \frac{K_I}{s} + K_D s = \frac{K_D s^2 + K_p s + K_I}{s},$$

This places a pole at the origin but provides two zeros for adjustment of the dynamic response. The two zeros may be real or complex depending on the gains used, but in either event both zeros will be in the left half plane. Thus, the design problem is to place the two zeros appropriately, adjust the loop gain, and determine the required values of the three adjustable gains. Clearly, this can be done with Bode or root locus methods.

With the advent of microprocessors, many applications are implementing compensators digitally. The PID controller is implemented readily with the microprocessor and no doubt will be widely used.

8.7   AN ANALYTIC APPROACH TO CASCADE COMPENSATION

In the preceding sections, the design of cascade compensation has been presented as a trial-and-error approach using graphical techniques. In brief, the designer starts with the Bode diagram of the basic system $G(s)$. He then alters the gain curve by inserting new *corners*, and changes the phase curve as required by the added zeros and poles. When the modified curves provide acceptable phase margin, gain margin, etc., the transfer function of the compensation, $G_c(s)$, is determined by reading the values of the zeros and poles from the Bode diagram. When the root locus is used instead of the Bode diagram, the approach differs in detail but the concept and philosophy are very similar. In terms of the transfer function algebra,

$$G_{eq}(s) = G_c(s)G(s), \tag{8.8}$$

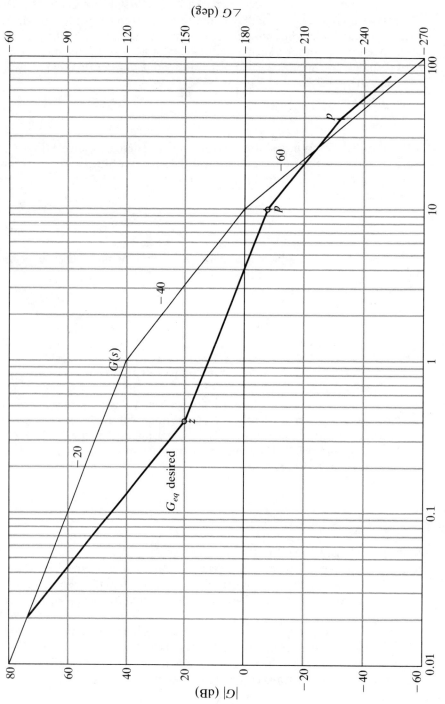

FIGURE 8.12 Example 8.6: (a) Bode Design of Cascade Compensation for the System of Examples 8.3 and 8.5 and (b) Step Response.

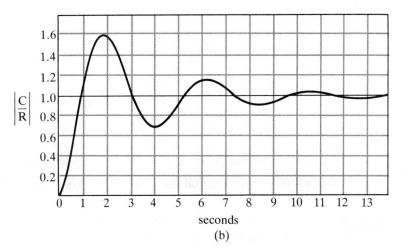

FIGURE 8.12   (*Continued*)

where $G(s)$ is known initially and $G_c(s)$ is determined by graphical trial and error. An alternative approach[g] is to *choose* the desired $G_{eq}(s)$ and solve for $G_c(s)$, i.e.,

$$G_c(s) = \frac{G_{eq}(s)}{G(s)},$$

where the choice of $G_{eq}(s)$ is based on the system specifications. It may or may not consider the nature of the plant transfer function. In choosing $G_{eq}(s)$, one may use any convenient method. Here we illustrate the approach using the Bode diagram as a convenient vehicle.

 EXAMPLE 8.6

Consider the system of Example 8.5:

$$G(s) = \frac{100}{s(s + 1)(.1s + 1)}$$

with the same specifications used in Example 8.4. The phase margin is approximately 45° and $T_s = 2.5$ s. To meet specifications, we choose a -20 dB/decade asymptote providing a gain crossover at $\omega = 4$. We place a zero at $\omega = 0.4$ and one pole at $\omega = 10$ in order to provide a substantial phase margin, a low-frequency pole at $\omega = 0.02$ to maintain K_v at 100, and a high-frequency pole at $\omega = 40$ to provide a -60 dB/decade asymptote and thus make the compensator physically realizable. Then, as shown in Figure 8.12(a),

$$G_{eq}(s) = \frac{100(0.4s + 1)}{s(50s + 1)(0.1s + 1)(0.025s + 1)}.$$

---

[g] This approach is conceptually similar to the Truxal-Guillemin method, which chooses the *closed-loop* transfer function from specifications and solves algebraically for the open-loop transfer functions.

Since

$$G(s) = \frac{100}{s(s + 1)(0.1s + 1)},$$

$$G_c(s) = \frac{G_{eq}(s)}{G(s)} = \frac{\dfrac{100(2.5s + 1)}{s(50s + 1)(0.1s + 1)(0.025s + 1)}}{\dfrac{100}{s(s + 1)(0.1s + 1)}} = \frac{(2.5s + 1)(s + 1)}{(50s + 1)(0.025s + 1)}.$$

The step response of the system is shown in Figure 8.12(b).

8.8   FEEDBACK COMPENSATION

In many systems, it is desirable to use one or more compensators in feedback rather than in cascade. Possible arrangements are shown in Figure 8.13. The decision to use a feedback scheme rather than[h] a cascade compensator is sometimes a matter of convenience, sometimes a matter of necessity, and for some problems it can be shown that a feedback compensator will do a better job. For example, use of velocity feedback for damping is very effective and consequently quite popular. Instrument servo motors with builtin tachometers are shelf items, so the use of velocity feedback can be very convenient. Conversely, when designing controls for hydraulic and pneumatic systems, as well as some process controls, it is difficult to devise an appropriate cascade component for a compensator, but sensors are available to measure a variety of variables and the resulting signals can be fed back to adjust a valve. Finally, in systems subjected to both input commands and load disturbances, a cascade compensator designed to damp transients due to a command input may not be as effective in suppressing transients due to a load disturbance as a feedback compensator designed specifically for that purpose.

The purposes for which the feedback compensator is designed are, in general, the same as those for which a cascade compensator would be used. In theory, either approach can be used and either should be capable of satisfying the same set of specifications. In practice, there may be reasons for preferring one over the other for a specific application.

One point of view that is helpful in developing techniques for analysis and design is to realize that either method, cascade or feedback compensation, ultimately replaces a forward transfer function $G(s)$ with a new transfer function $G_{eq}(s)$. The design process, then can be thought of as converting a $G(s)$ that is not suitable to a $G_{eq}(s)$ that is more satisfactory for the proposed system.

---

[h] On occasion, both may be used to adjust system performance. Discussion of such cases is beyond the scope of this text.

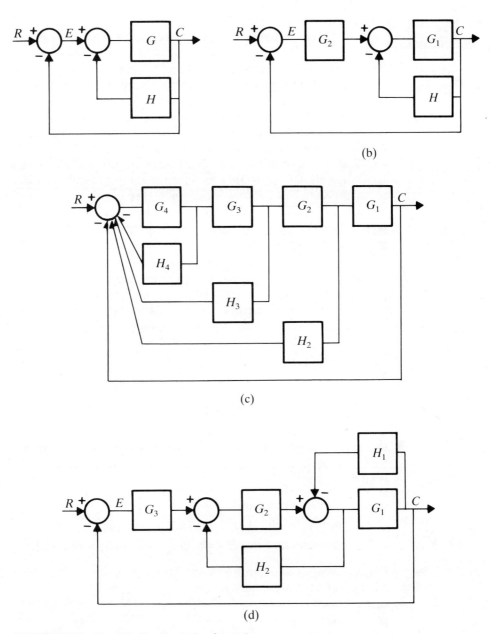

(b)

(c)

(d)

FIGURE 8.13   Possible Feedback Configurations.

8.9   ROOT LOCUS STUDIES: VELOCITY AND ACCELERATION FEEDBACK

Because the root locus method is basically a one parameter method its usefulness in studying feedback compensation is limited,[i] but very beneficial where it can be used. As will be shown, we can obtain not only numerical solutions to specific problems, but can sometimes obtain insight into entire classes of problems. Tachometer feedback is commonly used in practice; root locus studies help to point out its capabilities. Acceleration feedback is less commonly used but we shall also analyze its capabilities and limitations. With suitable constraints we can include both in some studies. Unfortunately, the important and more general case of state feedback usually involves too many variables for root locus studies, as is also the case with most feedback compensation problems using filters in the feedback path.

 EXAMPLE 8.7

Consider the system of Figure 8.14. The general case is indicated in Figure 8.14(a); the velocity is measured and the signal amplified and fed back around the plant. For the purposes of this study, we neglect the gain and dynamics (if any) of the error detector. For the second-order system of Figure 8.14(b), the characteristic equation is

$$s^2 + (P + Kk)s + K = 0, \tag{8.9}$$

which partitions to

$$\frac{kKs}{s^2 + Ps + K} = -1. \tag{8.10}$$

Note that the denominator is the characteristic equation of the uncompensated system so its factors are the roots of the uncompensated system. The resulting root locus, for k variable and selected values of K, is shown in Figure 8.15. Observe that as k is increased:

1. ω_n remains constant
2. ζ increases; critical damping and overdamping are available
3. σ increases (in a negative direction) and thus the settling time decreases.

Clearly, compensation with velocity feedback is an excellent way to improve the response of a second-order system.

 EXAMPLE 8.8

For the third-order system of Figure 8.14(c), the characteristic equation is

$$s(s + 1)(s + 7) + kKs + K = 0, \tag{8.11}$$

---

[i] As previously noted, root locus studies for cascade compensation are also limited.

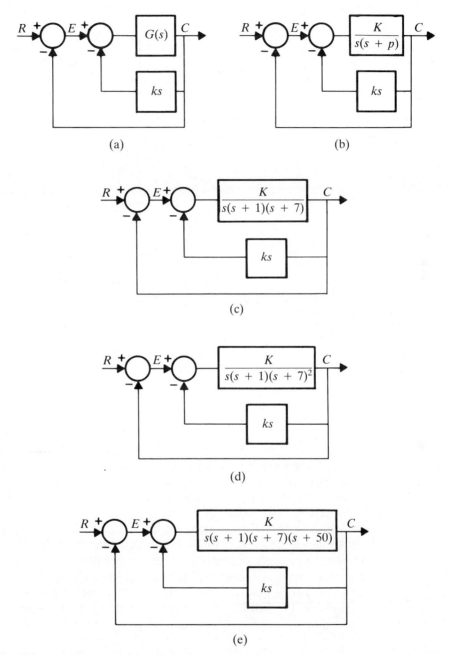

FIGURE 8.14 Block Diagrams for Velocity Feedback: (a) General Case, (b) Second-Order Plant, (c) Specific Third-Order Plant, (d) Specific Fourth-Order System, and (e) Another Fourth-Order Plant.

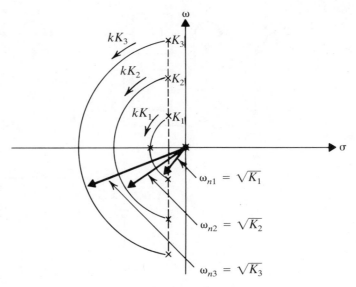

FIGURE 8.15   Example 8.7: Root Loci for Velocity Feedback of the Second-Order System of Figure 8.14(b).

and after partitioning we get

$$\frac{kKs}{s(s+1)(s+7)+K} = -1. \tag{8.12}$$

Again, the denominator is the characteristic equation of the uncompensated system and thus the poles of Eq. (8.12) are the roots of the uncompensated system. The root loci for several values of K are shown in Figure 8.16.

In Figure 8.16, the root locus for the uncompensated system has been drawn. In constructing the loci for the feedback gain k, note that the number of asymptotes is reduced from three to two. Also note that the asymptote centroid has been moved to the left and stays in the same location for all[j] values of K. Both of these features permit the velocity feedback to move the roots to the left, thus increasing both ζ and ω_n. Since there are only two asymptotes, increasing the feedback gain is guaranteed to drive the complex roots into the left half plane, i.e., *is guaranteed to stabilize the system*—not only this specific third-order system, but *all* minimum phase third-order systems.

Consider the root loci of Figure 8.16 in close detail. It appears that the root loci starting on the complex poles never return to the real axis (when the uncompensated

---

[j] For the uncompensated system, the sum of the poles is $0 + 1 + 7 = 8$, and the asymptote centroid is $\sigma = -8/3$. We note, on expansion of the characteristic equation to $s^3 + 8s^2 + 7s + K = 0$, that the sum of the roots is also equal to the sum of the poles, i.e., 8. Since the gain K may be given any value without changing the coefficient of the s^2 term, the sum of the poles of Eq. (8.12) will always be 8, therefore the asymptote centroid is invariant with k and K.

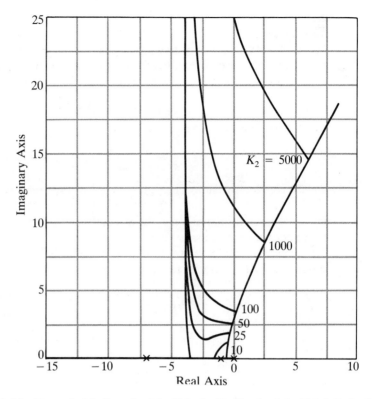

FIGURE 8.16   Example 8.8: Root Loci for Velocity Feedback of the Third-Order System of Figure 8.14(c).

system is unstable) so the velocity feedback does not overdamp such systems, or at least there will always be complex roots. This is true, yet the system can be adjusted to have a step response that is essentially exponential. Note that all three roots start on the poles at $k = 0$. As k is increased, the real root moves toward the zero at the origin, and for large k the real root dominates while the residue at the complex roots gets smaller so that the velocity feedback provides a response that is essentially overdamped.

From the preceding analysis of a specific third-order system with velocity feedback, we can draw some rather general conclusions about all minimum phase third-order systems:

1.   Velocity feedback is capable of stabilizing any third-order system.[k]

2.   Velocity feedback can provide an essentially overdamped response for any third-order system.

---

[k] Providing the asymptote centroid of the uncompensated system is in the left half plane.

EXAMPLE 8.9

When velocity feedback is used to stabilize or damp a fourth (or higher) order system the results may be excellent, but some limitations can be shown to exist. Consider the system of Figure 8.14(d). The characteristic equation is:

$$s(s + 1)(s + 7)^2 + kKs + K = 0, \tag{8.13}$$

which partitions to

$$\frac{kKs}{s(s + 1)(s + 7)^2 + K} = -1 \tag{8.14}$$

and the root loci are shown in Figure 8.17. We note from Figure 8.17 that there are now three asymptotes, because of the added pole. Of course the asymptotes at $\mp 60°$ cross into the right half plane. When the original gain K is set at the stability limit $K = K_1$, the velocity feedback does stabilize the system providing a moderate value of k is used; if k is made too large, the roots go into the right half plane again. If the main gain is raised to $K = K_2$, velocity feedback does *not* stabilize the system. The root locus does

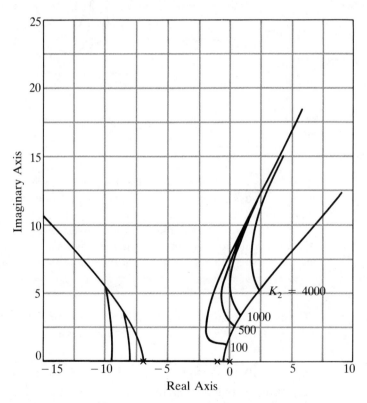

FIGURE 8.17   Example 8.9: Root Loci for Velocity Feedback for a Fourth-Order System ($P_4 = 7$).

not cross the imaginary axis but stays completely in the right half plane. Thus, if we choose to use velocity feedback with a fourth (or higher) order system, we are not guaranteed that it will do the job. We are not even guaranteed that it will stabilize the system. However, it should be realized that the example of Figure 8.17 was deliberately chosen to illustrate the point that velocity feedback is not *guaranteed* to work!

More realistic fourth-order systems would probably have a fourth pole that is much larger, as in Figure 8.14(e). For this system, the characteristic equation is

$$s(s + 1)(s + 7)(s + 50) + kKs + K = 0 \qquad (8.15)$$

and the partitioned equation is

$$\frac{kKs}{s(s + 1)(s + 7)(s + 50) + K} = -1. \qquad (8.16)$$

It is clear from the root loci of Figure 8.18 that velocity feedback is capable of

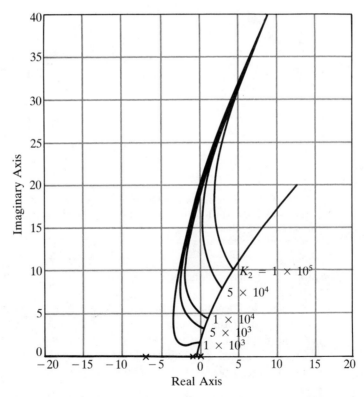

FIGURE 8.18 Example 8.9: Root Loci for Velocity Feedback for a Fourth-Order System ($P_4 = 50$).

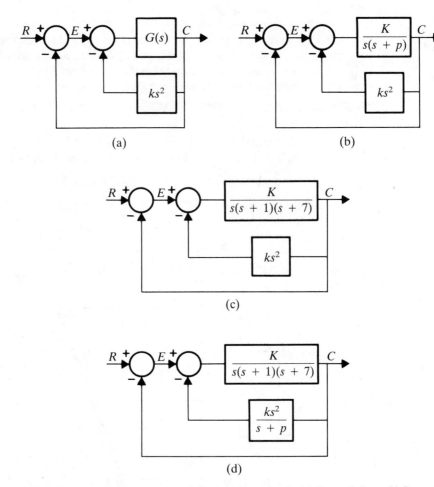

FIGURE 8.19 Block Diagrams for Acceleration Feedback: (a) General Case, (b) Second-Order System, (c) Third-Order Case, and (d) Third-Order System with *Approximate* Acceleration Feedback.

stabilizing the system and providing reasonable damping unless the forward gain K is very high.

Acceleration feedback is sometimes suggested as a means of compensating a system. We shall analyze several cases using root loci. The systems to be analyzed are shown in Figure 8.19. In general, as in Figure 8.19(a), the acceleration of the output is measured, fed back, and subtracted from the error signal. (Positive feedback may be useful in some applications but is not considered here.)

 EXAMPLE 8.10

For the second-order system of Figure 8.19(b), the characteristic equation is

$$s^2 + Ps + kKs^2 + K = 0 \qquad (8.17)$$

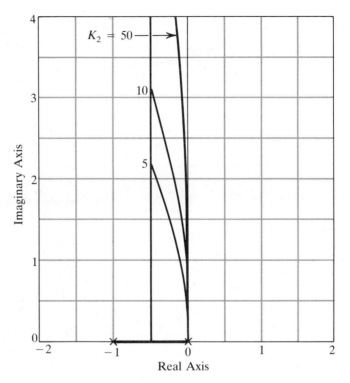

FIGURE 8.20   Example 8.10: Root Loci for Acceleration Feedback for the Second-Order System of Figure 8.19(b).

and the partitioned equation is

$$\frac{kKs^2}{s^2 + Ps + K} = -1. \tag{8.18}$$

The root loci are shown in Figure 8.20. It is clear that the acceleration feedback is reducing ζ, reducing ω_n, and reducing σ. The damping is reduced, oscillating frequencies are reduced (bandwidth reduced), and settling time increased. These are not desirable characteristics except for very special problems.

Since the effect of acceleration feedback on a second-order system is the reverse of what we usually want, an obvious question is, can acceleration feedback stabilize an unstable system? To answer this, let us look at the system of Figure 8.19(c).

 EXAMPLE 8.11

The characteristic equation is

$$s(s + 1)(s + 7) + kKs^2 + K = 0 \tag{8.19}$$

and the partitioned equation is

$$\frac{kKs^2}{s(s + 1)(s + 7) + K} = -1. \tag{8.20}$$

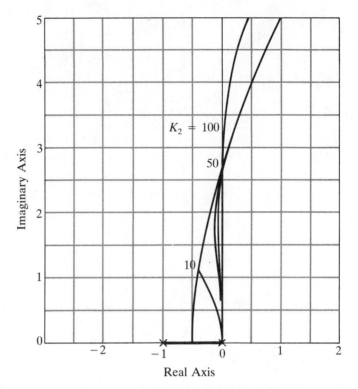

FIGURE 8.21 Example 8.11: Root Loci for Acceleration Feedback for the Third-Order System of Figure 8.19(c).

The root loci are shown in Figure 8.21. As seen, the root loci from the poles *do* cross the imaginary axis into the left half plane, so the system would be stabilized, though with little damping available.

The preceding analysis implies that acceleration feedback, used by itself, is not a very satisfactory means of compensation. Obviously the benefits are limited, so careful consideration should be given to any problem where such feedback is thought to be convenient and/or potentially useful. However, one should also consider the fact that commercially available accelerometers do not have a transfer function $H(s) = ks^2$! All of them have at least one pole and usually two poles because it is not physically possible to build an ideal accelerometer. The devices have specified ranges over which they provide a "good" acceleration signal. An accelerometer that is good over a wide bandwidth has poles that are remote, i.e., far into the left half s plane.

EXAMPLE 8.12

Let us consider a more realistic case as in Figure 8.19(d). The characteristic equation of this system is

$$s(s + 1)(s + 7)(s + P) + kKs^2 + K(s + P) = 0 \qquad \textbf{(8.21)}$$

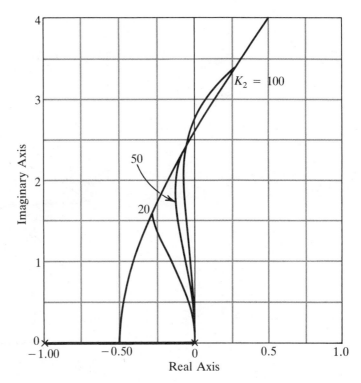

FIGURE 8.22   Example 8.12: Root Loci for Approximate Acceleration Feedback for the Third-Order System of Figure 8.19(d) where the Accelerometer Pole is Large.

and after partitioning

$$\frac{kKs^2}{(s + P)[s(s + 1)(s + 7) + K]} = -1. \tag{8.22}$$

Note that the accelerometer pole is a *factor* in the denominator. This means that when we study this system on the s plane we can locate this pole anywhere we wish in order to study its effect. The other poles remain fixed in locations determined by the gain of the uncompensated system. When P is a large number, the root loci are as shown in Figure 8.22. Observe that for a large value of P the root locus is almost exactly the same as for the system of Figure 8.19(c) (i.e., Figure 8.21). The reason is clear; when P is remote, its angle contribution to the angle summation is negligible at points near the origin, so these segments of the root loci are essentially unchanged by the presence of the remote pole. This also means that the accelerometer *is* a good one for the range of frequencies associated with the dominant complex roots of this system.

If our purpose is to stabilize and damp the system by feeding back a measurable signal, we do not necessarily require that the signal measured with the accelerometer be

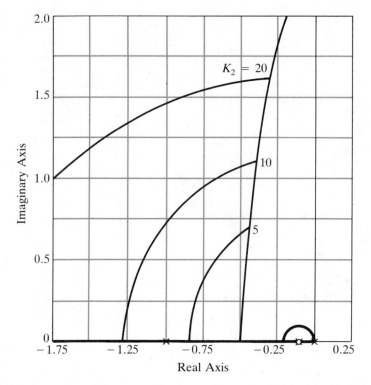

FIGURE 8.23 Example 8.17: Root Loci for Approximate Acceleration Feedback for a Third-Order System where the Accelerometer Pole is Small.

a good acceleration signal. Thus, we ask if the accelerometer pole is not large, what effect does this have on the root loci? Clearly, if the accelerometer pole is made smaller, its contribution to the angle summation increases and the shape of the loci change. As the limit is approached, i.e., as the pole is brought close to the origin, the result is as shown in Figure 8.23. Note that the root loci in the complex plane are quite similar to those of Figure 8.15 for velocity feedback. In the case of Figure 8.23, dominant complex roots can be located with a small value of k, for which there will also be a real root quite close to the origin. Fortunately, the accelerometer pole P becomes a zero of the closed-loop function so the contribution of this small root may be completely negligible.

It appears from Figure 8.23 that use of an acceleration measurement for feedback compensation may work quite well if a small pole is associated with the accelerometer. It also appears that the results are essentially the same as if a velocity signal were fed back. One then asks why use the accelerometer? There are several kinds of answers to this question. In some applications, it is not feasible to measure a velocity signal, but an accelerometer might be a practical measuring device. Under other circumstances, cost or weight or space limitations might favor the accelerometer. From the point of view of dynamic performance, there seems to be little difference in the response to step inputs,

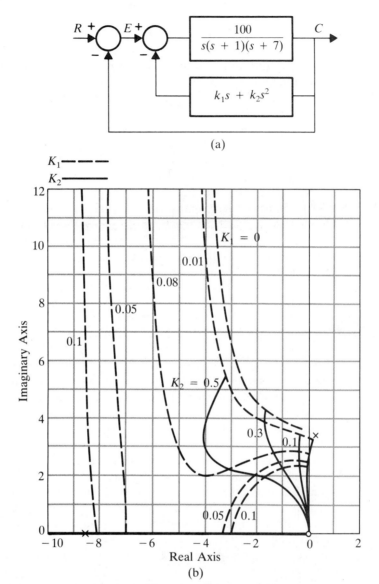

FIGURE 8.24 (a) Block Diagram for Root Locus Study of Velocity and Acceleration
Feedback.

Characteristic equation:

$$s(s + 1)(s + 7) + 100(k_1 s + k_2 s^2) + 100 = 0$$

From which:

$$\text{(I)} \quad \frac{100 k_1 s}{s^3 + (8 + 100 k_2)s^2 + 7s + 100} = -1$$

$$\text{(II)} \quad \frac{100 k_2 s^2}{s^3 + 8s^2 + (7 + 100 k_1)s + 100} = -1.$$

(b) Two Parameter Root Locus Family for System of Figure 8.24(a)

---

but if the sysrem is to be subjected to ramp inputs, velocity feedback reduces the error coefficient and thus impacts steady-state accuracy while the acceleration feedback does not.

It is physically possible to use both velocity and acceleration feedback and to have a variable forward gain also. For a third-order system, this would constitute complete state feedback, which will be considered later in this chapter. The effects of three variables on root locations cannot be studied conveniently with the root locus method, but if we consider the forward gain to be fixed the effects of both velocity and acceleration feedback can be studied. Consider the system of Figure 8.24(a). Let $k_2 = 0$, then the denominator of Eq. (I) from the figure caption is the characteristic equation of the uncompensated system and its roots are the poles of Eq. (I). We then draw the root loci for Eq. (I), which are functions of k_1 only. For specific values of k_1, we locate root points on the k_1 loci and substitute these values in Eq. (II) to obtain the k_2 loci. It is clear from Figure 8.24(b) that use of both feedback signals provides much more control of root locations that does either by itself.

8.10   BODE DIAGRAM METHODS FOR FEEDBACK COMPENSATION

The block diagrams of Figure 8.13 indicate a variety of configurations that might be used in feedback compensation of control systems. For purposes of terminology we note that the basic system (all H blocks removed) is usually called the *major loop*. Each closed loop containing an H block is called a *minor loop*. The purpose of such minor loops is primarily that of stabilizing the system and providing damping of system response to the command input R. As has been stated previously, cascade compensation and feedback compensation are entirely equivalent when the system is linear and only basic specifications are considered; that is, they are equivalent in that either can be used and there is no mathematical reason for a preference. However, when the system contains nonlinearities, when load disturbances may occur, or when noise may enter the system at an intermediate point, feedback compensation may be demonstrably better than cascade compensation. It is therefore important that we consider the design of feedback compensators. Their advantages will be discussed at a later point.

A basic point of view (one that is convenient though not necessary) is that the use of a feedback loop around one or more components effectively replaces the transfer function of the forward elements $G(s)$, which are not desirable for the system, with a new transfer function that is more suitable for our purposes. Consider the simple block diagram of Figure 8.13(a), which has a forward block $G(s)$ and a feedback compensator $H(s)$. By block diagram reduction of the minor loop, we see that $G(s)$ is replaced by

$$G_{eq}(s) = \frac{G(s)}{1 + GH(s)}. \tag{8.23}$$

Theoretically, we may feedback around the entire forward transfer function and thus replace it completely with a new one. Whether this is possible depends on the availability of signals and of summing junctions. Whether or not it is practical to use such compensation depends on a number of factors, among which are design problems in stabilizing the minor loop and the complexity of the resulting $H(s)$ compensator. Other factors depend on the specific system and on the purposes for which a feedback compensator was chosen. In either case, since the feedback compensator replaces $G(s)$ with $G_{eq}(s)$, the desired $G_{eq}(s)$ must be specified. In general, this may be done on the Bode diagram by first drawing the asymptotes for the major loop, then locating the poles of $G(s)$ that are to be changed by $H(s)$. Procedures beyond that point will be demonstrated after the design tools are developed.

Let us consider the basic equation in the frequency domain:

$$G_{eq}(j\omega) = \frac{G(j\omega)}{1 + G(j\omega)H(j\omega)}. \tag{8.24}$$

In general, there will be a range of frequencies in which $|G(j\omega)|$ is large and some other range in which it is small. This is also true for $|H(j\omega)|$, and the ranges of frequencies will probably overlap but surely will not be identical. We observe, then, that over some range of frequencies

$$|G(j\omega)H(j\omega)| \gg 1.0, \tag{8.25a}$$

and in this range Eq. (8.24) reduces to

$$G_{eq} = \frac{G(j\omega)}{G(j\omega)H(j\omega)} = \frac{1}{H(j\omega)}. \tag{8.25b}$$

Over some other range of frequencies,

$$|G(j\omega)H(j\omega)| \ll 1.0 \tag{8.25c}$$

and

$$G_{eq} = \frac{G(j\omega)}{1} = G(j\omega). \tag{8.25d}$$

For convenience, and because $G(j\omega)H(j\omega)$ is a vector quantity, we simplify equations (8.25a) and (8.25c) to the simpler conditions

$$|G(j\omega)H(j\omega)| > 1.0,$$
$$|G(j\omega)H(j\omega)| < 1.0. \tag{8.26}$$

Equation (8.26) is easy to interpret graphically on the Bode diagram and, although an approximation is thereby introduced, the resulting designs usually do not deviate significantly from the less approximate results.

The graphical interpretation of Eq. (8.26) is developed as follows:

$$20\log_{10}|GH| = 20\log_{10}|G| + 20\log_{10}|H|$$
$$= 20\log_{10}|G| - 20\log_{10}|1/H|; \tag{8.27}$$

thus, if we plot the asymptotes for G and the asymptotes for $1/H$ on the Bode diagram, whenever the G curve is *above* the $1/H$ curve, $|GH| > 1$. Whenever the G curve is *below* the $1/H$ curve, $|GH| < 1$. At the intersection of these curves, $|GH| \cong 1$.

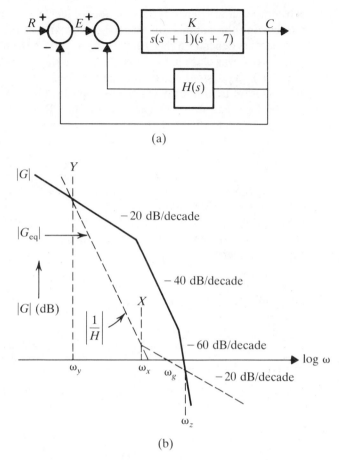

FIGURE 8.25 Bode Design of a Feedback Compensator: (a) Block Diagram and (b) Bode Diagram Showing Design Construction.

To illustrate the methods involved, consider the system of Figure 8.25(a). To determine the H function, we first draw the Bode diagram for $|G|$, as shown in the sketch of Figure 8.25(b). We do *not* draw the phase curve because the criteria of Eqs. (8.25) are magnitude criteria. Next, the $|1/H|$ function is constructed to satisfy specifications. For this illustration, let the requirements be simply to stabilize the system with reasonable phase margin and without reducing K_v. Therefore, we draw in an asymptote with slope of -20 dB/decade, which crosses the 0-dB axis to the left of the uncompensated gain crossover as shown. This segment must be long enough to generate a substantial positive phase shift and the gain crossover must be sufficiently to the left of the original gain crossover so that a substantial amount of this introduced phase shift becomes phase margin. To satisfy the error coefficient restriction, a second asymptote is introduced. Its slope is chosen at -40 dB/decade for this illustration. It connects with the previously introduced segment providing a corner at X, and it is extended to cross the original low frequency -20 dB/decade asymptote at Y. These two introduced asymptotes design the compensator. We define them to be the Bode diagram of $1/H$, and we read off the transfer function of H as follows:

$$\frac{1}{H} = \frac{(\omega_g)^2[(s/\omega_x) + 1]}{s^2}, \tag{8.28}$$

$$H = \frac{(\omega_g)^{-2}s^2}{(s/\omega_x) + 1}. \tag{8.29}$$

Note that H has two zeros and only one pole so it may not be physically realizable, in which case we introduce another pole at much higher frequency to satisfy the realizability requirement without noticeably changing conditions at the gain crossover frequency. Then the feedback compensator becomes

$$H = \frac{(\omega_g)^{-2}s^2}{[(s/\omega_x) + 1][(s/P) + 1]}. \tag{8.30}$$

From Figure 8.25(b), we also read off the equivalent transfer function of the major loop noting as follows:

1. At frequencies lower than ω_y, $|1/H| > G$. Therefore, $|GH| < 1$ and $G_{eq} = G$.
2. At frequencies between ω_y and ω_z, $|G| > |1/H|$. Therefore, $|GH| > 1$ and $G_{eq} = 1/H$.
3. At frequencies greater than ω_z, $|GH| < 1$ and $G_{eq} = G$.

Then from Figure 8.25(b),

$$G_{eq} = \frac{K_v\left(\dfrac{s}{\omega_x} + 1\right)}{s\left(\dfrac{s}{\omega_y} + 1\right)\left(\dfrac{s}{\omega_z} + 1\right)^2}, \tag{8.31}$$

which is an approximation, of course, because of the approximations made in constructing the diagram. The corner frequencies x and z are not exactly correct, because at these points $|G| \equiv |1/H|$ and the basic approximations have their greatest error.

It is always wise to check G_{eq} of the minor[1] loop (algebraically if possible). In this case, the exact equation for G_{eq} is

$$G_{eq} = \frac{K/s(s + 1)(s + 7)}{1 + [(Ks^2/\omega_g^2\omega_x P)/s(s + 1)(s + 7)(s + \omega_x)(s + P)]}$$

$$= \frac{K(s + \omega_x)(s + P)}{s(s + 1)(s + 7)(s + \omega_x)(s + P) + (Ks^2/\omega_g^2\omega_x P)}. \tag{8.32}$$

---

[1] For this example, G_{eq} for the system and G_{eq} for the minor loop are the same because the feedback is connected around the entire forward transfer function.

Note that the denominator is now fifth order and is not in factored form. If any of its factors are in the right half plane, then:

1. The minor loop is unstable.
2. The G_{eq} is nonminimum phase.

Item 1 is an undesirable condition and normally the minor loop design would be modified to correct it. (Note that it is possible for the system to be stable even though the minor loop is unstable). Item 2 restricts our interpretation of Eq. (8.31). Normally, we would simply add the phase curve for G_{eq} to Figure 8.25(b) and evaluate stability from the phase margin, but if G_{eq} is nonminimum phase we cannot construct the phase angle curve as usual and, in addition, we could not interpret stability because phase margin has no meaning when the transfer function is nonminimum phase (pole in right half plane).

The steps in designing the transfer function of a feedback compensator, as illustrated, are simple, straightforward, and very effective. However, they are based on approximations and may give rise to an unstable minor loop. Therefore, such designs should be checked carefully.

Since tachometer feedback is commonly used to stabilize systems, we will use the Bode diagram design method to set the gain of the tachometer channel. In the process of doing this, additional basic concepts will be discussed.

 EXAMPLE 8.13

Consider the block diagram of Figure 8.26(a), which shows velocity feedback around a second-order system. Here, the problem is not stabilization but improvement of damping. The Bode diagram of Figure 8.26(b) shows the asymptotic curve for G and, although the phase curve is not shown, it is clear that there is little phase margin.

The tachometer transfer function is

$$H(s) = ks; \tag{8.33}$$

therefore

$$\frac{1}{H(s)} = \frac{1/k}{s}. \tag{8.34}$$

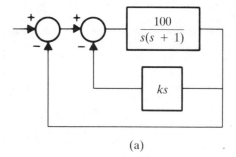

(a)

FIGURE 8.26 Example 8.13: (a) Block Diagram and (b) Bode Design of Velocity Feedback for a Second-Order System.

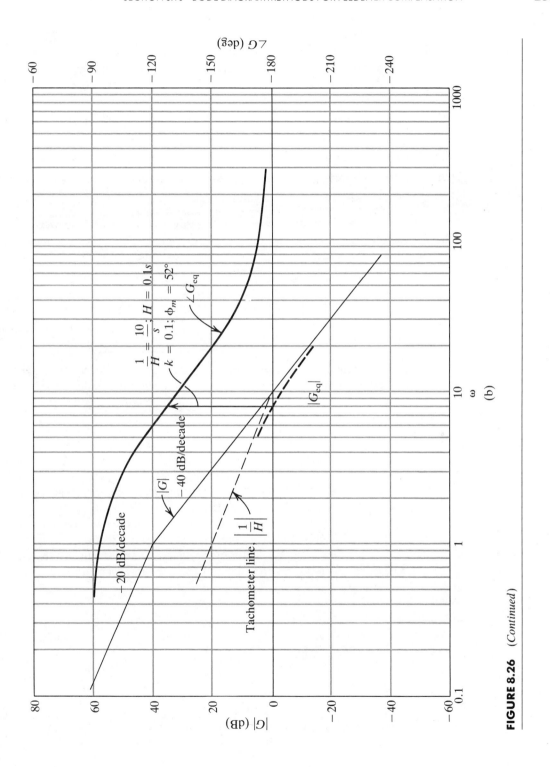

FIGURE 8.26 (*Continued*)

We draw a straight line with slope of -20 dB/decade to represent Eq. (8.34). This $1/H$ line of Figure 8.26(b) must lie *below* the G line if it is to represent G_{eq} over some selected frequency range. The gain $1/k$ is determined by the value of ω at which this line crosses the 0-dB axis. We choose the location of this line so that the compensated system has the amount of phase margin we want, or to meet some other specification. For the system of Figure 8.26(a), let us assume that a phase margin of about $50°$ is desired. The basic procedure is trial and error:

1. Draw the -20 dB/decade line and extend it until it crosses the G curve.
2. For the resulting G_{eq}, construct the phase curve and determine the phase margin.
3. Move the -20 dB/decade line and repeat.

With experience, the procedure converges rapidly to the desired answer—or it indicates that the desired answer is not possible. In Figure 8.26(b), the tachometer line has been drawn to intersect $|G|$ at the gain crossover. Thus,

$$\frac{1}{H(s)} = \frac{10}{s}; \qquad H(s) = 0.1s; \tag{8.35}$$

$$G_{eq} = \frac{10}{s(0.1s + 1)}. \tag{8.36}$$

The angle of G_{eq} is shown, $|G_{eq}|$ has been corrected at the gain crossover, and $\phi_m = 52°$. The tachometer gain, which was the quantity to be determined by the design, is $k = 0.1$.

When the system is of order three or higher, several possibilities exist, depending on the availability of a convenient place to add in the tachometer signal. Consider Figure 8.27(a) and (b), where $G(s)$ is third order in both diagrams, but the velocity signal is summed with the forward path at different locations. For Figure 8.27(a), the entire $G(s)$ is replaced by a G_{eq}. The graphical construction may indicate that the pole at 10 remains a pole, or it may not so indicate, but in the exact solution it will not be a pole of G_{eq}. On the other hand, for the system of Figure 8.27(b), the block $10/(0.1s + 1)$ is not in the minor loop and remains exactly a factor of the forward transfer function regardless of the value of k. This fact also restricts the graphical design procedure.

 EXAMPLE 8.14

Consider the Bode diagram of Figure 8.28. Two trial designs are shown. For each, the phase angle of G_{eq} was drawn assuming that the poles of the graphical design are real and correct numerically. From the diagram,

$$G_{eq} \; \#1 = \frac{7}{s[(s/7) + 1][(s/10) + 1]}, \tag{8.37}$$

$$G_{eq} \; \#2 = \frac{5}{s[(s/10) + 1]^2}. \tag{8.38}$$

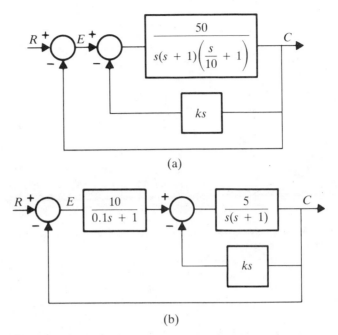

FIGURE 8.27 Third-Order Systems with Velocity Feedback: (a) Around All Forward Dynamics and (b) Around Part of Dynamics.

From the algebra, with $H_1(s) = s/7$,

$$G_{eq} \#1 = \frac{50/s(s + 1)(0.1s + 1)}{1 + (50s/7)/s(s + 1)(0.1s + 1)}$$

$$= \frac{50}{s(s + 1)(0.1s + 1) + 50s/7}$$

$$= \frac{6.14}{s[(s/9.024)^2 + (1.22s/9.024) + 1]}. \qquad (8.39)$$

We note that the poles will actually be complex if the indicated value of k is used, and the resulting error coefficient is less than the approximate 7. For $\zeta = 0.61$, the resonance peak is very small so the true gain crossover will not differ much from the approximate one and the indicated phase margin should be nearly correct. With $H(s) = s/5 = 0.2s$,

$$G_{eq} \#2 = \frac{50/s(s + 1)(0.1s + 1)}{1 + [50(0.2s)/s(s + 1)(0.1s + 1)]}$$

$$= \frac{50}{s(s + 1)(0.1s + 1) + 10s}$$

$$= \frac{4.545}{s[(s/10.49)^2 + 0.1s + 1]}. \qquad (8.40)$$

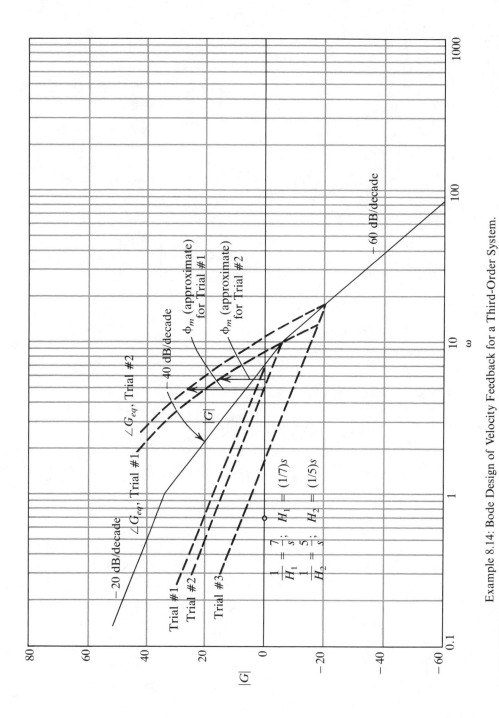

Example 8.14: Bode Design of Velocity Feedback for a Third-Order System.

For this trial also, K_v is lower than the graphical result, and with $\zeta = 0.524$, the complex poles will have a resonance peak of about 2 dB, but the gain curve will not rise above 0 dB, so the phase margin shown by the graphical design is very close to correct.

■ EXAMPLE 8.15

For the system of Figure 8.27(b), we may still use the Bode diagram of Figure 8.28. The value of $|G|$ uncompensated is unchanged and trials #1 and #2 are made in exactly the same way graphically and also *interpreted* the same *graphically*. There are two differences to be considered when designing for the system of Figure 8.27(b):

1. Trial #3 (Figure 8.28) would be a permissible design for the system of Figure 8.27(a), but is *not possible* for Figure 8.27(b) because the feedback loop does not include the pole at 10. Since the feedback $H(s)$ does not affect that pole, the plot of $1/H(s)$ is not permitted to pass below that point.[m]

2. The *algebraic* interpretation of results is different.

The algebraic solution for G_{eq} when the system of Figure 8.25(b) is used follows. For trial #1 with $H(s) = s/7$,

$$G_{eq} = \frac{10}{0.1s + 1} \frac{5/s(s + 1)}{1 + (5s/7)/s(s + 1)}$$

$$= \frac{29.2}{s(0.583s + 1)(0.1s + 1)}, \tag{8.41}$$

which may be compared with Eqs. (8.37) and (8.39). For trial #2 with $H(s) = 0.2s$,

$$G_{eq} = \frac{10}{0.1s + 1} \frac{5/s(s + 1)}{1 + 5(0.2s)/s(s + 1)}$$

$$= \frac{25}{s(0.5s + 1)(0.1s + 1)} \tag{8.42}$$

and Eq. (8.42) may be compared with Eqs. (8.38) and (8.40).

The combination of velocity and acceleration feedback is also handled easily with the asymptotic methods. Consider the block diagram of Figure 8.29(a) and the Bode diagram of Figure 8.29(b). Note that

$$H(s) = k_1 s + k_2 s^2 = s(k_2 s + k_1) = k_1 s\left(\frac{k_2}{k_1}s + 1\right) \tag{8.43}$$

and therefore,

$$\frac{1}{H(s)} = \frac{1/k_1}{s[(k_2 s/k_1) + 1]}. \tag{8.44}$$

---

[m] The feedback gain can be increased to the value of trial #3, of course, but the graphical construction is not applicable to it and it is not correct to read off G_{eq}.

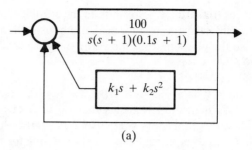

(a)

FIGURE 8.29 (a) Block Diagram for Valocity and Acceleration Feedback and (b) a Bode Diagram for a Trial Design.

To stabilize the system with about 50° phase margin, we want the new gain cross-over to be in the -20 dB/decade slope generated by the velocity feedback. Location of one pole of G_{eq} is determined by the k_1/k_2 ratio, the other by crossing of the -60 dB/decade asymptote with the accelerometer line. Thus, considerable freedom is available.

 EXAMPLE 8.16

One trial is shown in Figure 8.29(b). The tachometer line is chosen and from its intersection with the 0-dB axis, $k_1 = 1/5 = 0.2$. The pole is placed at a somewhat higher frequency to ensure the desired phase margin and $P = 8 = k_1/k_2$, so $k_2 = k_1/8 = 0.2/8. = .025$; thus,

$$G_{eq} = \frac{5}{s[(s/8) + 1][(s/25) + 1]} \tag{8.45}$$

by inspection of the diagrams. Using algebra,

$$H(s) = k_1 s + k_2 s^2$$
$$= 0.2s + 0.025s^2,$$

$$G_{eq} = \frac{G}{1 + GH}$$

$$= \frac{100/s(s + 1)(0.1s + 1)}{1 + 100(0.025s^2 + 0.2s)/s(s + 1)(0.1s + 1)}$$

$$= \frac{100}{s(0.1s^2 + 3.6s + 21)}. \tag{8.46}$$

Manipulating to Bode form,

$$G_{eq} = \frac{1000}{s(s^2 + 36s + 210)} = \frac{1000/210}{s[(s/7.33) + 1][(s/28.68) + 1]}. \tag{8.47}$$

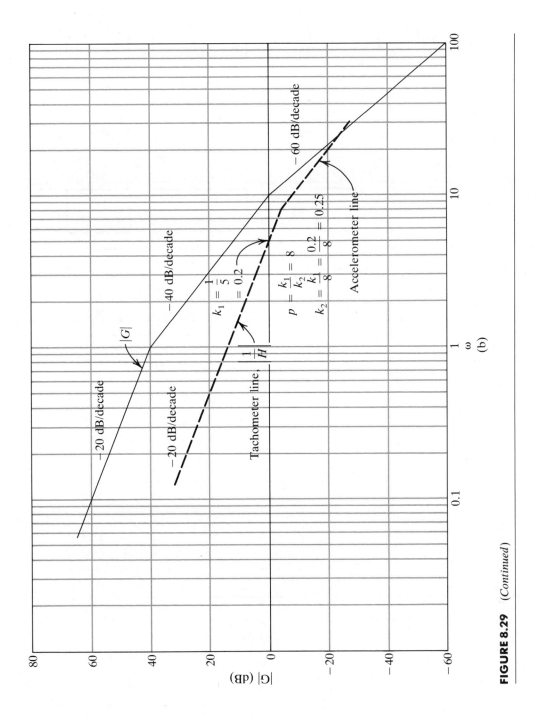

FIGURE 8.29   (*Continued*)

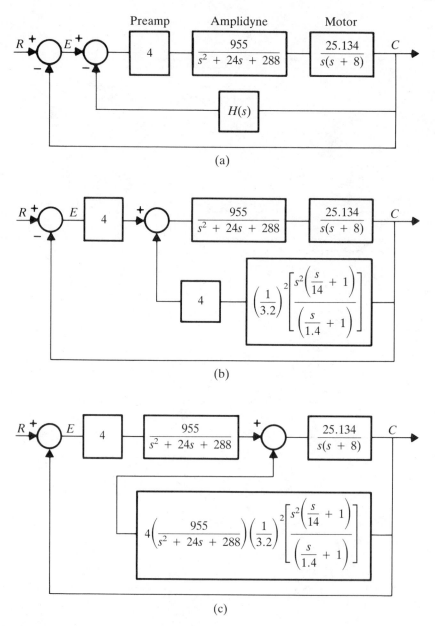

FIGURE 8.30   Block Diagrams of an Azimuth Positioning System: (a) Feedback Compensation Around All of the Forward Dynamics, (b) an Alternative Realization, (c) Another Possible Realization, (d) Another Possible Realization, and (e) Still Another Connection.

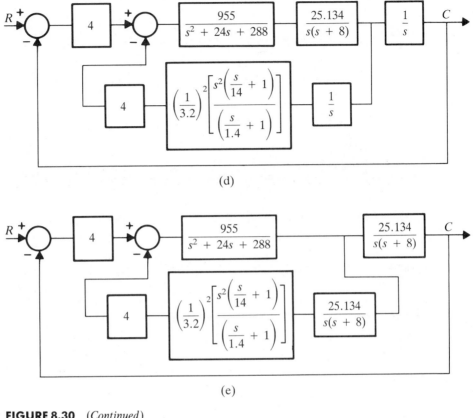

(d)

(e)

FIGURE 8.30 (*Continued*)

For a more general case of feedback compensation, consider the system of Figure 8.30. We shall explore several ways to design feedback compensation for this system. The case illustrated in Figure 8.30(a) suggests designing a minor loop that encloses all of the forward dynamics. This has the conceptual advantage that we can design $1/H(s)$ to give exactly what we want. The principal disadvantage of this approach is that the minor loop design may be unstable because the loop transfer function is of higher order.

A preliminary design is carried out in Figure 8.31 and results in

$$H(s) = \left(\frac{1}{3.2}\right)^2 \frac{s^2[(s/14)+1]}{[(s/1.4)+1]}. \tag{8.48}$$

Since this transfer function has more zeros than poles, it is not physically realizable and, if we wished to implement the connection shown in Figure 8.30(a), two poles should be added. Alternative realizations of the same design are obtained by block diagram manipulation and are shown in Figures 8.30(b), (c), (d), and (e). All of these have exactly

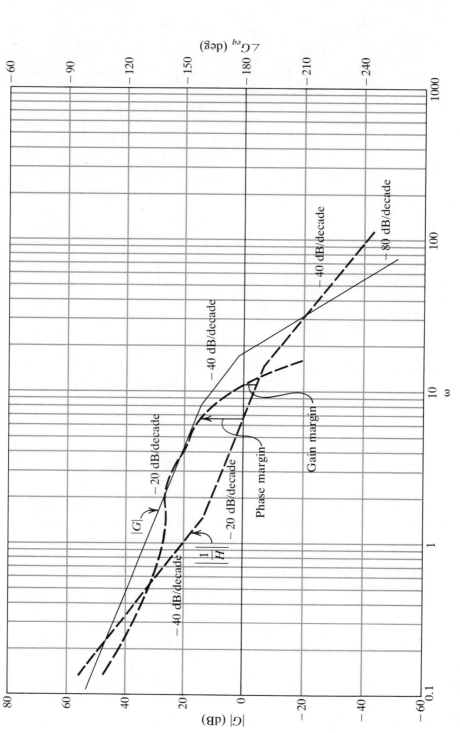

FIGURE 8.31 Bode Diagram for System of Figure 8.30(a).

the same loop transfer function for the minor loop, which is

$$GH = 4\left(\frac{955}{s^2 + 24s + 288}\right)\left[\frac{25.134}{s(s + 8)}\right]\left(\frac{1}{3.2}\right)^2\left\{\frac{s^2[(s/14) + 1)]}{[(s/1.4) + 1]}\right\}, \qquad (8.49)$$

and all have the same G_{eq}. Note that if the configuration of Figures 8.30(c) or (e) is used, the required feedback compensator is realizable.

The minor loop dynamics is fifth order, so the stability of the minor loop must be checked. Any available method may be used. Let us try Routh. The characteristic equation of the minor loop is

$$s(s + 8)(s^2 + 24s + 288)(s + 1.4) + (4)(955)(25.134)\left(\frac{1}{3.2}\right)^2\left(\frac{1.4}{14}\right)s^2(s + 14) = 0.$$

This becomes

$$s^5 + 33.4s^4 + 1462.4s^3 + 16102.4s^2 + 3225.6s = 0.$$

The Routh array is

s^5	1.0	1462.4	3225.6
s^4	33.4	16102.4	0
s^3	980.292	3225.6	0
s^2	15992.5	0	
s^1	3225.6		
s^0	0		

Thus, the minor loop has no roots in the right half plane, so G_{eq} has no poles in the right half plane. Therefore, the Bode diagram of Figure 8.31 may be completed by adding the phase curve as shown. Since the phase margin and gain margin are rather small, some redesign of $H(s)$ might be in order.

The conceptual approach of designing a feedback compensator to encompass *all* of the forward dynamics is a good starting point for simple systems, such as the fourth-order example just considered. It is a reasonable first trial for any order system, but for higher order systems it is quite likely that the minor loop may be unstable, and the resulting modifications needed to stabilize the minor loop require an undesirably complex $H(s)$. Alternative approaches are indicated in the block diagram of Figure 8.32. It is not uncommon to use multiple minor loops in compensating systems. The loops may be nested, as in Figure 8.32(a), or multiple, as in Figure 8.32(b), and of course other configurations are possible and practical depending on the specific system and the specifications. Such designs may result in excellent systems, but the procedures used in design involve considerable trial and error. A major difficulty is that each loop in a multiple-loop design contributes only a portion of the required compensation and it is difficult to assign realistic specifications to each loop.

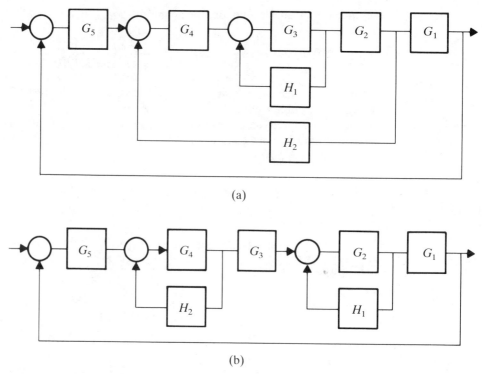

(a)

(b)

FIGURE 8.32 Alternative Configurations for Feedback Compensation: (a) Nested Minor Loops and (b) Multiple Minor Loops.

EFFECT OF FEEDBACK COMPENSATION ON NONLINEARITIES AND PARAMETER VARIATIONS

Nonlinearities occur in all physical systems, and many parameters such as gain, resistance, magnetic field strength, viscosity, etc., change due to aging, temperature, and other conditions associated with the operation of the system and its environment. In addition, if a system is mass produced, many parameters vary due to tolerances on the components. The very nature of feedback tends to reduce the effects of such variations, due to the basic relationships of Eqs. (8.25) and (8.26), i.e.,

$$\text{if} \quad GH > 1, \quad G_{eq} \cong 1/H,$$

$$\text{if} \quad GH < 1, \quad G_{eq} \cong G. \tag{8.48}$$

In brief, if a feedback compensator is used and it is connected around a nonlinear element or an element subject to parameter variation, G_{eq} tends to act as $1/H$, reducing the net effect of the nonlinearity (or parameter) on system dynamics. Note, however, that this suppression is not optimized by the servo compensator since it has been designed to augment the stability of the system. It is possible to design the feedback loop to optimize its effect on a nonlinear element, or to minimize the impact of parameter variation, but the objectives of such a design are not necessarily compatible with desired system performance. The designer can utilize both capabilities if this seems desirable, i.e., design a loop specifically to "clean up" a component by suppressing nonlinear characteristics or parameter variations, then design another stabilizing loop.

| **8.12** | EXTERNAL DISTURBANCES |

Most systems are subjected to disturbances other than the command input. Such disturbances may enter the loop at locations other than the input. Two common types of disturbances are *noise*, which often enters the electronics, and *load disturbances*, such as wind loads on an antenna, force or torque loads on a machine tool, gravitational forces, etc. Regardless of the nature of the compensation, disturbances cause a transient excursion of the output and may also cause a steady-state error. If there is a choice between cascade compensation and feedback compensation, which is more effective against such external disturbances? In general, if the disturbance causes a steady-state error, neither is effective, but a well-designed feedback compensator does not permit as much transient excursion as an equivalent cascade compensator.

The steady-state result is easily explained. Since the basic steady-state accuracy of a feedback control is determined by the system type number and error coefficient, these must be the same with either design. This restricts the nature of the feedback transfer function $H(s)$, which cannot transmit the output variable (for this would reduce the type number) nor can it transmit the first derivative[n] of output (for this would reduce the error coefficient). Therefore, for a disturbance such as wind load inducing a constant load torque, the feedback through $H(s)$ is zero, and the error caused by the torque loading is the same as if the minor loop were not present.

Under dynamic operating conditions, the transient response to the load disturbance or noise is different for each kind of compensation even though they are "equivalent" in the sense that the overall *forward* transfer functions are the same for both designs. This can be shown in a general way by consideration of the block diagrams as shown in Figure 8.33. Figure 8.33(a) shows the cascade-compensated loop and (c) shows the feedback compensation scheme. Figures 8.33(b) and (d) show rearrangements for $R = 0$. From Figure 8.33(b), we see that the feedback signal must pass through the entire system dynamics, thus introducing maximum time delay before a corrective signal is formed. In Figure 8.33(d), however, it is seen that there is a feedback path through HG_2, the minor loop dynamics, which clearly introduces less delay than for the cascade-compensated case. Thus, corrective action is initiated earlier and some of the transient excursion is suppressed.

---

[n] An exception would be a system subjected to only step inputs.

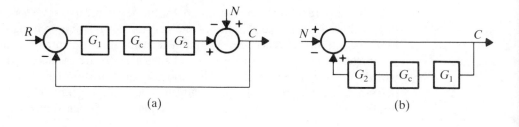

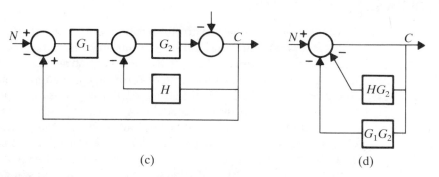

FIGURE 8.33 Comparison of Cascade and Feedback Compensation for Disturbance Rejection: (a) Cascade Compensation and (b) Equivalent Diagram, Cascade Compensation; (c) Feedback Compensation and (d) Equivalent Diagram, Feedback Compensation.

8.13   COMPENSATION BY STATE VARIABLE FEEDBACK

In previous sections, we have seen that compensation using feedback can be very effective. A very effective type of feedback compensation, where feasible, is to feedback signals proportional to all (or some) of the states. Figure 8.34(a) shows the concept in block diagram form, for which the plant equation is

$$\dot{X} = \mathbf{A}X + \mathbf{B}u$$

but

$$u = K(R - k^T X) = -Kk^T X + KR.$$

Substituting,

$$\dot{X} = \mathbf{A}X - BKk^T X + BkR$$

$$= (\mathbf{A} - BKk^T)X + BkR$$

so that the effect of the state feedback is to alter some terms in the **A** matrix.

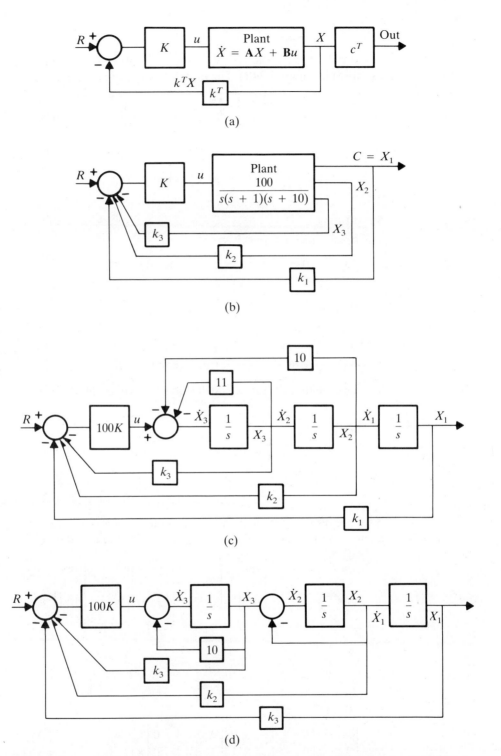

FIGURE 8.34 Block Diagrams for State Variable Feedback: (a) Basic Concept, (b) A Third-Order Example, (c) The Third-Order System with Plant Equations Arranged for One Set of States, and (d) Third-Order System with Different States.

Consider the specific system of Figure 8.34(b) with a particular definition of state as shown in Figure 8.34(c). The matrix form of the equations is as follows:

$$
\begin{bmatrix} \dot{X}_1 \\ \dot{X}_2 \\ \dot{X}_3 \end{bmatrix} = \begin{bmatrix} 0 & 1 & 0 \\ 0 & 0 & 1 \\ 0 & -10 & -11 \end{bmatrix} \begin{bmatrix} X_1 \\ X_2 \\ X_3 \end{bmatrix} + \begin{bmatrix} 0 \\ 0 \\ 100 \end{bmatrix} U
$$

$$
U = K[R - k^T X] = KR - K[k_1, \quad k_2, \quad k_3] \begin{bmatrix} X_1 \\ X_2 \\ X_3 \end{bmatrix}.
$$

Substituting,

$$
\begin{bmatrix} \dot{X}_1 \\ \dot{X}_2 \\ \dot{X}_3 \end{bmatrix} = \begin{bmatrix} 0 & 1 & 0 \\ 0 & 0 & 1 \\ 0 & -10 & -11 \end{bmatrix} \begin{bmatrix} X_1 \\ X_2 \\ X_3 \end{bmatrix} - \begin{bmatrix} 0 \\ 0 \\ 100 \end{bmatrix} K[k_1, \quad k_2, \quad k_3] \begin{bmatrix} X_1 \\ X_2 \\ X_3 \end{bmatrix} + \begin{bmatrix} 0 \\ 0 \\ 100 \end{bmatrix} KR.
$$

Combining,

$$
\begin{bmatrix} \dot{X}_1 \\ \dot{X}_2 \\ \dot{X}_3 \end{bmatrix} = \begin{bmatrix} 0 & 1 & 0 \\ 0 & 0 & 1 \\ -100k_1 K & -10 - 100k_2 K & -11 - 100k_3 K \end{bmatrix} \cdot \begin{bmatrix} X_1 \\ X_2 \\ X_3 \end{bmatrix} + \begin{bmatrix} 0 \\ 0 \\ 100 \end{bmatrix} KR.
$$

Thus, we see that the use of state feedback modifies the **A** matrix. From this, we form

$$
[s\mathbf{I} - \mathbf{A}] = \begin{bmatrix} s & -1 & 0 \\ 0 & s & -1 \\ +100k_1 K & +11 + 100k_2 K & s + 10 + 100k_3 K \end{bmatrix},
$$

and from this, $\det[s\mathbf{I} - \mathbf{A}]$ is

$$
s^3 + (11 + 100k_3 K)s^2 + (10 + 100k_2 K)s + 100k_1 K = 0.
$$

A similar result is obtained for any other choice of states. Consider Figure 8.34(d). The plant is the same as for Figure 8.34(c) but with the new definition of state variables:

$$
\begin{bmatrix} \dot{X}_1 \\ \dot{X}_2 \\ \dot{X}_3 \end{bmatrix} = \begin{bmatrix} 0 & 1 & 0 \\ 0 & -1 & 1 \\ 0 & 0 & -10 \end{bmatrix} \begin{bmatrix} X_1 \\ X_2 \\ X_3 \end{bmatrix} + \begin{bmatrix} 0 \\ 0 \\ 100 \end{bmatrix} U
$$

$$
U = K[R - k^T X]
$$

$$
= kR - K[k_1, \quad k_2, \quad k_3] \begin{bmatrix} X_1 \\ X_2 \\ X_3 \end{bmatrix}.
$$

Substituting

$$
\begin{bmatrix} \dot{X}_1 \\ \dot{X}_2 \\ \dot{X}_3 \end{bmatrix} = \begin{bmatrix} 0 & 1 & 0 \\ 0 & -1 & 1 \\ 0 & 0 & -10 \end{bmatrix} \begin{bmatrix} X_1 \\ X_2 \\ X_3 \end{bmatrix} - \begin{bmatrix} 0 \\ 0 \\ 100 \end{bmatrix} K[k_1, \quad k_2, \quad k_3] \begin{bmatrix} X_1 \\ X_2 \\ X_3 \end{bmatrix} + \begin{bmatrix} 0 \\ 0 \\ 100 \end{bmatrix} KR.
$$

Combining,

$$\begin{bmatrix} \dot{X}_1 \\ \dot{X}_2 \\ \dot{X}_3 \end{bmatrix} = \begin{bmatrix} 0 & 1 & 0 \\ 0 & -1 & 1 \\ -100Kk_1 & -100Kk_2 & -10 - 100Kk_3 \end{bmatrix} \begin{bmatrix} X_1 \\ X_2 \\ X_3 \end{bmatrix} + \begin{bmatrix} 0 \\ 0 \\ 100 \end{bmatrix} KR.$$

From this, we form $[s\mathbf{I} - \mathbf{A}]$:

$$[s\mathbf{I} - \mathbf{A}] = \begin{bmatrix} s & -1 & 0 \\ 0 & s+1 & -1 \\ +100Kk_1 & +100Kk_2 & s + 10 + 100Kk_3 \end{bmatrix}$$

and $\det[s\mathbf{I} - \mathbf{A}]$ is

$$s^3 + (11 + 100Kk_3)s^2 + (10 + 100Kk_2 + 100Kk_3)s + 100Kk_1 = 0.$$

This characteristic equation for the system of Figure 8.34(d) differs from that for Figure 8.34(c) because the states are defined differently, but both have the same basic characteristics; i.e., the feedback gains appear in all coefficients and appear linearly. Since this provides independent adjustment of all coefficients, and since all coefficients are functions of the roots, we can place the roots wherever we wish and thus have complete control of the dynamic modes of the system.

For calculation purposes, i.e., to calculate the feedback gains required for a chosen set of roots, the simplest procedure is to obtain *both*[o] characteristic equations and equate coefficients. This provides N algebraic equations in N unknown gains. For example, Figure 8.34(d) has provided the third-order characteristic equation,

$$s^3 + (11 + 100Kk_3)s^2 + (10 + 100Kk_2 + 100Kk_3)s + 100Kk_1.$$

Suppose we desire roots at $s = -1 \mp j1$ and $s = -15$. We obtain the polynomial with these roots as

$$(s + 1 + j1)(s + 1 - j1)(s + 15) = s^3 + 17s^2 + 32s + 30.$$

Equating coefficients of like powers,

$$100Kk_1 = 30$$

$$10 + 100Kk_2 + 100Kk_3 = 32$$

$$11 + 100Kk_3 = 17.$$

From which

$$Kk_1 = \frac{30}{100} = 0.3$$

$$Kk_3 = \frac{17 - 11}{100} = \frac{6}{100} = 0.06$$

$$Kk_2 = \frac{32 - 10 - (100)(0.06)}{100} = \frac{16}{100} = 0.16.$$

---

[o] One using the chosen roots, one with the unknown feedback gains.

For the example above (or for any third-order polynomial), the choice of desired root locations is not difficult. For higher order systems, however, the choice of "desired" roots may be a more serious problem. Of course, the choice of *dominant* roots is not hard. The difficulty in choosing additional roots lies in the fact that we usually have no clearly defined reasons for a choice. We want all such roots sufficiently far into the left half plane so that their contribution to the transient response is small compared to that of the dominant roots. However, the specific locations chosen determine:

1. the numerical values of the feedback gains
2. whether a given state is fed back as negative or positive feedback
3. the system error coefficient (accuracy)
4. the sensitivity characteristics of the design.

Therefore, the desired root locations should be chosen with some care.

Since accuracy is always an important consideration, and is usually specified quantitatively, it is important to understand the relationships between the static error coefficients and the state feedback. In general, the closed-loop transmission function is

$$\frac{C(s)}{R(s)} = \frac{a_m s^m + a_{m-1} s^{m-1} + \cdots a_1 s + a_0}{b_n s^n + b_{n-1} s^{n-1} + \cdots b_1 s + b_0} = \frac{K_{c1} \prod\limits_{i=0}^{m}(s + Zi)}{\prod\limits_{j=0}^{n}(s + P_j)}.$$

Also, we define the error to be

$$E(s) = R(s) - C(s) = R(s)\left[1 - \frac{C(s)}{R(s)}\right],$$

and the steady-state error (using the final value theorem) is

$$E_{ss} = \lim_{s \to 0} sR(s)\left[1 - \frac{C(s)}{R(s)}\right].$$

We can use this expression to derive the error coefficients of type-zero, -one, and -two systems and relate them to state feedback gains. For a type-zero system with step input,

$$E_{ss} \lim_{s \to 0} s\left(\frac{1}{s}\right)\left[1 - \frac{C(0)}{R(0)}\right] = 1 - \frac{K_{C1} \prod\limits_{i=0}^{m}(Zi)}{\prod\limits_{i=0}^{n}(P_j)},$$

but the steady-state error for a unit step is known to be

$$E_{ss} = \frac{1}{1 + K_P} = 1 - \frac{K_{C1} \prod\limits_{i=0}^{m}(Zi)}{\prod\limits_{i=0}^{n}(P_j)}.$$

Solving for K_p,

$$
K_p = \frac{K_{C1} \prod_{i=0}^{m} (Z_i)}{\prod_{j=0}^{n} (P_j) - K_{C1} \prod_{i=0}^{m} (Z_i)}
$$

$$
= \frac{a_0}{b_0 - a_0}.
$$

For the system of Figure 8.34,

$$
a_0 = 100K
$$

$$
b_0 = 100Kk_1
$$

$$
K_P = \frac{100K}{100Kk_1 - 100K} = \frac{100K}{100K(k_1 - 1)} = \frac{1}{k_1 - 1}.
$$

Note that for $k_1 \neq 1$, K_P is a finite number. This means that the system will have a steady-state error for a step input, i.e., the original type-one system has been converted[P] to a type-zero system by using $k_1 \neq 1$. However, if we use $k_1 = 1$, then $K_P = \infty$ and the system remains type one. In practice, we can achieve this by observing that, in all of the state equations for the arrangements of Figure 8.34, the feedback gains are all multiplied by the forward gain K so that the system has four adjustable gains, but only three *independent* adjustments. If the system is slightly rearranged, as in Figure 8.35, setting $k_1 = 1.0$ and moving the summing junction to the right of the adjustable gain, the main gain K becomes an independent adjustment and is equivalent to feeding back X_1 through a gain Kk_1. The system remains type one, and therefore the position error coefficient $K_P = \infty$.

Although we can maintain the type number of a type-one system as indicated, the use of state feedback affects the value of the error coefficient K_v. Consider the basic equation for steady-state error with a ramp input:

$$
E_{ss} = \lim_{s \to 0} s \left(\frac{1}{s^2} \right) \left[1 - \frac{C(s)}{R(s)} \right]
$$

$$
= \lim_{s \to 0} \left[\frac{1 - C(s)/R(s)}{s} \right]
$$

$$
= \frac{1 - C(0)/R(0)}{0}.
$$

In order to have a finite error, or zero error, it is necessary that

$$
\frac{C(0)}{R(0)} = 1.
$$

[P] If $k_1 \neq 1$, any system type is converted to type zero.

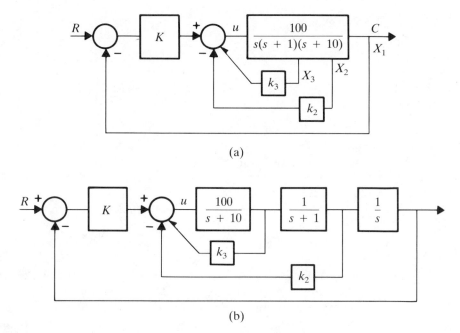

FIGURE 8.35   Rearrangement of System Gains to Preserve the Integration of a Type-One System: (a) General Scheme and (b) a Special Choice of States.

Then the steady-state error evaluates to

$$E_{ss} = \frac{0}{0}$$

and we apply L'Hospital's rule, noting that for a unit ramp input to a type-one system,

$$E_{ss} = \frac{1}{K_v}.$$

Then

$$\frac{1}{K_v} = -\lim_{s \to 0} \frac{d}{ds} \left[\frac{C(s)}{R(s)} \right].$$

Since $C(s)/R(s)\big|_{s=0} = 1.0$ for systems that are type one or higher, it is convenient to divide the right side of the equation by it, obtaining

$$\frac{1}{K_v} = -\lim_{s \to 0} \frac{d}{ds} \left[\frac{C(s)}{R(s)} \right] \Big/ \frac{C(s)}{R(s)}$$

$$= -\lim_{s \to 0} \frac{d}{ds} \log_e \left[\frac{C(s)}{R(s)} \right].$$

Since

$$\log_e \frac{C(s)}{R(s)} = \log_e K_{C1} + \log_e \prod(s + Z_i) - \log_e \prod(s + P_j)$$

$$= \log_e K_{C1} + \sum_{i=0}^{m} \log_e(s + Z_i) - \sum_{j=0}^{m} \log_e(s + P_j)$$

$$\frac{d}{ds} \log_e \frac{C(s)}{R(s)} = \sum_{i=0}^{m} \frac{1}{Z_1} - \sum_{j=0}^{n} \frac{1}{P_j} = -\frac{1}{K_v}.$$

From theory of equations,

$$\sum \frac{1}{Z_1} = \frac{a_1}{a_0},$$

$$\sum \frac{1}{P_j} = \frac{b_1}{b_0},$$

and since the system is type one, $a_0 = b_0$. Thus,

$$-\frac{1}{K_v} = \frac{a_1 b_0 - b_1 a_0}{a_0^2} = \frac{a_1 - b_1}{a_0},$$

$$-K_v = \frac{a_0}{a_1 - b_1} = \frac{b_0}{a_1 - b_1}.$$

For the system of Figure 8.35(b), the closed-loop transfer function is

$$\frac{C(s)}{R(s)} = \frac{100K}{s^3 + (11 + 100k_3)s^2 + (10 + 100k_2 + 100k_3)s + 100K},$$

from which

$$b_0 = 100K$$

$$b_1 = 10 + 100k_2 + 100k_3$$

$$a_1 = 0;$$

so

$$-K_v = \frac{100K}{-(10 + 100k_2 + 100k_3)}$$

$$K_v = \frac{100K}{(10 + 100k_2 + 100k_3)}.$$

All three of the adjustable gains—K, k_2, and k_3—contribute to K_v, but if the three roots are specified, the root values determine the required values of the gains and the resulting value of K_v may not meet specifications for accuracy. While some adjustment of desired root locations is often acceptable, such adjustment may not be sufficient to obtain the specified error coefficient. As has been shown

$$-\frac{1}{K_v} = \sum_{i=0}^{m} \frac{1}{Z_1} - \sum_{j=0}^{m} \frac{1}{P_j},$$

where the Z's and P's are the *closed-loop* zeros and poles. The third-order system considered here does not have zeros. Thus, if we introduce a zero (with accompanying pole of course), the value of K_v can be adjusted with more freedom. Consider the diagram of Figure 8.36(a). The closed-loop transfer function is

$$\frac{C(s)}{R(s)} = \frac{[100K/(1+k_4)]s + [100K/(1+k_4)]Z}{s^4 + [1 + 10(\alpha + k_4)Z + 100k_3/(1+k_4)]s^3}$$
$$+ \left[\frac{10(\alpha + k_4) + 100k_3}{1 + k_4} + 10(\alpha + k_4)Z + \frac{100(k_3 + k_2)}{1 + k_4}\right]s^2$$
$$+ \left[\frac{10(\alpha + k_4)Z + 100k_3 Z + 100k_2 Z + 100K}{1 + k_4}\right]s + \frac{100KZ}{1 + k_4}.$$

From this,

$$a_0 = b_0 = \frac{100KZ}{1 + k_4},$$

$$a_1 = \frac{100K}{1 + k_4},$$

$$b_1 = \frac{10(\alpha + k_4)Z + 100k_3 Z + 100k_2 Z + 100K}{1 + k_4}.$$

Then

$$K_v = \frac{-100KZ}{100K - [10(\alpha + k_4)Z + 100k_3 Z + 100k_2 Z + 100K]}$$
$$= \frac{+100K}{10(\alpha + K_4) + 100k_3 + 100k_2}.$$

Note that there are five independent variables in the expression for K_v. They are k_1, k_2, k_3, k_4, and α. Four of these are required to place the roots at designated locations. The fifth may be used to set K_v to its specified value. For computational purposes, equating coefficients provides four equations in five unknowns; the expression for K_v provides a fifth equation.

Instead of using a cascade lag network to generate a closed-loop zero, if an additional feedback path is used and a filter placed in that path, the pole of the filter becomes a closed-loop zero. A possible configuration is shown in Figure 8.36(b). For this system, the closed-loop transfer function is

$$\frac{C}{R}(s) = \frac{100Ks + 100KP}{s^4 + (11 + P + 100k_3)s^3 + (10 + 11P + 100k_3 P + 100\alpha + 100k_3 + 100k_2)s^2}$$
$$+ (10P + 100\alpha + 100k_3 P + 100k_2 P + 100K)s + 100KP.$$

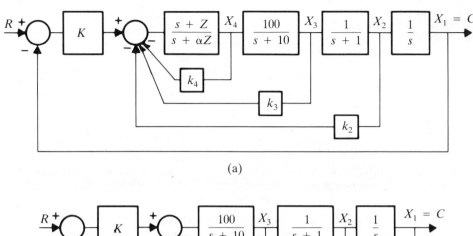

(a)

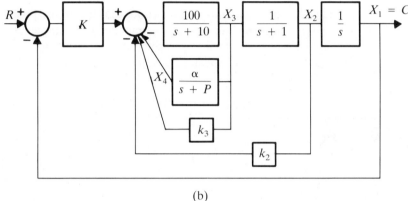

(b)

FIGURE 8.36 Variations on State Variable Feedback: (a) Use of a Lag Filter to Introduce a Closed-Loop Zero, Thus Allowing Adjustment of K_v and (b) Use of a Feedback Pole to Generate the Closed-Loop Zero.

From this,

$$a_0 = b_0 = 100KP$$

$$a_1 = 100K$$

$$b_1 = 10P + 100\alpha + 100k_3P + 100k_2P + 100K.$$

then

$$K_v = \frac{-a_0}{a_1 - b_1}$$

$$= \frac{-100KP}{100K - (10P + 100\alpha + 100k_3P + 100k_2P + 100K)}$$

$$= \frac{100kP}{10P + 100\alpha + 100k_3P + 100k_2P}$$

and again there are five variables: K, P, α, k_2, and k_3, which can be used to set four roots and K_v.

8.14 SUMMARY

The design of feedback control systems includes stabilization and adjustment of dynamic (transient) response. The design process is commonly called *compensation*. Compensators may be designed using Bode, root locus, or state variable methods. The compensators may be placed in cascade with the forward path, or in one or more feedback paths.

When Bode diagram methods are used, the philosophy is to reshape the open-loop phase and gain curves until acceptable phase margin, gain margin, and bandwidth have been obtained. A number of cases have been discussed with examples given.

When Root locus methods are used, the best procedure is to obtain computer-generated curves for the root locus (or two parameter root locus). A suitable point is selected for the dominant complex roots and the required parameters of the compensator are read from the diagram. A number of examples illustrate the procedures.

State variable feedback can be used to place the roots (poles) of the characteristic equation at specified locations. The matrix methods for accomplishing this have been presented and examples given.

REFERENCES

1. Ziegler, J. G.; and Nichols, N. B. "Optimum Settings for Automatic Controllers." Vol. 64 *Trans. ASME.* (1942).

2. Ziegler, J. G.; and Nichols, N. B. "Process Lags in Automatic Control Circuits." Vol. 65 *Trans. ASME.* (1943).

BIBLIOGRAPHY

Bower, J. L.; and Schultheiss, P. M. *Introduction to the Design of Servomechanisms.* New York: John Wiley and Sons (1958).

Eveleigh, V. W. *Introduction to Control Systems Design.* New York: McGraw-Hill Book Co. (1972).

Thaler, G. J.; and Brown, R. G. *Analysis and Design of Feedback Control Systems.* New York: McGraw-Hill Book Co. (1961).

PROBLEMS

8.1 An ideal instrument servo is to be damped by velocity feedback. Maximum overshoot to a step input is to be less than 25%. Design the value of k.

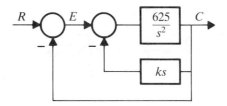

a. Use the root locus method.
b. Use the Bode diagram method.
c. Compare results.
d. Simulate, using a step input. Are specs met?

8.2 The system of Problem 8.1 is to be damped using a cascade compensator (one zero, one pole). Repeat parts (a), (b), (c), and (d) of Problem 8.1.

8.3 Design the velocity feedback (i.e., find k) so that the system is stable and "well-damped." Use *both* Bode and root locus methods; design with one, check results with the other. Record:
a. value of k.
b. phase and gain crossovers
c. phase margin
d. ζ and ω
e. error coefficient of compensated system.

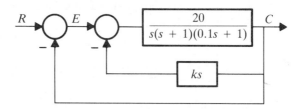

8.4 A phase-locked loop may be represented (for small signals) as a single-loop, unity feedback system with

$$G = \frac{10^{12}}{s^2}.$$

Design a compensator that will stabilize the loop and provide a settling time of less than 5 microseconds. (Note: The gain may be changed if needed.)

8.5 Design k to stabilize the system and provide at least $30°$ phase margin. After determining your value for k, check the phase margin by block diagram manipulation and another Bode diagram.

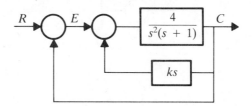

8.6   Design the tachometer feedback (find k) such that the phase margin is approximately 30°.

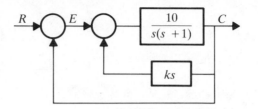

8.7   Given:

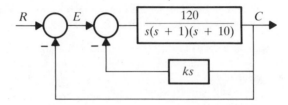

a.   Determine k to provide approximately 25° phase margin.
b.   What is K_v for your compensated system?

8.8   A third-order, type-one servo is to be stabilized with tachometer feedback. The system may be described as shown below.
a.   Design the feedback to provide approximately 45° phase margin.
b.   Record the value of k.
c.   Feedback around *all* of the time constants turns out to be impossible. Define the transfer function needed for H. [see Fig. P8.8(b)].

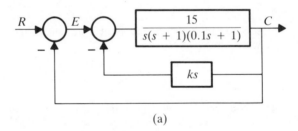

(a)

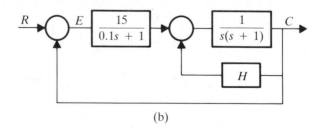

(b)

8.9 Find values of α and β to place roots at $s = -1 \mp j2$.

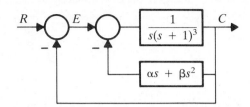

8.10 For the following system, find values of A and B to place roots at $s = -1 + j5$. Where is the third root?

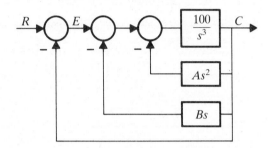

8.11 Given:

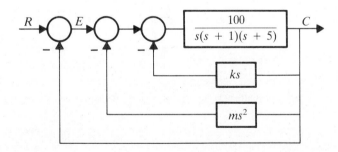

Place complex roots at $s = -5 \mp j5$. by determining the required values of k and m.

8.12 Design G_c to stabilize the system and provide at least $20°$ phase margin without reducing the error coefficient. Record:

    a. your compensator transfer function
    b. phase and gain margins
    c. estimated settling times.

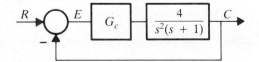

8.13 The error coefficient, $K_v = 300$, is not to be reduced. Design G_c to provide at least 40° phase margin and a bandwidth of approximately 100 rad/s. You may use as many sections of filter as you wish.

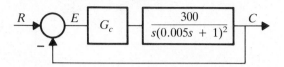

8.14 The servo is to be designed to have a settling time of approximately 10 microseconds. Choose a suitable value of k, then design τ_1 and τ_2 to provide about 40° phase margin.
a. Record your values for K, τ_1, and τ_2.
b. Calculate the settling time (*show work*).
c. Calculate the steady-state error if the system is subjected to a ramp input of 1 rad/s.

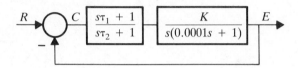

8.15 G_c is to be a lag (low-pass) filter. Design G_c to provide about 40° phase margin.
a. Record your transfer function for G_c.
b. Calculate the settling time.

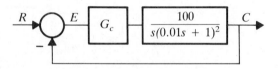

8.16 An instrument servo has a motor-load transfer function of

$$G = \frac{15}{s(0.05s + 1)(0.01s + 1)}.$$

Specifications require a phase margin of 50° and an error constant of $K_v = 2000s^{-1}$. Design a cascade compensator to achieve these specifications (transfer functions only). Draw block diagram for the compensated system with transfer functions shown.

8.17 Given a single-loop, unity feedback system with

$$G = \frac{1000.}{s(0.01s + 1)(0.002s + 1)}.$$

a. Draw the Bode diagram for the uncompensated system.
b. From the diagram, evaluate (and record) the phase margin and gain margin. Is the system stable?
c. Design a cascade compensator that will provide at least 35° phase margin, under the

following constraints:
1. K_v must not be reduced
2. The open-loop bandwidth must not be changed by more than 10%.

8.18 Design a feedback compensator for the system of Problem 8.17.

8.19 Given a single-loop, unity feedback control system with forward transfer function

$$G(s) = \frac{100}{s^2(0.1s + 1)}.$$

  a. Design a cascade compensator that will stabilize the system and keep M_{pt} less than 1.5.
  b. Explain why you chose the kind of compensation you used.
  c. Evaluate ζ, ω_n, and settling time for the compensated system.
  Note: Bode gain may not be reduced.

8.20 Given a single-loop feedback control system with Bode diagram (open loop) as shown, design a lag-lead compensator to:
  a. keep error coefficient unchanged
  b. provide open-loop gain crossover at $\omega = 35$
  c. provide at least 30° phase margin. Record compensator.

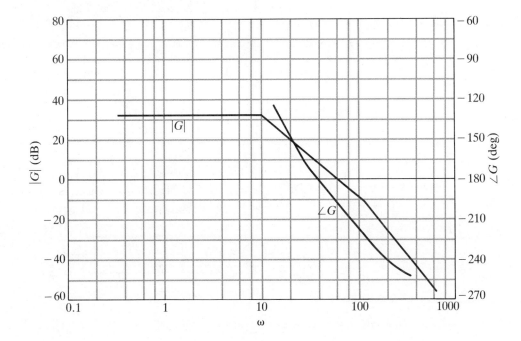

8.21 Design G_c as a phase lead compensator to provide a phase margin of 25° to 30° without reducing K_v. Record:
  a. your compensator transfer function (P, Z)
  b. new gain crossover frequency

    c.   new phase crossover frequency

    d.   phase margin

    e.   gain margin.

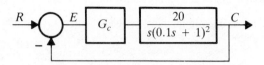

8.22   The purpose of the compensator is to obtain complex roots at $s = -1 \pm j6$.

    a.   Locate any Z, P pair that forces the root locus through the desired point.

    b.   Evaluate k and K_v for the Z, P pair of (a).

    c.   Find a Z, P pair that keeps K_v at its original value of 10. (*Note*: Solution of this part is trial and error.)

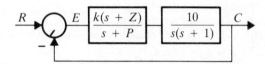

8.23   Design G_c to provide a pair of complex roots at $s = -5 \mp j10$ without reducing the error coefficient of the system. After designing G_c, check to see if the specified roots are dominant. (Do not try to take the inverse transform).

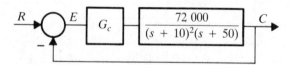

8.24   Keep K_v unchanged. Design G_c as a phase lead filter to obtain a phase margin of at least 50°. Be sure to record the Z, P, and gain of your filter.

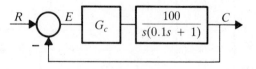

8.25   Design G_c to place complex roots at $s = -0.2 \pm j1$.

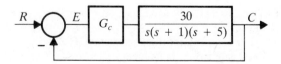

8.26 Keep K_p unchanged. Design G_c to stabilize the system and give 50° phase margin.
  a. Before starting the design, specify whether you are using a phase lead or lag, and *why*.
  b. After design, evaluate the settling time.

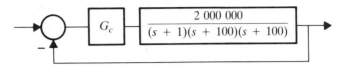

8.27 Determine P and Z to place roots at $s = 0 \pm j10$.

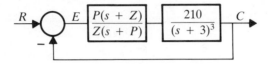

8.28 Given:

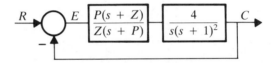

  a. Find P and Z to give roots on the imaginary axis at $\pm j5$.
  b. What are the locations of the remaining two roots?

8.29 Given:

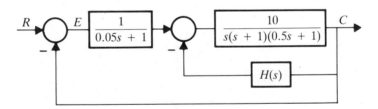

Design $H(s)$ to stabilize the system without reducing K_v.

8.30 Determine k_1 *and* k_2 to place complex roots at $s = -1 \pm j2$.

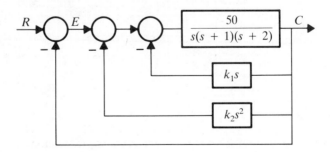

8.31   Given:

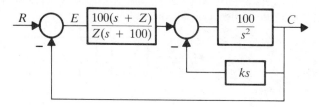

Find k and Z to place roots at $s = -7.5 \mp j7.5$.

8.32   Given:

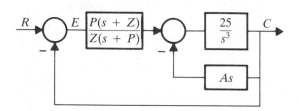

Determine A, P, and Z to place roots at $-1 \mp j2$.

8.33 $|G|$ is given. Design H to stabilize and provide at least $30°$ phase margin. Record all numerical values needed for H.

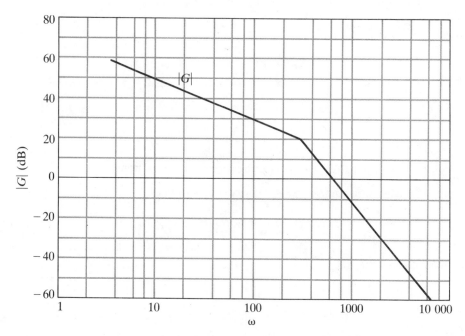

8.34 G is given by the Bode diagram. Design G_c to stabilize and provide at least $30°$ phase margin. What is the settling time of the compensated system?

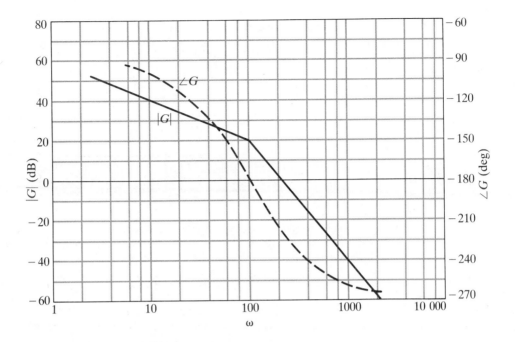

8.35 G is given by the Bode diagram. Design G_c as a lead filter (one zero, one pole), with zero pole spacing exactly one decade. Stabilize system and obtain as much phase margin as possible. *Record Z, P, ϕ_m, and gain crossover frequency.*

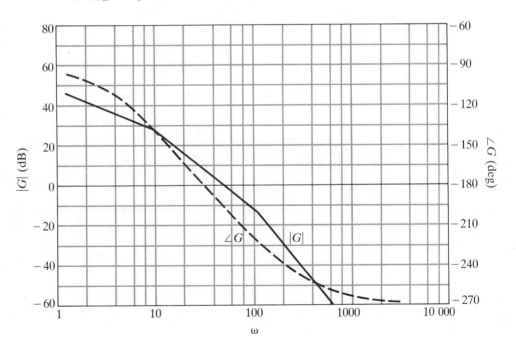

8.36 Design G_c such that the peak overshoot to a step input is less than 25% (but close to it), and the settling time is approximately 0.1 s. Be sure to explain what you are doing!

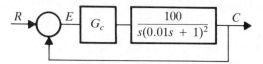

8.37 Given:

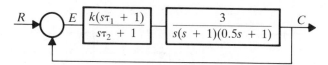

a. Draw the Bode diagram for the uncompensated system.
b. Stabilize the system using the lead compensator. The Bode gain is not to be reduced and the phase margin must be at least 30°. What are your values for k, τ_1, and τ_2?
c. Repeat the design, but use a lag compensator.
d. Compare the two designs—i.e., compare the behavior of the two systems.

8.38 A single-loop, unity feedback servo has a transfer function

$$G(s) = \frac{300}{s\left(\dfrac{s}{3.5} + 1\right)\left(\dfrac{s}{25} + 1\right)}.$$

a. Design a single-section lag compensator to stabilize and provide a phase margin of about 30°. Write down the transfer function of your compensator.
b. Estimate the settling time of the compensated system. Explain how you arrived at this estimate.

8.39 The gain of the system in Problem 8.38 is reduced so that $G(s)$ becomes

$$G(s) = \frac{30}{s\left(\dfrac{s}{3.5} + 1\right)\left(\dfrac{s}{25} + 1\right)}.$$

a. Design a single-section lead compensator to stabilize the system and provide a phase margin of about 30°. Write down the transfer function of your compensator.
b. Estimate the settling time of the compensated system. Explain how you arrived at this estimate.

8.40 For each of the following forward transfer functions, draw the root loci, locate the roots approximately, and sketch the *root relocation zones* for a single-section cascaded lead compensator and also a single-section cascaded lag compensator. For each case, state whether or not it is possible to stabilize the system with a single section of lead compensation, and if it is possible estimate the maximum value of ζ that can be obtained.

a. $G = \dfrac{100}{s^2}$ b. $G = \dfrac{100}{s(s+1)(s+2)}$

c. $G = \dfrac{100}{s^3}$ d. $G = \dfrac{100}{(s+5)(s+10)^2}$

8.41 Given:

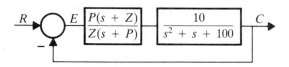

Find the root relocation zones.

8.42 Given:

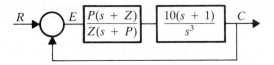

Draw the root relocation zones on the s plane.

8.43 Given:

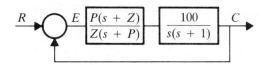

The compensator is a lead network. Draw the root relocation zone on the s plane.

8.44 Given:

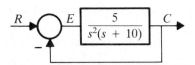

Determine the root relocation zones for both lead and lag compensation. Show them on the s plane.

8.45 A unity feedback system, as shown below, is to have a settling time of 10 microseconds and a peak overshoot to a step input of less than 30%.
   a. Determine a value for K_v that makes the specs possible.
   b. Design a cascade compensator to meet specs without reducing K_v.

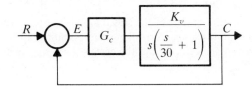

8.46   Problem 8.45 is given different accuracy specs so that $K_v = 10^{10}$. The dynamic specs are not changed, but the compensation is to be accomplished with a feedback filter, H_c, which will not reduce K_v. Design the feedback compensator.

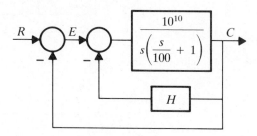

8.47   The error coefficient is not to be reduced. Design G_c to give at least 30° phase margin and a bandwidth of about 100 rad/s. Use several sections of compensator if you wish.

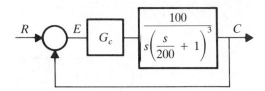

8.48   Given:

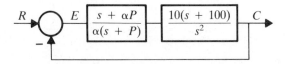

For this cascade-compensated system, construct the root relocation zones.

8.49   Find the root relocation zones on the s plane for the system given below (show derivation).

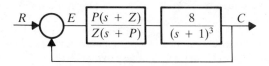

8.50   Design the compensator to satisfy the following specifications:
   a.   The closed-loop system is to have a step response in which the maximum overshoot does not exceed 35%.
   b.   The settling time of the closed loop is not to exceed 500 milliseconds.
   First write an explanation of your interpretation of these specifications and list your step-by-step procedure for carrying out the design. Then do the design and list the numerical values of τ_1 and τ_2.

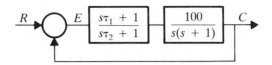

8.51   For the system of Problem 8.21, a PID controller is to be used for G_c. Design suitable gains for the PID.

8.52   For the system of Problem 8.36, a PI controller is used to compensate the system. Use the Bode diagram to design the gains of the PI.

8.53   For the system of Problem 8.23, a PID controller is used to convert the system to type one and to adjust performance. If it is desired to have $K_v \approx 20$, what should be the controller gains? The system must be stable and have a phase margin of no less than 30°.

8.54   State feedback is to be used with the plant of Problem 8.26, instead of a cascade compensator. It is desired to have closed-loop roots at $s = -2, -20$, and -40. Choose suitable states and design the required feedback gains.

8.55   For the plant of Problem 8.47, compensation is to be achieved with state variable feedback, but the states are not available for measurement. Design an observer. Then set feedback gains to place roots at $s = -2 \mp j2, -20$, and -40.

9 Digital and Sampled Data Control Systems

9.1 SOME BACKGROUND

The physical nature of most dynamic systems is such that their states vary as continuous functions of time. Thus, any signals generated by such systems are continuous signals, the sensors we use for measurement usually produce continuous signals, and control schemes developed for such systems have historically used these continuous signals.

At times it has been advantageous to use *samples* of pertinent continuous signals in data processing as well as in control systems. Sampling is necessary in some applications such as radar range determination or the measurement of phase difference between periodic signals. In other applications, an economic advantage is gained, as in telemetering systems where one communication link is used for many measurements. Of course, whenever signals are to be processed digitally, sampling is necessary because digital numbers are obtained by conversion of the amplitude of a sample of a continuous signal via an analog-to-digital (A/D) converter.

Regardless of the reason for sampling signals, such sampled signals must be processed to be useful. When used in control systems, the sampled (or digital) signals are fed into the controller, which then develops the continuous signal needed to drive the plant. To understand the operation of sampled data/digital controls, the fundamental nature of the sampling process must be considered.

A single sample contains specific, but limited, information. A sequence of samples obviously contains more information than a single sample, and it can be shown that a sequence of samples can contain all of the information that was in the original continuous signal. To retrieve this information, the sample sequence must be processed and this processing takes time. Thus, while it is theoretically possible to sample a continuous signal and then reproduce this signal exactly by processing the samples,

321

there is an unavoidable time delay caused by the processing. This is the *best* we can do. In practice, such time delays may be permissible when processing communication data, since the desired result is recovery of information; when sampling is used in a control system, where stability is of prime importance, time delay is destabilizing. Sampled data control systems may discard some of the information in the sampled data in order to obtain better stability and dynamic performance.

9.2   THE SAMPLING PROCESS: METHODS AND RATES

The basic problem in sampling data is the measurement of the magnitude of a variable. When the variable is an available electrical signal, the measurement is usually easy and the process may be represented as in Figure 9.1, which shows a sampling switch, the continuous input signal, and the sampled output. The switch operates at a rate determined by the designer (or perhaps the process!). Usually, the sampling rate is *constant*, i.e., samples are taken every T seconds, where T is a constant and is called the *sampling period*. When digital control is used, the sampling rate is almost always constant, since many algorithms require that T be constant. Nonconstant sampling rates are possible and may be useful in special applications.

The samples in Figure 9.1 are shown as vertical lines, and are to be interpreted as *impulses*. This is the normal definition and is used in developing mathematical analyses. In any physical system, the impulse may be approximated with a short duration pulse. Systems sometimes use finite duration pulses, others use variable duration pulses, and many use pulse-width modulation. However, because almost all physical plants are designed for continuous input and output signals, it is not practical to use impulse samples as inputs, and the interval between samples is "filled" with a continuous signal, which is generated by a *hold* circuit. The simplest hold circuit is commonly called *zero-order hold* (ZOH) and it simply maintains its output at the level of the preceding sample until the next sample is taken, as illustrated in Figure 9.1. This is also what happens when using an A/D converter—a sample is taken and converted to a digital number, which is retained in the output of the A/D converter until the next sample is taken. When sampled data are used in feedback controls, the hold circuit used is almost exclusively a ZOH because other types of higher order hold, e.g., first-order or second-order hold, introduce additional time delay and adversely affect stability, which makes them undesirable for control system use.

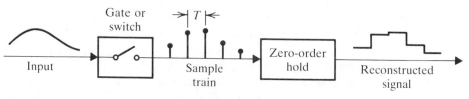

FIGURE 9.1 Sampling a Continuous Signal.

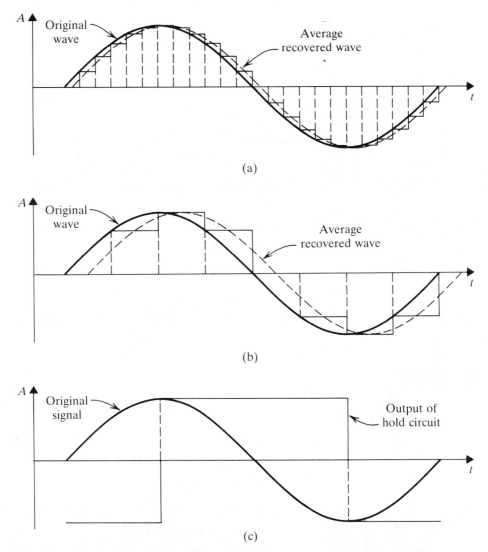

FIGURE 9.2 Illustration of Effect of Sampling Rate: (a) *High Sampling Rate*: Reproduction of Sine Wave is Good. Average of Staircase Wave is Clearly a Sine Wave Delayed About One Half Sampling Period. (b) Medium Sampling Rate: Reproduction of Sine Wave is Fair. Average of Staircase Wave is Again a Sine Wave Delayed About One Half Sampling Period. (c) Two Samples per Cycles, Shannon's Limit.

The time delay introduced by a ZOH is approximately one half of the sampling period. This is illustrated in Figures 9.2(a) and (b). A transfer function may be derived for the ZOH by considering Figure 9.3, which shows the time behavior of the hold system: at $t = 0$, a unit impulse set the hold level at $g(t) = 1.0$, which is held for T

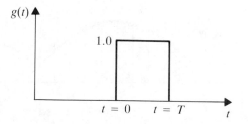

FIGURE 9.3   Definition of a ZOH and its Response to a Unit Impulse at $t = 0$:

$$g(t) = 1.0\ u(t) - 1.0\ u(t - T)$$

$$g_H(s) = \frac{1.0}{s} - \frac{1.0\ e^{-sT}}{s} = \frac{1.0 - e^{-sT}}{s}.$$

seconds, at which instant it is reset to zero. This operation might be continued with another sample at $t = T$. Note that, from Figure 9.3,

$$g(t) = 1.0u(t) - 1.0u(t - T). \tag{9.1}$$

Transforming,

$$G(s) = \frac{1.0}{s} - \frac{1.0e^{-sT}}{s} = \frac{1.0 - e^{-sT}}{s}. \tag{9.2}$$

9.3   THE SAMPLING PROCESS: MATHEMATICAL ANALYSIS

Let a continuous signal be defined by

$$f(t) = \text{a continuous time function}$$
$$\delta_T(t) = \text{a unit amplitude train of impulses with sampling period } T.$$

Then a sampled time function may be defined:

$$f^*(t) = f(t)\delta_T(t) = \text{train of impulse samples.} \tag{9.3}$$

A mathematical definition of the impulse train is

$$\delta_T(t) = \sum_{K=-\infty}^{\infty} \delta(t - KT) = \frac{1}{T}\sum_{K=-\infty}^{\infty} \exp\left(j\frac{K2\pi t}{T}\right) = \frac{1}{T}\sum_{K=-\infty}^{\infty} e^{jK\omega_s t}, \tag{9.4}$$

where $\omega_s = 2\pi/T = $ sampling frequency.
Using Eq. (9.4),

$$f^*(t) = \sum_{K=-\infty}^{\infty} f(KT)\delta(t - KT) = \frac{1}{T}\sum_{K=-\infty}^{\infty} f(t)e^{jK\omega_s t}. \tag{9.5}$$

Transforming Eq. (9.5),

$$F^*(s) = \sum_{K=-\infty}^{\infty} f(KT)e^{-KTs} = \frac{1}{T}\sum_{K=-\infty}^{\infty} F(s + jK\omega_s). \tag{9.6}$$

Using the first term, i.e.,

$$F^*(s) = \sum_{K=-\infty}^{\infty} f(KT)e^{-KTs},$$

leads to the Z transformation. Using the second term and letting

$$s = 0 + j\omega, \qquad F^*(s) = \frac{1}{T}\sum_{K=-\infty}^{\infty} F[j(\omega + K\omega_s)]$$

leads to frequency domain analysis. We will consider both approaches.

9.3A   Frequency Domain Analysis

When a continuous signal is sampled, frequencies are generated by the sampling process. These frequencies are called *complementary* or *alias frequencies*, and are in addition to those frequencies inherent to the continuous input signal. To demonstrate this, consider that a sine wave is being sampled:

$$v(t) = A \sin \omega_a t = A\left[\frac{e^{j\omega_a t} - e^{(-j\omega_a t)}}{2j}\right], \tag{9.7}$$

$$v^*(j\omega_a) = \frac{A}{2Tj}\sum_{K=-\infty}^{\infty} v[j(\omega_a + K\omega_s)]. \tag{9.8}$$

Expanding the series, starting at $K = 0$, and arranging the terms in the sequence $K = 0$, $+1, -1, +2, -2, \ldots,$

$$v^*(j\omega_a) = \frac{A}{2Tj}\{v(j\omega_a) + v[j(\omega_a + \omega_s)] + v[j(\omega_a - \omega_s)]$$

$$+ v[j(\omega_a + 2\omega_s)] + v[j(\omega_a - 2\omega_s)] + \cdots\}. \tag{9.9}$$

The frequency spectrum of Eq. (9.9) is shown in Figure 9.4(a). Note that all components have the same amplitude. If the continuous input signal is *band limited*, i.e., contains frequencies only within known limits, the spectrum is shown in Figure 9.4(b), and if the input signal in *not* band limited the spectrum is shown in Figure 9.4(c). To recover the original signal, we must filter out all of the alias frequencies. It is apparent that this can be done if the original signal is band limited, and cannot be done if the original signal is not band limited. A fundamental requirement, then, is that

$$\omega_s \geq 2\omega_b, \tag{9.10}$$

where ω_b is the bandwidth of the signal. This, in essence, is Shannon's theorem, and the value of the sampling frequency determined from Eq. (9.10) is also known as the Nyquist rate. It should be noted that this limit applies to recovery of information from the sampled signal and is not the primary limitation for sampled data control systems. When sampling is done within the feedback loop, the effects of time delay and alias signals on stability require that higher sampling rates be used. A common figure of merit is $\omega_s = 10\omega_b$, where ω_b is the open-loop, 0-dB bandwidth.

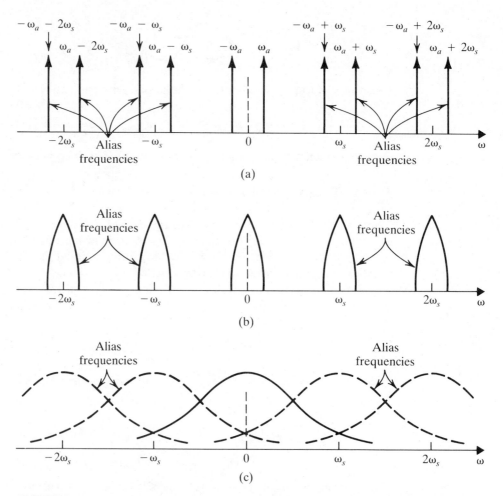

FIGURE 9.4 Frequency Spectra Due to Sampling: Spectrum of (a) a Sampled Sine Wave, (b) a Band-Limited Signal, and (c) a Wideband Signal.

From the preceding discussion, one infers that the sampling rate should be made as high as possible. This seems to be true for nondigital systems. When the signal is digitized, however, the word length must be increased if the higher rate is to be effective, and a practical limit is soon reached.

Equations (9.7), (9.8), and (9.9) consider a signal of one frequency, ω_a. As mentioned above, Figure 9.4(a) portrays the spectrum of this sampled signal and Figures 9.4(b) and (c) show the spectra for a band-limited and nonband-limited signal. To apply this to feedback controls, consider the block diagram of Figure 9.5. For this system, the equations are:

$$E = R - C, \tag{9.11}$$

$$C = G_H G E^*. \tag{9.12}$$

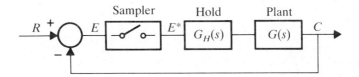

FIGURE 9.5 A Single-Loop, Unity Feedback Sampled Data System.

Equation (9.12) defines the continuous output C in terms of the s domain transfer function and the sample train E^*. We can solve these equations for the signals at the sampling instants if we discard intersample information by defining:

$$E^* = R^* - C^*, \tag{9.13}$$

$$C^* = (G_H G E^*)^*, \tag{9.14}$$

and noting that

$$(G_H G E^*)^* \equiv (G_H G)^* E^*, \tag{9.15}$$

from which

$$\frac{C^*}{R^*} = \frac{(G_H G)^*}{1 + (G_H G)^*}. \tag{9.16}$$

Thus, the characteristic equation is

$$1 + (G_H G)^* = 0. \tag{9.17}$$

The frequency spectrum of $G_H G(j\omega)$ replaces the frequency ω_a in Eq. (9.9). By summing the series of Eq. (9.9), the open-loop frequency response of the system is obtained and the Nyquist criterion may be applied to determine stability. Such methods are implemented most readily with the Z transform, but it is instructive to observe that when $\omega_s \gg \omega_b$ the series converges rapidly; often, all terms except the $K = 0$ term may be neglected (i.e., sampling does not affect stability).

 EXAMPLE 9.1

Let $G(s) = 4/s(s + 1)$, and assume that no hold is used ($G_H = 1.0$). Let $\omega_s = 2\pi/T = 62.8$ ($T = 0.1$). Then

$$G^*(j\omega) = \frac{1}{0.1} \left\{ \frac{4}{j\omega(j\omega + 1)} + \frac{4}{j(\omega - 62.8)[j(\omega - 62.8) + 1]} \right.$$

$$\left. + \frac{4}{j(\omega + 62.8)[j(\omega + 62.8) + 1]} \right\}.$$

Choose $\omega = 1.0$:

$$G^*(j1) = \left\{ \frac{40}{j(j1 + 1)} + \frac{40}{-j61.8[j(-61.8) + 1]} + \frac{40}{j(63.8)[j(63.8) + 1]} \right\}$$

and it is clear that the second and third terms are negligible compared to the first term.

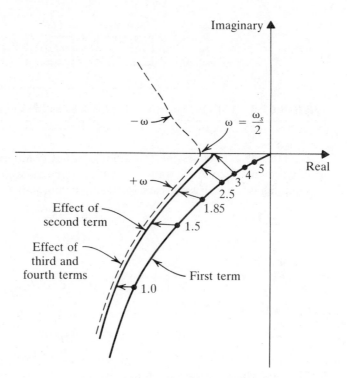

FIGURE 9.6 Polar Plot Showing Series Convergence and Stability Interpretation.

If the same system is sampled at a lower rate, e.g., $\omega_s = 2\pi/T = 6.28$ and $T = 1.0$, the result is (at $\omega = 1.0$):

$$G^*(j1) = \frac{4}{j1(j1 + 1)} + \frac{4}{-j5.28(-j5.28 + 1)} + \frac{4}{j7.28(j7.28 + 1)}$$

and the second and third terms are not negligible. Figure 9.6 gives a polar plot for the case where $\omega_s = 6.28$. This shows the importance of rapid convergence of the series (and thus of high sampling rates). It is interesting to note that the vector for $\omega = \omega_s/2$ always lies on the negative real axis. Thus, evaluation of the series at this one frequency may often be a sufficient test for stability, and additional evaluation for several frequencies somewhat lower than $\omega_s/2$ should always be sufficient for a minimum-phase system. Use of the Z transformation can simplify the calculations.

As a consequence of these relationships, design in the s domain via Bode diagram methods is practical as long as $\omega_s \geq 10\omega_b$. For lower sampling rates, additional terms in the series must be evaluated and/or the Z transfer function of the compensated system must be derived from the s transfer function of the compensated system. While such methods can be used, they are not recommended except for very simple systems.

It should be noted that Eq. (9.3) through (9.9) all relate to sampled *signals*, not transfer functions. In developing the equations for the system of Figure 9.5, it is seen that only one sampled signal, $E*$, exists physically. If both continuous and sampled signals are considered, there are too many variables and the equations cannot be solved simultaneously. This dilemma is solved by considering all signals *at the sampling instants only*. This eliminates the continuous signal equations as shown in Eqs. (9.13) through (9.17) and, in doing so, a starred or sampled transfer function is defined as the ratio of the value of the output (continuous) signal at the sampling instant to the value of the sampled input signal. Thus, $G(s)$ can be expanded in a series as in Example 9.1, or Z transformed as in Example 9.2 in the next section. These examples are for a very simple system.

In general, the rules for manipulating transfer functions in the s domain do not apply exactly to sampled data systems. The correct sampled transfer function depends on the number of samplers used and their location in the system. Consider the following cases as shown in Figure 9.7. From Figure 9.7(a) it is clear that the Z transfer function of a single block with input signal sampled is obtained by taking the Z transform of $G(s)$. This is true whether the output is actually sampled or not; however, use of such a $G(Z)$ depends on the way the block is used in the system. Figure 9.7(b) shows that, for a cascade connection of blocks with intermediate samplers, $G(Z)$ for the cascade combination is the product of the Z transfer functions of the individual blocks, and this applies to any number of blocks in cascade. On the other hand, if blocks are cascaded without intermediate samplers, the s transfer functions must be multiplied and the resulting $G(s)$ transformed to $G(Z)$.

The basic rule is: for any signal path through a system, the $G(s)$ transfer functions of all cascaded blocks between samplers must be multiplied before applying the Z transform. When in doubt, the basic methods of Eqs. (9.11) through (9.17) may be applied.

9.3B Z Transform Analysis

The first term in Eq. (9.6), i.e.,

$$F*(s) = \sum_{K=-\infty}^{\infty} f(KT)e^{-KTs}, \qquad (9.6)$$

may be used to define the Z transformation. Let us define

$$Z = e^{sT}; \qquad (9.18)$$

then Eq. (9.6) converts to

$$F(Z) = \sum_{K=-\infty}^{\infty} f(KT)Z^{-K}, \qquad (9.19)$$

which defines an infinite series in Z. To be useful, we need to convert the series to closed form. Explicit closed forms must be obtained for each function (or functions) that generate $f(KT)$. The closed forms are found either by inspection of the specific series or

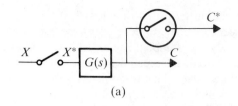

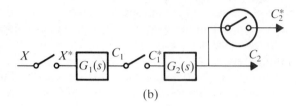

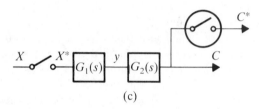

FIGURE 9.7 Transfer Functions for Sampled Data Systems:

a. Single block with continuous output: $\qquad C = G(s)X^*$
   Introduce a fictitious sampler at C: $\qquad C^* = G^*(s)X^*$

   Then: $\qquad\qquad\qquad\qquad\qquad\qquad \dfrac{C^*}{X^*} = G^*(s) \to G(Z).$

b. Sampled system with two actual samplers and two cascaded blocks. The Z transfer functions multiply:

$$C_1^* = G_1^* X^* \qquad C_2^* = G_2^* C_1^* \qquad C_2 = G_2(s)C_1^*$$

$$\frac{C_1^*}{X^*} = G_1^*(s) \qquad \frac{C_2^*}{C_1^*} = G_2^*(s)$$

$$\frac{C_2^*}{X^*} = G_1^*(s)G_2^*(s) \to G_1(Z)G_2(Z)$$

c. System with two cascaded blocks not separated by a sampler. Transfer function is $G_1G_2(Z) \neq G_1(Z)G_2(Z)$:

$$C = G_2(s)y \qquad\qquad C^* = (G_1(s)G_2(s)X^*)^*$$
$$y = G_1(s)X^* \qquad\qquad\quad = [G_1(s)G_2(s)]^*X^*$$

$$C = G_2(s)G_1(s)X^* \qquad \frac{C^*}{X^*} = [G_1(s)G_2(s)]^* \to G_1G_2(Z).$$

TABLE 9.1 Z Transfer Functions

Comment	s Form	Z Form
Step or pole at origin	$\dfrac{A}{s}$	$\dfrac{AZ}{Z-1}$
Ramp or two poles at origin	$\dfrac{1}{s^2}$	$\dfrac{TZ}{(Z-1)^2}$
Parabola	$\dfrac{1}{s^3}$	$\dfrac{T^2 Z(Z+1)}{2(Z-1)^3}$
Real pole	$\dfrac{B}{s+P}$	$\dfrac{BZ}{Z-e^{-PT}}$
Complex poles	$\dfrac{Ka}{(s+\alpha)^2 + a^2}$	$\dfrac{KZe^{-\alpha T}\sin\alpha T}{Z^2 - 2e^{-\alpha T}Z\cos\alpha T + e^{-2\alpha T}}$
Imaginary poles	$\dfrac{Ka}{s^2 + a^2}$	$\dfrac{KZ\sin\alpha T}{Z^2 - 2Z\cos\alpha T + 1}$

Note: Poles in the s plane map to poles in the Z plane on a one-for-one basis: $-P \to e^{-PT}$. There is no rule for mapping zeros from the s plane to the Z plane. When transforming from s to Z new zeros are generated on the Z plane. Normally, the Z transfer function has N poles and $N-1$ zeros.

by application of the complex convolution theorem. Results for commonly encountered cases have been collected in tables such as Table 9.1.

 EXAMPLE 9.2

Consider the system of Example 9.1. Expand $G(s)$ in partial fractions to obtain forms existing in Table 9.1:

$$G(s) = \frac{4}{s(s+1)} = \frac{4}{s} - \frac{4}{s+1}.$$

From the table,

$$G(Z) = \frac{4Z}{Z-1} - \frac{4Z}{Z-e^{-T}} = \frac{4Z(1-e^{-T})}{(Z-1)(Z-e^{-T})}.$$

Let $T = 1.0$, then $e^{-T} = 0.368$. Hence,

$$G(Z) = \frac{4(1-0.368)Z}{(Z-1)(Z-0.368)} = \frac{2.528Z}{(Z-1)(Z-0.368)}.$$

This is the closed form of the series of Eq. (9.19) as applied to the system of Example 9.1.

Because of the basic definition, $Z = e^{sT}$, Z and s are related functionally—each is a complex variable—and a map of a function on one plane can be related to a map of the same function on the other plane. It is useful to relate certain lines on the s plane to corresponding lines on the Z plane:

1. The real axis of the s plane maps to the real axis on the Z plane as follows:

    a. At the origin of the s plane, $s = 0 + j0$, $Z = e^{sT} = 1.0 + j0$.

    b. At $s = -\infty + j0$, $Z = 0 + j0$.

    Thus, the negative real axis of the s plane maps into the segment $0 + j0 \le Z \le +1 + j0$. For $\sigma > 0$, $e^{sT} > +1.0$ and $Z = +|e^{\sigma T}| + j0$. Thus, the positive real axis of the s plane maps into the positive real axis of the Z plane for $+1.0 + j0 \le Z \le +\infty + j0$.

2. The imaginary axis of the s plane maps into the unit circle on the Z plane. Note that, for sampled data, the sampling frequency is

$$f_s = \frac{1}{T},$$

$$\omega_s = 2\pi f_s = \frac{2\pi}{T}; \qquad T = \frac{2\pi}{\omega_s}.$$

For $s = 0 + j\omega$,

$$Z = e^{j\omega T} = \exp\left[j\omega\left(\frac{2\pi}{\omega_s}\right)\right] = \exp\left(j2\pi\frac{\omega}{\omega_s}\right)$$

$$= \exp\left[j\pi\frac{\omega}{(\omega_s/2)}\right].$$

By inspection $|Z| = 1.0$ for all ω and $\angle Z = \pi\omega/\omega_s/2$, which are the parametric equations of a circle of unit radius with center at the origin of the Z plane. Figure 9.8 illustrates these results. Note that the unit circle maps frequencies from $-\omega_s/2 \le \omega \le +\omega_s/2$.

3. The primary strip; effect of alias frequencies.

The sampling process generates alias frequencies around multiples of the sampling frequency. This can be represented on the s plane by dividing the imaginary axis into segments as shown in Figure 9.9. The unit circle on the Z plane is the map of the segment $-\omega_s/2 < \omega < \omega_s/2$. Horizontal lines at $-\omega_s/2$ and $+\omega_s/2$ bound the region of interest for a system band limited to $\omega_b < \omega_s/2$, and the strip so defined is called the *primary strip*. On the horizontal boundary.

$$s = \mp\sigma \mp j\frac{\omega_s}{2};$$

therefore,

$$Z = e^{sT} = \exp\left(\mp\sigma j\frac{\omega_s}{2}\right)T = e^{\mp\sigma T}e^{\mp j\pi},$$

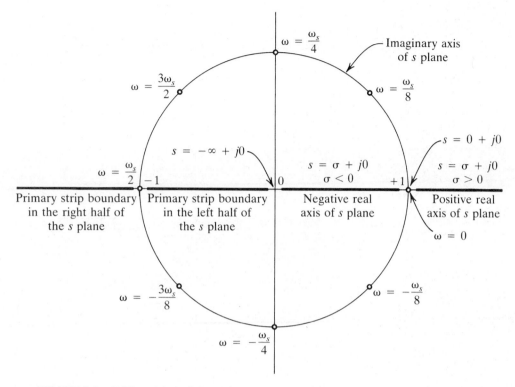

FIGURE 9.8   *Z*-Plane Map of the *s* Plane.

from which it is seen that the boundaries of the primary strip map onto the negative real axis of the *Z* plane. The boundaries in the left half of the *s* plane map on the *Z* plane between the origin and -1.0 and, those in the right half of the *s* plane map on the *Z* plane between -1 and $-\infty$. This is indicated in Figure 9.9.

  In like manner, it is readily shown that the imaginary axis segments map on top of the unit circle; also, all strip boundaries map on top of each other. The left half of the *s* plane maps inside of the unit circle. Clearly, the unit circle is the stability boundary in the *Z* plane. The Routh test[a] cannot be used because it applies to the imaginary axis of a rectangular coordinate system. The Nyquist criterion is applied by mapping the unit circle on the *Z* plane through the *Z* transfer function. The result is a polar plot such as that of Figure 9.6, but the curve obtained is the sum of the *entire* series rather than the sum of a finite number of terms. The root locus method may also be used on the *Z* plane; note that plotting procedures are unchanged, but the stability limits are the points at which the locus crosses the unit circle.

---

[a] Array tests for sampled data systems are Jury's tests and the Schur-Cohn criteria. They are not discussed in this text.

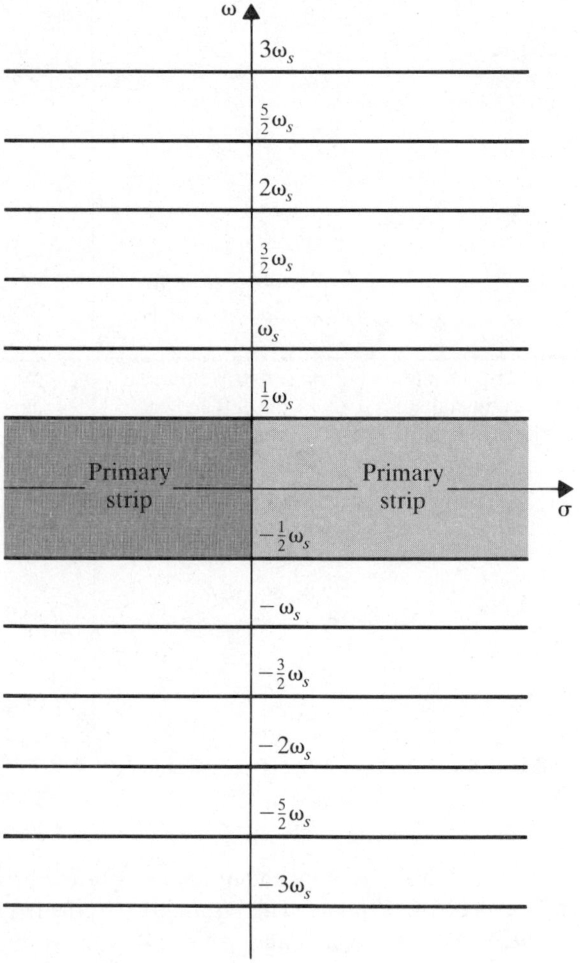

FIGURE 9.9   Effects of Aliasing on the *s* Plane.

 EXAMPLE 9.3

For the system of Example 9.2, the Z transfer function is

$$G(Z) = \frac{4(1 - e^{-T})Z}{(Z - 1)(Z - e^{-T})}.$$

The root locus is shown in Figure 9.10(a). If a ZOH is added to the system,

$$G(s) = \left(\frac{1 - e^{-sT}}{s}\right)\left[\frac{K}{s(s + a)}\right] = \frac{K(1 - e^{-st})}{s^2(s + a)}$$

$$G(Z) = K(1 - Z^{-1}) \times Z\text{-transform of } \left[\frac{1}{s^2(s + a)}\right]$$

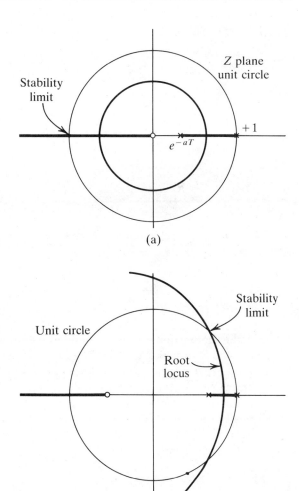

FIGURE 9.10 Example 9.3, Root Loci on the Z Plane:

a. root locus for

$$G(Z) = \frac{4(1 - e^{-T})}{(Z - 1)(Z - e^{-T})}$$

b. root locus for

$$G(Z) = \frac{A(Z - Z_1)}{(Z - 1)(Z - e^{-aT})}.$$

Expanding in fractions,

$$\frac{1}{s^2(s+a)} = \frac{1/a}{s^2} - \frac{1/a^2}{s} + \frac{1/a^2}{s+a},$$

from which

$$G(Z) = \frac{A(Z - Z_1)}{(Z - 1)(Z - e^{-aT})},$$

where

$$A = \frac{K}{a^2}(aT - 1 + e^{-aT}) \qquad Z_1 = -\left[\frac{1 - e^{aT}(1 + aT)}{aT - 1 + e^{-aT}}\right].$$

The root locus is shown in Figure 9.10(b).

9.4 FEEDBACK SYSTEM ANALYSIS IN THE Z DOMAIN

When Z transfer functions are used, considerable labor is required to convert from the s variable to the Z variable. Also, because the entire left half plane maps inside the unit circle on the Z plane, pole, zero, and gain values must be defined with a larger number of significant figures. It is therefore advisable to use the computer whenever possible.

After obtaining proper Z transfer functions, the closed-loop roots can be found by various methods. It is of interest to study dynamic behavior as a function of root location on the Z plane. Figure 9.11 shows the unit circle on the Z plane, with possible root points indexed with letters A, B, C, etc. Figure 9.12 shows typical time responses

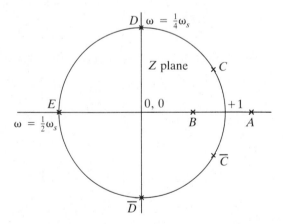

FIGURE 9.11 Unit Circle on the Z Plane with Possible Root Points Indexed for Illustrations.

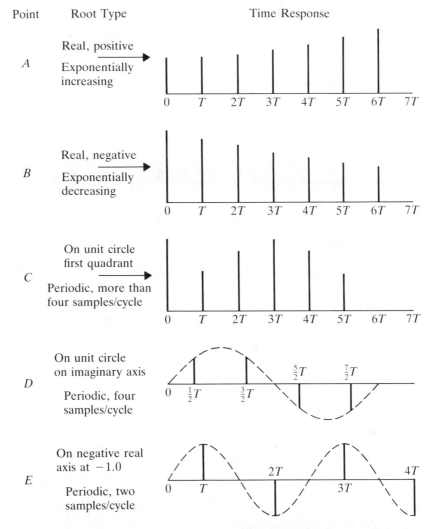

FIGURE 9.12   Time Response Due to Roots at Locations Indexed in Figure 9.11.

due to roots at the points designated on Figure 9.11. It is clear that well-behaved sampled data control systems must have all roots inside the unit circle, and preferably in the right half of the Z plane.

Sampled data control systems are used for the same basic purposes as comparable nonsampled systems. Thus, the same types of command signals and load disturbances are expected and a primary requirement is, again, steady-state accuracy. (Note that for digital systems perfect accuracy is not possible because of truncation to finite word length). In general, the final value theorem can be used. In the Z domain, the final value theorem is:

$$F(t = \infty) = \lim_{Z \to 1} \left(\frac{Z - 1}{Z}\right)[F(Z)]. \tag{9.20}$$

The basic command signals are the step, ramp, and parabola as for continuous systems. In the Z domain, these are

$$\text{Step: } R(Z) = \frac{Z}{Z - 1}, \tag{9.21a}$$

$$\text{Ramp: } R(Z) = \frac{TZ}{(Z - 1)^2}, \tag{9.21b}$$

$$\text{Parabola: } R(Z) = \frac{T^2 Z(Z + 1)}{2(Z - 1)^3}. \tag{9.21c}$$

■ EXAMPLE 9.4

Consider the system of Example 9.3 with ZOH:

$$G(Z) = \frac{A(Z - Z_1)}{(Z - 1)(Z - e^{-aT})}.$$

For the single-loop, error-sampled system,

$$E(Z) = \frac{R(Z)}{1 + G(Z)},$$

and for a unit step input,

$$E_{ss}(t) = \lim_{Z \to 1} \frac{Z - 1}{Z} E(Z) = \lim_{Z \to 1} \frac{Z - 1}{Z} \left[\frac{Z/(Z - 1)}{1 + G(Z)}\right].$$

$$= \lim_{Z \to 1} \frac{1}{1 + [A(Z - Z_1)/(Z - 1)(Z - e^{-aT})]} = 0$$

For a unit ramp input,

$$E_{ss}(t) = \lim_{Z \to 1} \frac{Z - 1}{Z} \left[\frac{TZ/(Z - 1)^2}{1 + G(Z)}\right] = \lim_{Z \to 1} \frac{T/(Z - 1)}{1 + [A(Z - Z_1)/(Z - 1)(Z - e^{-aT})]}.$$

$$= \lim_{Z \to 1} \frac{T(Z - e^{-aT})}{(Z - 1)(Z - e^{-aT}) + A(Z - Z_1)} = \frac{T(1 - e^{-aT})}{A(1 - Z_1)}$$

It should be noted that error coefficients can be defined for type-zero, -one, or -two systems. This is left to the student as an exercise. It should also be noted that gain adjustment can be used to reduce the steady-state error (for those cases where steady-state error exists), provided the stability limit is not exceeded.

<table>
<tr><td>**9.5**</td><td>DESIGN OF COMPENSATION: SOME SIMPLE CASES</td></tr>
</table>

In this text, only cascade compensation of error-sampled systems is considered. Some possible cases are given by the block diagrams of Figure 9.13. If a single sampler is used, then the cascade compensator must be an analog filter, but may be designed in either

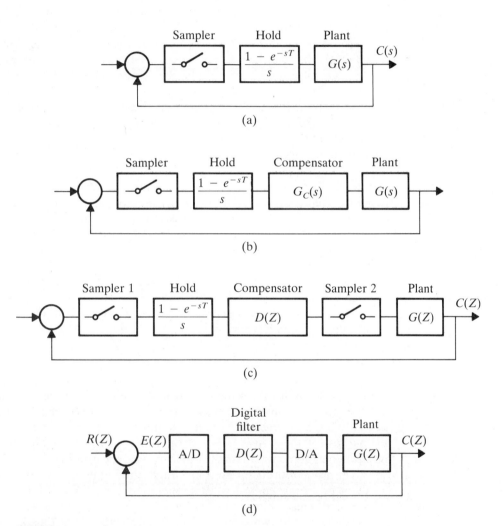

(a)

(b)

(c)

(d)

FIGURE 9.13 Block Diagrams for Cascade-Compensated Sampled Data Systems: (a) Sample and Hold, No Compensator; (b) Sampler, Hold, and Analog Compensator; (c) Sample, Hold, Compensator [Designed as a Discrete Compensator $D(Z)$; Realization as Analog, Discrete, or Digital Depending on Nature of Sampler 2]; and (d) Digital Compensator.

the s domain or the Z domain. If two samplers are used, the compensator would be designed in the Z domain and realized as a continuous, discrete, or digital filter. When the microprocessor is used and a digital compensator is designed, the initial A/D requires one sampler and the output D/A requires the second sampler. All numbers in the microprocessor are held, of course, and the total time delay in the digital compensator is due to the sampling period of the A/D plus the computation time of the particular algorithm.

In practice, control systems may be completely analog, or they may be sampled for reasons other than digital control, or they may be sampled so that the microprocessor (or other digital processor) can be used. For many systems that require sampling, the sampling rate is fixed or at least bounded by the hardware to be used. Any resulting compensator design must be able to accommodate the time delay due to the sampling period. On the other hand, if sampling is done to make digital control possible, then the limits on sampling rate are presumably set by the conversion time of the A/D, the microprocessor clock rate, available word length, or, if a central processing unit is used instead of a dedicated microprocessor, by the time-sharing allocations. In any of these cases, the design of a continuous (unsampled) compensator will be the best possible design and may be used as a reference for comparison. All discrete and/or digital compensators designed for the system must provide a less desirable result because of the inherent time delays. In addition to using the continuous system design as a reference for comparison, it may also be used for the actual design of the digital compensator by using a transformation, as will be demonstrated.

▮ EXAMPLE 9.5

Consider a phase-locked oscillator used for getting digital data from a disk memory. The block diagram is that of Figure 9.13(b). The plant is a voltage-controlled oscillator with output *frequency* proportional to applied voltage and output phase is the *integral* of the frequency. With typical numbers,

$$G_{plant} = \frac{5 \times 10^6}{s}.$$

Since the disk rotates at constant speed, data cells are equally spaced and a maximum data frequency is known. For design purposes, consider the input to the loop to be a sine wave[b] of this frequency. Assume that the oscillator generates a sine wave[b] also, and the purpose of the loop is to adjust the oscillator frequency and phase so that the positive going zero crossovers of the waves are synchronized.

The *phase* of the input is the integral of the frequency, and so is a *ramp* of phase. Steady-state error is to be zero so a type-two system is needed. This means that the compensator must contain an integrator. If only the integrator was added, the open-loop transfer function would consist of two poles at the origin; thus, a lead filter must be

---

[b] Actual wave shapes are quite different, of course.

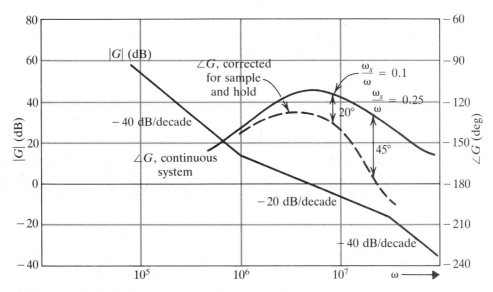

FIGURE 9.14 Bode Diagram for the Phase-Locked Loop.

used for stability and damping. The open-loop transfer function is

$$G_{\text{open}} = G_H G_C G_P + \frac{1 - e^{-sT}}{s} \frac{k(s+a)}{s(s+b)} \frac{K_{\text{osc}}}{s}.$$

For a particular design, $T = 74$ ns from which $\omega_s = 2\pi/T = 8.48 \times 10^7$ rad/s. The closed-loop settling time must be less than 1.0 μs, which should be achievable with an open-loop bandwidth (zero crossover) of 5×10^6 rad/s. Note that the sampling rate is more than 10 times the bandwidth, so only the primary term in the series is needed and an s plane design should be adequate if the sampling delay is accounted for. A Bode diagram design is given in Figure 9.14.

 EXAMPLE 9.6

Consider the block diagram of Figure 9.13(c). This figure implies that the compensator is to be designed as a discrete compensator but may be realized as either an analog filter if the second sampler is not used in the physical system, or as a pulsed analog filter if the second sampler is actually used. The design procedure is the same in either case, but the resulting filters are different.

To carry out the design, one obtains the Z transform of plant and hold, then designs the cascade compensator obtaining a filter transfer function $D(Z)$. Thus, the loop transfer function of the compensated system is

$$G_{\text{loop}}(Z) = D(Z)G_{HP}(Z).$$

If compensation is to be realized as an analog filter without the second sampler, then we must take the inverse Z transform of $G_{\text{loop}}(Z)$. If the second sampler is used so that $D(Z)$ is a pulsed filter, then we need take the inverse Z transform of $D(Z)$ only. For a system with a type-two plant:

$$G_{HP}(s) = \left(\frac{1 - e^{-sT}}{s}\right)\left(\frac{k}{s^2}\right)$$

$$G_{HP}(Z) = (1 - Z^{-1})\left[\frac{kT^2 Z(Z + 1)}{2(Z - 1)^3}\right] = \frac{kT^2(Z + 1)}{2(Z - 1)^2}.$$

Choosing to design with the Bode diagram, the bilinear transformation is used:

$$Z = \frac{1 + W}{1 - W},$$

where $W = u + jv$. Then,

$$G_{HP}(W) = \frac{(kT^2/2)(1/1 - W)}{4W^2/(1 - W)^2} = \frac{kT^2}{8}\left(\frac{1 - W}{W^2}\right),$$

$$G_{HP}(jv) = \frac{kT^2}{8}\frac{(1 - jv)}{(jv)^2}.$$

Let $k = 100$, $T = 0.031416$, and $T^2 = 0.00987$; then,

$$G_{HP}(jv) = \frac{0.0123(1 - jv)}{(jv)^2}.$$

Figure 9.15 shows the design on the Bode diagram, from which

$$G_C G_{HP}(jv) = \frac{0.0123(1 - jv)(1 + 10jv)}{(jv)^2(1 + jv/1.1)}$$

or, more generally,

$$G_C G_{HP}(W) = \frac{0.0123(1 - W)(1 + 10W)}{W^2(1 + W/1.1)}$$

$$= \frac{0.0123(1 - W)}{W^2}\frac{1 + 10W}{1 + W/1.1}$$

Next, transform back to the Z domain using the bilinear transform $W = (Z - 1)/(Z + 1)$:

$$G_C(Z)G_{HP}(Z) = \frac{1 + 10[(Z - 1)/(Z + 1)]}{1 + (Z - 1)/1.1(Z + 1)}\frac{50T^2(Z + 1)}{(Z - 1)^2}$$

$$= \frac{50T^2(Z + 1)}{(Z - 1)^2}\frac{5.24(Z - 0.818)}{(Z + 0.0476)}.$$

Realization as an analog system without second sampler requires inversion of the entire function to the s domain. If realization is to be as a pulsed filter only, $G_C(Z)$ needs

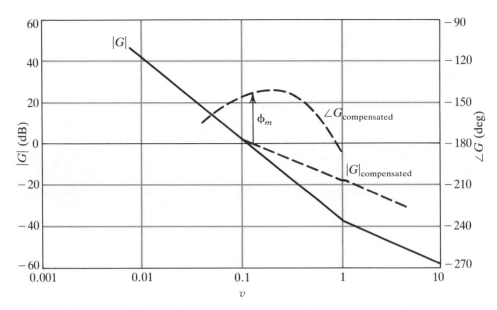

FIGURE 9.15 Bode Diagram for:

$$G(W) = \frac{0.0123(1 - W)(1 + W/0.1)}{W^2(1 + W/1.1)}.$$

to be inverted; and if realization is to be as a digital filter in a microprocessor, one need only write the software for $G_C(Z)$.

  EXAMPLE 9.7

A compensator for the system of Example 9.6 may be designed using the root locus on the Z plane. We know that $G_{HP}(Z) = K(Z + 1)/(Z - 1)^2$. This is used to plot the root locus of the uncompensated system as shown in Figure 9.16. Clearly, the system is unstable and a lead network is required. By trial and error, one arrives at $G_C(Z) = K(Z - 0.8)/(Z - 0.1)$. The gain is then adjusted to place the roots at an acceptable location on this root locus. The compensated transfer function is then

$$G_C G_{HP}(Z) = \frac{kK(Z + 1)(Z - 0.8)}{(Z - 1)^2(Z - 0.1)}.$$

If the compensator is to be realized as an analog filter, using the partial fraction expansion after reinserting the factors canceled in taking the Z transform of plant and hold, one obtains

$$G_C(s) = \frac{0.00875(s + 12.72)(s + 469.57)}{(s + 73.3)}.$$

Another pole is required for physical realizability and a suggested value is $(s + 5000)$.

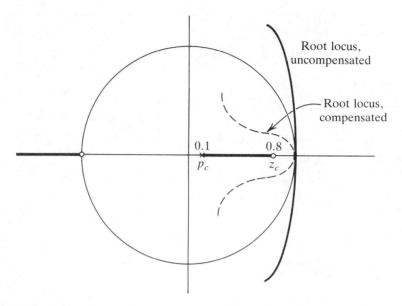

FIGURE 9.16 Design of Cascade Compensation with Root Loci on the Z Plane.

Another method is to use the *matched Z* transform, which maps poles onto the s plane in the standard fashion, e.g., $Z - 0.1 \Rightarrow s - 1/T \ln 0.1 = s + 73.31$, and arbitrarily maps zeros using the same rule, e.g., $z - 0.8 \Rightarrow s - 1/T \ln 0.8 = s + 7.1$, so that

$$G_C(s) = \frac{s + 7.1}{s + 73.31}.$$

Still another method uses the bilinear transformation:

$$Z = -\left(\frac{Ts/2 + 1}{Ts/2 - 1}\right),$$

which transforms $Z - 0.8/Z - 0.1$ to

$$G_C(s) = 1.63636\left(\frac{s + 7.07}{s + 52}\right).$$

 EXAMPLE 9.8

Example 9.5 illustrated design of an analog compensator for a sampled data system using conventional continuous system methods on the Bode diagram and accounting for the sample and hold by correcting the phase curve. If we wish to implement the filter as a digital filter using a microprocessor, how would we proceed? (Note that many compensator designs use this approach, because of the vast background of experience in analog system design). For linear systems, the transient response is completely defined by the frequency response—thus, if the digital system has the same frequency

response as the analog design, it will also have the same time response. Once $G_C(s)$ has been designed, the remaining problem is to design $G_C(Z)$ to duplicate the frequency response of $G_C(s)$. This is not easy to do. Most of the procedures for digitizing an analog filter do not preserve all of the desirable characteristics of the frequency response.

One good procedure is to *prewarp* the frequency response, then apply the bilinear transformation:

1. Design the analog filter, determining the required zeros and poles.

2. Prewarp the frequency response of the filter by shifting all zeros and poles as follows:

$$Z' = \frac{2}{T} \tan \frac{ZT}{2},$$

$$P' = \frac{2}{T} \tan \frac{PT}{2}.$$

3. Apply the bilinear transformation

$$s = \frac{2}{T} \left(\frac{Z - 1}{Z + 1} \right)$$

to obtain $G_C(Z)$.

Figure 9.17(a) gives the block diagram and (b) gives the Bode diagram design of the analog compensator for the system of Example 9.6, obtaining

$$G_C(s) = \frac{100}{5.6} \left(\frac{s + 5.6}{s + 100} \right).$$

Prewarping,

$$Z' = \frac{2}{0.031416} \tan \frac{(5.6)(0.031416)}{2}$$

$$= 0.098,$$

$$P' = \frac{2}{0.031416} \tan \frac{(100)(0.031416)}{2}$$

$$= 1.9457,$$

$$G'_C(s) = \frac{100}{5.6} \left(\frac{s + 0.098}{s + 1.7457} \right).$$

Using the bilinear transformation,

$$G_C(Z) = 17.414 \frac{Z - 0.9969}{Z - 0.94666}.$$

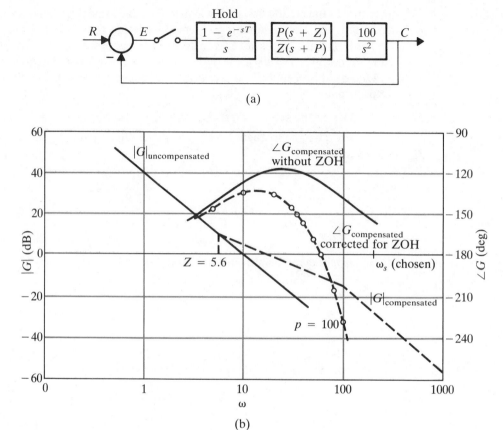

FIGURE 9.17   (a) Block Diagram for Bode Design; (b) D of Lead Compensator. Phase Curve Corrected for Sample and Hold.

9.6   SUMMARY

The sampling process is explained and analyzed mathematically in both the frequency domain and the Z transform domain. The results are applied to the analysis of sampled data feedback control systems. The use of these methods in design is presented and some simple examples are given.

BIBLIOGRAPHY

Franklin, G. F.; and Powell, J. D. *Digital Control of Dynamic Systems.* Reading, Mass.: Addison Wesley (1980).

Jury, E. I. *Sampled Data Control Systems.* New York: John Wiley and Sons (1958).

Kuo, B. C. *Digital Control Systems.* New York: Holt, Rinehart and Winston (1980).

Ogata, K. *Discrete Time Control Systems.* Englewood Cliffs, N.J.: Prentice Hall (1987).

Ragazzini, J. R.; and Franklin, G. F. *Sampled Data Control Systems.* New York: McGraw-Hill Book Co. (1958).

Tou, J. T. *Digital and Sampled Data Control Systems.* New York: McGraw-Hill Book Co. (1980).

PROBLEMS

9.1   If

$$G*(s) \triangleq \sum_{n=0}^{\infty} f(nT)e^{-nTs},$$

find the closed form solution if $G(s) =$

a. $\dfrac{1}{s + P}$

b. $\dfrac{1}{(s + P_1)(s + P_2)}.$

9.2   Find the Z transfer functions for each of the following where $T = 1.0$:

a. $\dfrac{K}{(s + a)(s + b)}$

b. $\dfrac{K(1 - e^{-sT})}{s(s + a)(s + b)}$

c. $\dfrac{(s + 3)}{(s + 1)(s + 2)}$

d. $\dfrac{(1 - e^{-sT})(s + 3)}{s(s + 1)(s + 2)}$

e. $\dfrac{K}{s(s + 1)(s + 2)}$

f. $\dfrac{K(1 - e^{-sT})}{s^2(s + 1)(s + 2)}$

g. $\dfrac{K(s + b)}{s^2}$

h. $\dfrac{K(1 - e^{-sT})(s + b)}{s^3}$

9.3   A second-order motor has a transfer function where $K = 2$:

$$G(s) = \frac{K}{s(s + 1)}.$$

a.   Obtain the open-loop Bode diagram and the root locus.
b.   Obtain the Z transform of $G(s)$. Plot the root locus on the Z plane. (Let $T = 1.0$.) Find the root locations.

9.4   The motor of Problem 9.1 is used in a closed-loop with ZOH, as shown below:

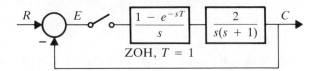

a.   Obtain the Bode diagram and correct the phase curve for the effect of the ZOH.
b.   Obtain the Z transfer function, plot the root locus, and mark the root locations.
c.   Simulate and obtain step response. Compare with the step response of the unsampled system.
d.   Using the simulation of part (c), try several values of the sampling period T. What is the effect of the sampling period on the step response? On system stability?

9.5   Given

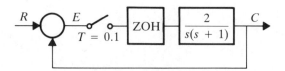

Is the system stable? What is the phase margin?

9.6   Given

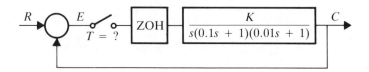

a.   For the *unsampled* system, find K for 45° phase margin.
b.   Choose T so that the sample/hold introduces $-10°$ phase shift at the gain crossover frequency found in Part (a).
c.   Simulate both sampled and unsampled systems and compare the step responses.

9.7   Given

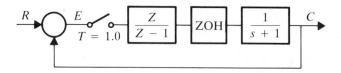

Find the step response.

9.8   A single-loop servo has a forward transfer function:

$$G(s) = \frac{20}{s(s + 1)(s + 5)}.$$

    a. Analyze for stability and phase margin.

    b. Assume that the system is error-sampled and has a ZOH. Set the sampling frequency at $5\omega_c$, where ω_c is the gain crossover frequency found in Part (a). Correct the phase curve of Part (a) and check for stability and phase margin.

    c. Design an analog compensator to increase the phase margin to approximately 45°.

    d. Simulate and obtain the step response.

9.9   Design a digital compensator for the system of Problem 9.8 by using the root locus on the Z plane. Obtain step response.

9.10   Use the bilinear transformation.

$$s = \frac{2}{T}\left(\frac{Z-1}{Z+1}\right)$$

to convert the analog filter found in Problem 9.8 to a digital filter. Simulate and obtain step response.

9.11   Repeat Problem 9.10 using the prewarped bilinear transformation.

9.12   Repeat Problem 9.10 using the pole-zero matching method.

10 Nonlinear Systems: Describing Functions and Limit Cycles

NATURE AND EFFECTS OF NONLINEARITIES

There are various ways to define a *nonlinear* system. Perhaps the best and simplest is to note that for linear systems the principle of superposition applies, then define a nonlinear system as any system for which this principle does *not* apply. Alternatively, we may note that a linear system may be modeled by a single linear differential equation or by a set of simultaneous equations, all linear. Then we may say that a nonlinear system is one for which the model is a single nonlinear equation, or a set of simultaneous equations at least one of which is nonlinear, or any set of equations which are not simultaneous but sequential. This last definition includes discontinuous (switched) systems for which the principle of superposition does not hold, even though all of the pertinent equations may be linear.

Commonly encountered physical phenomena that make a system nonlinear are

1. saturation
2. dead zone (or dead space, freeplay, etc.)
3. dry (Coulomb) friction or threshold
4. hysteresis
5. backlash.

These are usually unintentional nonlinearities, i.e., we do not want them in the system, they come with the hardware. However, there are other nonlinearities that result from intentional design:

6. relays
7. switches

8. multiplication of signals
9. deliberately adjustable gains (as in AGC circuits).

Such phenomena may affect steady-state accuracy, stability, and the dynamic response of a control system. They may also be the source of other dynamic phenomena that cannot exist in a linear system. For example, nonlinear systems may exhibit limit cycles that are steady-state oscillations of fixed amplitude and frequency. They may or may not be sinusoidal in wave shape—in fact, they may sometimes be relaxation-type oscillations. If forced with a periodic forcing function, the nonlinear system may exhibit jump resonance or subharmonic resonance. A jump resonance is a sudden discontinuous change in the amplitude of the periodic output when the input frequency passes through a critical value. A subharmonic resonance is the generation, in the periodic output, of a frequency lower (usually much lower) than the forcing frequency. In addition, the nonlinear system may respond differently to various input signals. Of course, we expect the response to different signal amplitudes to be different in wave shape because the principle of superposition does not apply, but in addition to this there are the effects of different *kinds* of forcing functions. For example, the system may not exhibit a limit cycle for a step input, but may go into a limit cycle for a ramp input. Also, as we will see later, if the system has an *unstable* limit cycle, it may respond as desired for small amplitude inputs, but be divergingly unstable for large amplitude inputs.

The design of control systems is concerned primarily with accuracy, but it must also take into account the constraints that the system must be stable and must have acceptable transient response. These considerations apply to essentially all control systems, and the presence of nonlinearities may affect all of these characteristics. Clearly, nonlinear friction, deadzone, backlash, and hysteresis will affect steady-state accuracy. If the loss of accuracy is not acceptable, one must either replace the guilty component or (usually) raise the loop gain, but raising the loop gain may cause loss of stability or unacceptable changes in the transient response. A saturation nonlinearity, on the other hand, does not usually affect accuracy, but it has substantial effect on the transient response when large disturbances are encountered—transient recovery from a large disturbance is much slower than from a small disturbance. A single saturating element can cause a limit cycle depending on the complexity of the system and the location of that saturating element in the system. If several elements can saturate, a variety of phenomena are possible. When nonlinearities are deliberately inserted into the system (such as switches, multipliers, quantizers, etc.), steady-state accuracy is not usually compromised, but the effects on stability and transient response can be significant. Furthermore, our ability to estimate these effects using linear theory—or nonlinear theory—is quite limited, so simulation studies are often needed.

Finally, one must recognize that nonlinear systems respond to forcing functions in a manner that can be significantly different from the response of linear systems for which the principle of superposition permits us to predict responses accurately. For example, a nonlinear system that responds in almost linear fashion to a step input may exhibit a limit cycle for a ramp input; or one that responds linearly to a step input may exhibit jump resonances when we try to measure its frequency response.

10.2   TOOLS AVAILABLE FOR ENGINEERS

In the engineering of control systems, we need the ability to analyze nonlinear control systems both qualitatively and quantitatively. We need to know whether the defects of nonlinearities can be neglected in a given application, or whether they must be considered quantitatively in order to meet performance specifications. We must be able to predict the occurrence of various types of nonlinear phenomena. If these phenomena are objectionable, we must know how to eliminate them or at least minimize the objectionable effects—and this, in general, requires sufficient knowledge for quantitative design. In still other cases, we may wish to *use* certain nonlinear phenomena such as limit cycles. To use a limit cycle effectively, we must know how to produce it, and we must be able to obtain a desired frequency and amplitude.

The tools available for analysis and design of nonlinear systems are, unfortunately, quite limited in nature, in number, in applicability, and in effectiveness. In the literature on nonlinear mechanics, one can find a number of techniques that have been developed through the years, but most of these are so specialized as to be virtually useless to the engineer. There are only three tools that have wide applicability:

1. the describing function method
2. the phase plane–phase space methods
3. simulation studies.

A few others, such as Lyaponuv's second method, are occasionally very useful, but cannot be considered generally applicable methods.

The engineering use of describing functions is normally applied to classes of problems quite different from those handled by phase space theory; the dominant theme is stability and one is usually concerned with either the elimination of limit cycles or the generation of acceptable limit cycles. Again, much of the work must be done by simulation. Stability calculations are carried out successfully in many cases, but the results must be verified; in other cases, the system is very complex and simulation studies become the primary tool and describing function theory merely provides background for intelligent planning of the simulations.

Phase plane theory originated as a graphical means of integrating nonlinear differential equations. The only graphical integration method presented here is the method of isoclines, which was chosen because the isocline and slope marker configurations are an independent aid to system analysis. Isoclines are used to explain the ability of derivative signals to "anticipate" and to damp; they are further developed to demonstrate the graphical meaning of eigenvectors. The concepts of dividing lines and switching lines are introduced and their use in analysis and design is demonstrated.

Phase plane analysis is used in the treatment of discontinuous systems—systems in which relays and switches are used in noncyclic operation so that describing function theory cannot be applied. After basic ideas are developed, including time optimal control, a few practical systems are treated in some detail, including servo positioning of read/write heads in disk memories, carriage control for the automatic typewriters used with word processors, and others.

Simulation with today's computers is an almost indispensable tool for analysis and design of nonlinear systems. Assuming that a suitable model is available, essentially any nonlinear system can be studied using simulation. Initial analysis by simulation, i.e., using the computer to study the various types of operation of a nonlinear system, is highly recommended. Also, economic considerations demand that any design—using any design method—should be checked thoroughly by simulation before hardware realization is undertaken. Finally, for many systems that cannot be analyzed readily or designed on paper, the use of simulation helps in system analysis and system development. It should be clearly understood that simulation studies are simply controlled experiments performed on a computer model; thus, adequate analysis results only if the experiments are carefully planned. In like manner, design using simulation is essentially analysis of a *modified* system, where the modifications are the changes and additions that the designer makes based on theory or intuition and, again, carefully planned experiments are needed. Thus, all simulation studies are trial-and-error procedures, and they result in satisfactory designs when the engineer's ideas are firmly based on theory and experience and the conduct of the studies is planned intelligently. Describing function theory is treated in this chapter, phase plane theory in Chapter 11, and simulation results are presented in both chapters.

10.3   THE DESCRIBING FUNCTION

Since the mathematical methods used to study the dynamics of linear systems do not work for nonlinear systems, we must look for other methods. However, it is natural to try to adapt familiar tools to new problems. Transfer function methods have been very useful; the invention of a *describing* function extends transfer function methods to nonlinear systems, and the describing function itself may be thought of as the transfer function of a nonlinearity.

Before attempting a mathematical definition of a describing function, let us note that one definition of a transfer function is the complex ratio of the output signal to the input signal. When we consider linear systems, the principle of superposition applies and the transfer function is an explicit linear function of the complex variable, but is *not* a function of signal amplitude. For nonlinear devices, however, the output/input ratio *is* a function of signal amplitude, the principle of superposition does *not* apply, and any function that describes the output/input ratio must be a function of signal amplitude and possibly a function of other variables. This situation is illustrated in Figure 10.1, where a transfer function is defined to be a function of the complex variable s (or of $j\omega$, for frequency response analysis). The describing function G_D is a function of input signal amplitude A and may also be a function of ω.

Consider the most common specific nonlinearity—saturation. Let us think of it as a saturating amplifier, as shown in Figure 10.2(a). For convenience, the saturation characteristic has been approximated (idealized) with straight line segments. Note that the nonlinearity is *single-valued*—for any instantaneous value of the input, there is *one*

Linear device

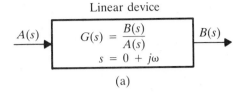

(a)

Nonlinear device

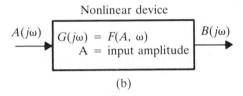

(b)

FIGURE 10.1 Comparison of Basic Concepts: (a) Transfer Function of a Linear Device and (b) Describing Function of a Nonlinear Device.

and *only one* value of the output. Many different kinds of nonlinearities are single-valued.

For any small value of the input signal such that $-a \leq \text{IN} \leq +a$, the amplifier is linear, i.e., the output signal is K times the input, where K is the gain. However, for $\text{IN} < -a$, the magnitude of the output is $-sat$ and, for $\text{IN} > +a$, the output is $+sat$, and the gain is not K, but is some number $k < K$ that is a function of the magnitude of the input signal. Thus, we can say that the gain associated with a saturation non-linearity is variable, or that the describing function of a saturation nonlinearity is a variable gain. This concept of *variable gain* can be quite useful (as we shall see) even without a quantitative definition, but it is more useful to define the describing function quantitatively.

To define a describing function quantitatively, we must restrict the type of nonlinearity to be considered and choose the input signal. As a result, a variety of describing functions have been defined. Most have been special purpose functions needed for specific problems and are not discussed here. The most commonly used describing function has been defined for use with frequency response analysis and the Nyquist criterion. It requires that:

1. The nonlinearity be symmetrical.

2. The input signal be a pure sinusoid.

By inspection of Figure 10.2(b), we see that if

$$\text{In} = A \sin \omega t, \tag{10.1}$$

where $|A| > a$, then the output signal will be distorted, i.e., the peaks will be clipped indicating that it is not a pure sine wave.

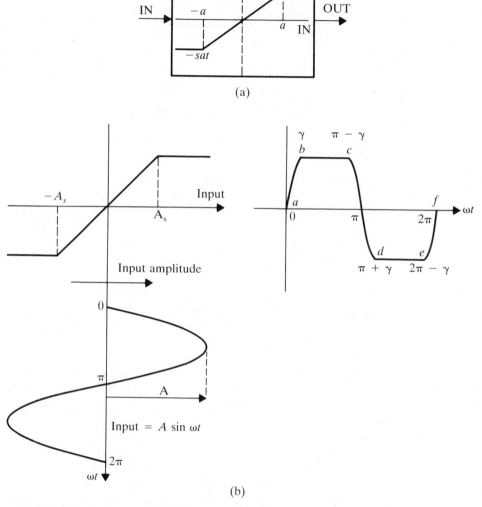

FIGURE 10.2   (a) Saturated Amplifier with Idealized Saturation Characteristic and (b) Construction for Derivation of a Describing Function.

The output wave shape can be represented by a Fourier series, so that in general,

$$\text{OUT}(t) = F^0 + \sum_{n=1}^{\infty} (A_n \cos n\omega t + B_n \sin n\omega t), \tag{10.2}$$

where

$$F^0 = \frac{1}{2\pi} \int_0^{2\pi} \text{OUT}(t)\, d(\omega t) \tag{10.3}$$

and

$$A_n = \frac{1}{\pi} \int_0^{2\pi} \text{OUT}(t) \cos n\omega t \, d(\omega t), \tag{10.4}$$

$$B_n = \frac{1}{\pi} \int_0^{2\pi} \text{OUT}(t) \sin n\omega t \, d(\omega t), \tag{10.5}$$

$$n = 1, 2, 3, 4, \ldots . \tag{10.6}$$

If the nonlinearity is symmetrical, then $F^0 \equiv 0$, and we linearize the relationship by truncating the series after the $n = 1$ term. The describing function is then defined to be the ratio of the fundamental frequency term in the output series to the input term. Thus, we have

$$\text{OUT}(t) = A_1 \cos \omega t + B_1 \sin \omega t = \sqrt{A_1^2 + B_1^2} \, \angle \phi, \tag{10.7}$$

$$\text{IN}(t) = A \sin \omega t = A \angle 0°, \tag{10.8}$$

$$G_D = \frac{\sqrt{A_1^2 + B_1^2} \, \angle \phi}{A}. \tag{10.9}$$

For the specific case of saturation, there is no phase shift so $A_1 = 0$ and

$$G_D = \frac{B_1}{A} \angle 0° \qquad \text{for} \quad |A| > |a|. \tag{10.10}$$

We note that this result is just a variable gain, as previously indicated, but it is now also a specific, quantitative relationship that is a function of A, the amplitude of the input signal, but is *not* a function of ω.

This procedure (i.e., using a pure sine wave input, determining the Fourier series for the output, and using only the fundamental frequency term of this series) can be used to derive a describing function for any symmetrical nonlinearity whether single- or multiple-valued. A table of typical, commonly encountered nonlinearities with their describing function is given in Table 10.1. The procedure can also be used when the nonlinear block contains energy storage elements (and G_D is then a function of ω also), but it is not practical to tabulate such cases.

■ EXAMPLE 10.1 DERIVATION OF A SINUSOIDAL DESCRIBING FUNCTION

Let the nonlinearity be saturation and, for convenience, we approximate it with three straight line segments as shown in Figure 10.2(b). When the input to the saturating element is a sine wave, the output wave reproduces the input unless the instantaneous amplitude of the input exceeds the saturation level. Thus, those segments of the output wave from a to b, from c to d, and from e to f are sinusoidal in nature. From b to c and from d to e, the output is constant at the saturation level. With this information, and assuming that the gain of the saturating element is *one*, we can write an equation for

TABLE 10.1 Describing Functions of Common Nonlinearities

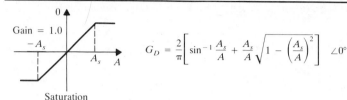

$$G_D = \frac{2}{\pi}\left[\sin^{-1}\frac{A_s}{A} + \frac{A_s}{A}\sqrt{1 - \left(\frac{A_s}{A}\right)^2}\right] \quad \angle 0°$$

Saturation

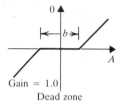

$$G_D = \frac{2}{\pi}\left[\frac{\pi}{2} - \sin^{-1}\frac{b/2}{A} - \frac{b/2}{A}\sqrt{1 - \left(\frac{b/2}{A}\right)^2}\right] \quad \angle 0°$$

Dead zone

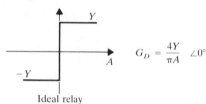

$$G_D = \frac{4Y}{\pi A} \quad \angle 0°$$

Ideal relay

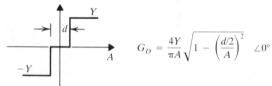

$$G_D = \frac{4Y}{\pi A}\sqrt{1 - \left(\frac{d/2}{A}\right)^2} \quad \angle 0°$$

Relay with dead zone

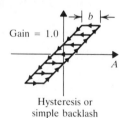

Hysteresis or
simple backlash

$$G_D = \frac{1}{\pi}\left\{\left[\frac{\pi}{2} + \sin^{-1}\left(1 - 2\frac{b/2}{A}\right)\right]^2 + 4\frac{b/2}{A}\left(1 - \frac{b/2}{A}\right)\right.$$
$$+ \left[\pi + 2\sin^{-1}\left(1 - 2\frac{b/2}{A}\right)\right]2\frac{b/2}{A} \cdot$$
$$\left.\left(1 - 2\frac{b/2}{A}\right)\sqrt{\frac{1 - \frac{b/2}{A}}{\frac{b/2}{A}}}\right\}^{1/2}$$

$$\angle G_D = \tan^{-1}\frac{4\frac{b/2}{A}\left(\frac{b/2}{A} - 1\right)}{\frac{\pi}{2} + \sin^{-1}\left(1 - 2\frac{b/2}{A}\right) + 2\frac{b/2}{A}\left(1 - 2\frac{b/2}{A}\right)\sqrt{\frac{1 - \frac{b/2}{A}}{\frac{b/2}{A}}}}$$

each segment:

$$\text{Input} = A \sin \omega t$$
$$\text{Output} = \text{Input} \qquad\qquad 0 \le \omega t \le \gamma$$
$$\text{Output} = A_s \qquad\qquad \gamma \le \omega t \le \pi - \gamma$$
$$\text{Output} = \text{Input} \qquad\quad \pi - \gamma \le \omega t \le \pi + \gamma$$
$$\text{Output} = -A_s \qquad\quad \pi + \gamma \le \omega t \le 2\pi - \gamma$$
$$\text{Output} = \text{Input} \qquad 2\pi - \gamma \le \omega t \le 2\pi.$$

Because of symmetry, only the positive half wave need be considered. Also because of symmetry, the device enters saturation at $\omega t = \gamma$ and leaves saturation at $\omega t = \pi - \gamma$.

To evaluate the coefficient of the fundamental frequency term in the Fourier series for the output wave, substitute in the appropriate equation:

$$A_1 = \frac{2}{\pi} \int_0^\gamma A \sin \omega t \cos \omega t \, d\omega t + \frac{2}{\pi} \int_\gamma^{\pi-\gamma} A_s \cos \omega t \, d\omega t + \frac{2}{\pi} \int_{\pi-\gamma}^\pi A \sin \omega t \cos \omega t \, d\omega t,$$

$$B_1 = \frac{2}{\pi} \int_0^\gamma A \sin^2 \omega t \, d\omega t + \frac{2}{\pi} \int_\gamma^{\pi-\gamma} A_s \sin \omega t \, d\omega t + \frac{2}{\pi} \int_{\pi-\gamma}^\pi A \sin^2 \omega t \, d\omega t.$$

Evaluating these integrals,

$$A_1 \equiv 0$$

$$B_1 = \frac{2A}{\pi} \left[\sin^{-1} \frac{A_s}{A} + \frac{A_s}{A} \sqrt{1 - \left(\frac{A_s}{A}\right)^2} \right].$$

Then the describing function is

$$G_D = \frac{\sqrt{A_1^2 + B_1^2}}{A} \angle \tan^{-1} \frac{A_1}{B_1} = \frac{2}{\pi} \left[\sin^{-1} \frac{A_s}{A} + \frac{A_s}{A} \sqrt{1 - \left(\frac{A_s}{A}\right)^2} \right] \angle 0°.$$

10.4 STABILITY AND LIMIT CYCLES

We will eventually use the describing function in its quantitative form to study system stability via the Nyquist criterion. First, however, it is both instructive and useful to study stability using the qualitative concept of a *variable gain* in conjunction with the root locus method. Note that for single-valued nonlinearities no phase angle is introduced; the nonlinearity may be considered a *pure gain*. Then the system has been linearized in the sense that we may evaluate roots for each possible value of gain, resulting in the well-known root locus of a linear system. For any specific system, we can study stability and determine the existence of limit cycles (if any), their frequency, and some other characteristics if needed. We shall demonstrate the techniques by examples.

■ EXAMPLE 10.2

Consider the third-order system of Figure 10.3(a), with root locus as in Figure 10.3(b). We note that the nonlinearity is saturation and we know that if saturation were not present the stability of the system would depend on the loop gain. The logical questions, therefore, are:

1. If the linear system is stable, can saturation make it unstable?
2. If the linear system is unstable, can the saturation stabilize it?

We can answer both of these questions, for this specific system, by root locus analysis.

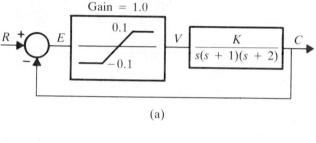

(a)

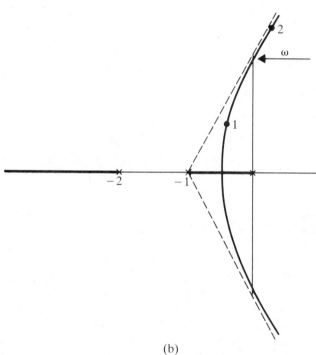

(b)

FIGURE 10.3 Stability Analysis of System with Saturation: (a) Third-Order Servo with Saturation Nonlinearity and (b) Root Locus.

Assume that the gain for linear (small signal) operation places the roots at point 1 on Figure 10.3(b). Then, for small signals into the amplifier, it does not saturate and the roots are at point 1, which is in the left half of the s plane, so the transient oscillations decay in amplitude while the roots remain fixed in value. For large signals into the amplifier, it saturates and the equivalent gain of the amplifier decreases. When this happens, the root point is no longer at point 1; it moves toward the *poles*. For this system, it is clear that for motion toward the poles the roots always remain in the left half s plane and, therefore, the transient oscillations decrease in amplitude. Reduction in signal amplitude, however, decreases the amount of saturation and permits the amplifier gain to increase, so that the roots move away from the poles and point 1 ultimately becomes the operating point. We conclude that the system is always stable for the given value of linear gain. Transient response is affected, but all oscillations eventually die out.

Next, consider the case when the linear gain is high enough to place roots at point 2. The linear system is unstable because there are roots in the right half plane. Can saturation stabilize the system? If the initial amplitude of the signal into the amplifier is small, then the gain is maximum and the roots are in the right half plane at point 2. Therefore, the oscillation will increase in amplitude, which saturates the amplifier, reduces the gain, and moves the root point away from point 2 toward the poles. This continues until the imaginary axis is reached at ω, where the root locus crosses it. At this point, the real part of the roots is zero so the signals do not change amplitude and the system continues to oscillate at constant frequency ω and at constant amplitude. This conclusion is also reached if we assume that the initial signal is of large amplitude, saturating the amplifier and reducing the gain so that the initial root point is, say, point 1. At point 1, the roots are in the left half plane so the amplitude of oscillation decreases. This increases the gain so the roots move on the root locus but *away* from the poles until point ω is reached. Thus, the intersection of the root locus with the imaginary axis is a limit approached by the operating point regardless of the initial conditions. It is clearly a stable equilibrium point, and operation at that point is a constant amplitude, constant frequency oscillation, which we call a *limit cycle*. Also, because the system seeks this condition for its steady state, we call it a *stable* limit cycle.

To illustrate the concepts and extend them, four values of gain are chosen and the saturation limits set at ± 0.1 as shown in Figure 10.4, where the limit cycles are shown on the E versus $\dot{E}$ plane. An input step of 1.0 unit is used for all four gain values, so that the amplifier saturates heavily in each case. For $K = 1$, the system is stable and well damped; there is a small overshoot, little oscillation, and no limit cycle. For $K = 4$, the system is still stable but is lightly damped so there is quite a bit of transient oscillation, which eventually damps out, and there is no limit cycle. For $K = 6$, the linear system is at the stability limit, so after the initial overshoot the system settles into a stable limit cycle. When the gain is raised to $K = 10$, the behavior is similar, but the amplitude of the limit cycle is increased. Note that the transient oscillations have an amplitude greater than that of the limit cycle because the step input is large.

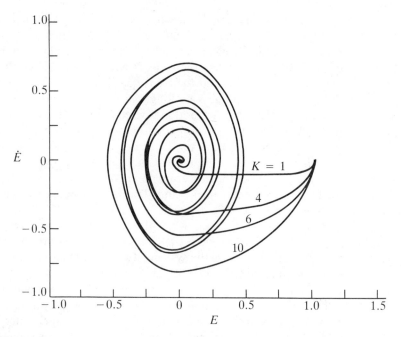

FIGURE 10.4   Example of Saturation Limit Cycles.

The preceding illustration, in addition to outlining an approach to stability analysis, has defined a *stable limit cycle* to be a steady-state condition of oscillation at fixed amplitude and frequency, said condition being one which the system *seeks* as an equilibrium. There can also be an *unstable limit cycle*, which is also a condition of oscillation at fixed amplitude and frequency, from which system operation *diverges* if perturbed, i.e., it is an unstable equilibrium condition. If the root locus method is used, then either (or both) kind of limit cycle is defined at an intersection of the root locus with the imaginary axis of the *s* plane. For example, if the system of Figure 10.3(a) has a dead zone nonlinearity rather than saturation, the limit cycle is unstable. Also, if we retain the saturation nonlinearity but have a plant

$$G(s) = \frac{K(s + 4)(s + 8)}{s(s + 0.4)(s + 0.8)(s + 50)},$$

the root locus crosses the imaginary axis twice. One crossing defines a stable limit cycle, the other an unstable limit cycle.

 EXAMPLE 10.3

To illustrate the unstable limit cycle, Figure 10.5 gives simulation results for the third-order servo when the nonlinearity has dead zone as well as saturation. System gain is

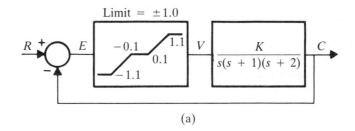

(a)

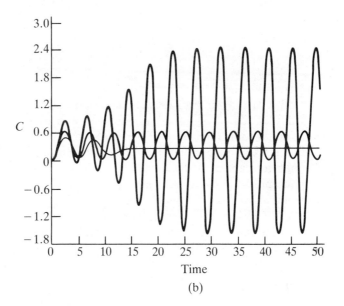

(b)

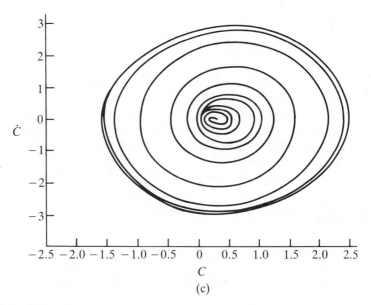

(c)

FIGURE 10.5 (a) Nonlinearity with both Dead Space and Saturation, (b) Transient Response, and (c) Limit Cycles on the Phase Plane.

kept constant at $K = 10$ and three steps of increasing amplitude are used to test the system. For the small step, the gain of the nonlinear block is small so its effective gain is small and the transient oscillations die out. For the intermediate step (carefully chosen by trial and error), the system is at the stability limit and the oscillation amplitude remains constant—this is the unstable limit cycle. For the larger step, the system is unstable and the oscillations increase in amplitude until they reach the stable limit cycle due to saturation. Note that *both* limit cycles are at the same frequency! Figure 10.5(b) shows the result on the C versus time plane and Figure 10.5(c) shows the limit cycles on the phase plane.

The frequency of the limit cycle is the value of ω at the ω axis intersection, and it is readily evaluated in various ways. The amplitude of the limit cycle, however, is not as readily evaluated quantitatively. If we try to do so using root locus concepts, we encounter some difficulties: When the system is operating in a limit cycle, the signal feeding the nonlinearity has a periodically varying magnitude; and if the gain of the nonlinear element is evaluated on an instantaneous basis, it too is periodic—an uninterpretable condition. Clearly, the gain of the nonlinear element must be evaluated in some *average* sense. This *average* value for the gain of the nonlinear element has to be such that it will provide a loop gain exactly equal to the root locus gain at the limit cycle point! However, it is difficult to correlate this requirement with the actual magnitudes of the signal into (and out of) the nonlinear element. The describing function, which we defined previously, provides a quantitative, average value for the gain of the nonlinear element and is reasonably accurate when frequency response methods are used, but it relates the amplitude of the input sinusoid to the amplitude of the Fourier fundamental of the output, and correlation of this number with the root locus gain is not obvious.

Stability analysis of nonlinear systems using the root locus method is simple and direct whenever the system contains only one nonlinearity that is single-valued. For any other situation, the root locus is seldom convenient, though analysis may be possible (as with hysteretic,[1] i.e., two-valued, nonlinearities). When one wishes to design the system, i.e., to adjust the stability and dynamic behavior, then the choice of design tools depends on the specific problems. The root locus method is very useful if we want to retain the limit cycle, but wish to adjust its frequency and amplitude. On the other hand, if we wish to eliminate the limit cycle and also adjust the transient response of the nonlinear system, then frequency response methods using the describing function may be preferable.

10.5   DESIGN OF LIMIT CYCLE OPERATION

To explain the design procedures, we use as an example a third-order saturating servo, as shown in Figure 10.3(a), with root locus as in Figure 10.3(b). The system will limit cycle with frequency $\omega = \sqrt{2}$, and the amplitude of the limit cycle *at the input to the amplifier* (i.e., signal E) will be whatever value is required to make the loop gain go to $K = 6$. No attempt is made to eliminate the limit cycle, so the design concerns itself with:

1. limit cycle frequency
2. limit cycle amplitude
3. damping of any transients.

10.5A Limit Cycle Frequency

This is determined solely by the root locus crossing of the $j\omega$ axis and is the value of ω at that point. Therefore, we can set the limit cycle frequency by designing a compensator to place the imaginary axis crossover at the desired ω; we need not be concerned with signal amplitudes or gains. (In practice, the signal amplitude is important and must be considered in the compensator design. At this point, we study only the simplest case.)

Assume that we want the limit cycle frequency to be $\omega = 3$, and we choose to use a cascaded filter to accomplish this. Let

$$G_c = \frac{s + Z}{s + P}$$

For this illustration, we ignore the gain and reshape the locus with Z and P. The filter may be placed between error detector and amplifier or it may be placed between amplifier and plant. The root locus of the compensated system must pass through the point $s = 0 + j3$, so at that point

$$\angle G_c G(j3) = -180°$$

or

$$-90° - \tan^{-1}3 - \tan^{-1}\frac{3}{2} - \tan^{-1}\frac{3}{P} + \tan^{-1}\frac{3}{Z} = -180°.$$

Thus,

$$\tan^{-1}\frac{3}{Z} - \tan^{-1}\frac{3}{P} = -180° + 90° + \tan^{-1}3.0 + \tan^{-1}\frac{3}{2} \simeq 38°$$

and the filter must introduce a net positive angle of $+38°$ degrees. The filter must therefore be a *lead* filter, but the pole and zero locations are not unique. Since an infinite number of zero-pole pairs will provide the same limit cycle frequency, we then expect each pair to provide a different amplitude of limit cycle, because each pair provides a different root locus gain at $s = 0 + j3$. We also expect the transient responses to be different. Figure 10.6 and Table 10.2 tabulate the characteristics obtained by simulation.

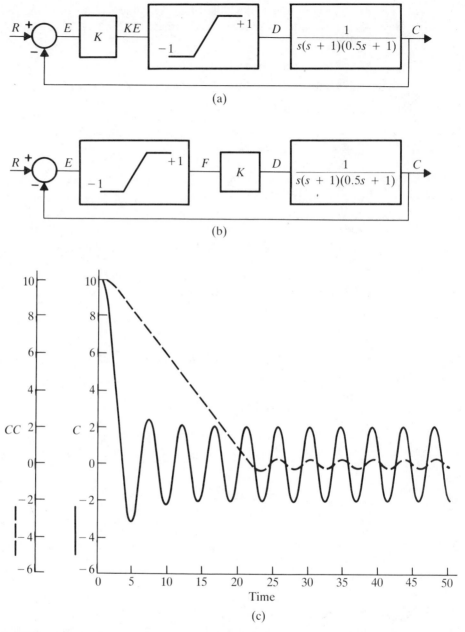

FIGURE 10.6 Limit Cycle Behavior in a Saturating Servo: (a) System A where Limiter Follows Gain, (b) System B where Gain Follows Limiter, and (c) Transient Response of the Saturating Systems.

TABLE 10.2 Simulation Results

SYSTEM A

Gain	ω_{lc}	E_{max}	Overshoot (%)	Settling Time (s)
$k = 6$	1.385	0.421	6.8	$\simeq 20$
10	1.385	0.433	6.8	$\simeq 20$
20	1.385	0.438	6.8	$\simeq 20$

SYSTEM B

Gain	ω_{lc}	E_{max}	Overshoot (%)	Settling Time (s)
6	1.4	2.52	37.0	$\simeq 7.0$
10	1.383	4.33	55.0	$\simeq 5.0$
20	1.383	8.77	86.0	$\simeq 3.0$

Notes: Limit cycle frequency predicted via Routh criterion is $\omega = \sqrt{2}$. By inspection of the block diagrams, system B should respond more quickly because the signal driving the plant (D) can be very large, but for system A, signal $D = \pm 1.0$ maximum.

10.5B Limit Cycle Amplitude and Damping

The limit cycle frequency and amplitude are both determined by the root locus conditions at the axis crossover of the locus. The limit cycle frequency is the value of ω at that point. The limit cycle amplitude depends on the root locus gain at that point. We note, however, that the root locus gain is the product of the gain of the nonlinear element and the gain of the linear part of the system. The linear gain is fixed, but the gain of the nonlinear element is a function of signal amplitude at the input to the nonlinear block. The root locus gain at the axis crossover is determined solely by the locations of the poles and zeros of the linear transfer function, and it is a fixed number for a given pole-zero configuration. If, then, we design the system with a large linear gain, the nonlinear gain must be relatively small and the amplitude of the limit cycle adjusts itself to provide the required nonlinear gain. For example, in the illustration of Figure 10.3, the root locus gain at the axis crossover is $K = 6$. If we set the linear gain, $k_1 = 1$, then $k(E) = 6$; if $k_1 = 60$, then $k(E) = 0.1$, etc., and in each case the signal amplitude at E must adjust so that the correct value of $k(E)$ is obtained. Thus, we can adjust the limit cycle amplitude to any value we wish if it is possible to adjust the linear gain, resulting in independent adjustment of limit cycle amplitude through loop gain adjustment.

Design of system damping is not analyzed as readily. From the preceding discussion, it is clear that we can use a variety of compensators to achieve precisely the same limit cycle frequency. For each possible compensator, we can adjust the amplitude of the limit cycle to exactly the same value by adjusting the linear gain and, clearly, we can choose any value for the amplitude.

Consider the tabulated simulation results in Figure 10.7 and Table 10.3, first noting the structural difference between systems A and B: the *gain distribution* is reversed, which affects behavior significantly. In both cases, the adjustable gain K and the saturating element are in cascade and constitute all of the loop gain except for the ratio P/Z. Thus, once the zero-pole pair is chosen, the gain of this *cascaded combination* must adjust to provide the proper root locus gain for the limit cycle. This fact is independent of the sequence of the two elements, but the signal amplitudes needed to satisfy this condition do indeed depend on the sequence.

We may summarize the tabulated result as follows:

1. The limit cycle frequency is essentially the same for all cases. The slight variations observed are due primarily to imprecision of the zero-pole pair values.

2. Limit cycle amplitude at E (and therefore at C) is smaller for system A by a factor of K! Clearly, this is due to the fact that signal D is the limiter output and cannot exceed 1.0; therefore, to achieve the required gain the signal at E must be much less than 1.0. Conversely, for system B, it is the signal at F that is limited to 1.0, but the signal at D is $K \times F$ and may be quite large. In fact, the signal at F *must be* 1.0 over an appreciable portion of the limit cycle, since the gain cannot be reduced if the amplifier does not saturate! So the signal at D must reach the large value $K \times F$ and it follows that the signal E must be $K \times$ (the limit cycle value for system A). Thus, it is clear that *structure*, in this case, the *distribution of gain*, is very important in adjusting limit cycle amplitude.

3. The transient responses of the two systems are quite different, as shown in Figure 10.7. The amplifier of system A saturates for small error signals and, since signal $D = 1.0$ in saturation, the plant must run "open loop" for a long time to correct a large error, but there is little overshoot of the reference ($R = 0$) value, and the system goes smoothly into the limit cycle. In system B, however, the large signal available to D drives the output into correspondence very rapidly. Note that both systems reach the steady-state limit cycle after about one overshoot, at least for the conditions of Figure 10.7.

    The tabulated results of Figure 10.7 show that the settling times of both systems, and the percent overshoot of both systems, are altered by the zero-pole pair. However, for system A, these variations are small due to the low limit on the signal D. For system B, the variations are substantial, and for different zero-pole pairs the settling time is substantially different.

The settling time of the system appears to depend *primarily* on the structure; i.e., on the specific location of the nonlinear block in the system; or perhaps we should say it depends on the gain distribution. Clearly, it does *not* depend on the linear transfer function since that is the same for both cases; likewise it cannot be interpreted from the

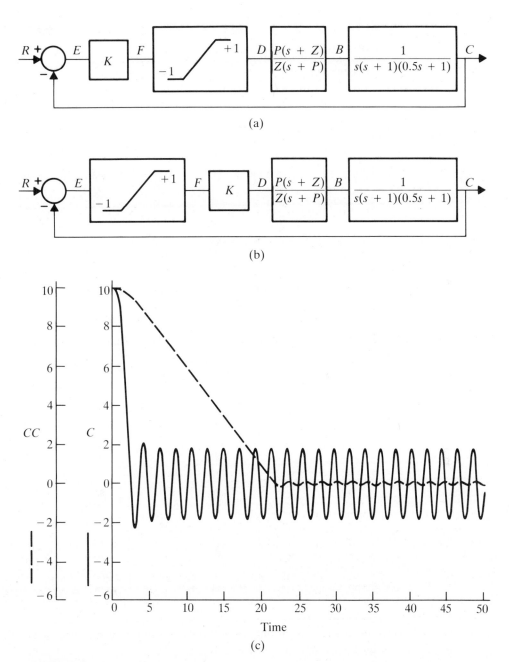

FIGURE 10.7   Limit Cycle Behavior in a Compensated Saturating Servo: (a) System A where Limiter Follows Gain, (b) System B where Gain Follows Limiter, and (c) Transient Responses of the Compensated Systems.

TABLE 10.3 Simulation Results

SYSTEM A

Gain	Z/P	ω_{lc}	E_{max}	Overshoot (%)	Settling Time (s)
10	1/4.53	2.96	0.185	3.2	12
20	1/4.53	2.93	0.198	3.2	12
30	1/4.53	2.96	0.200	3.2	12
20	2/9.1	2.93	0.126	4.07	13
30	2/9.1	2.93	0.13	4.07	13
40	2/9.1	2.90	0.132	4.07	13
20	3/24.43	2.96	0.100	4.42	14
40	3/24.43	2.90	0.107	4.42	14
60	3/24.43	2.90	0.108	4.41	14
20	3.5/53.9	2.90	0.099	4.56	14
40	3.5/53.9	2.85	0.104	4.56	14
60	3.5/53.9	2.85	0.105	4.56	14

SYSTEM B

Gain	Z/P	ω_{lc}	E_{max}	Overshoot (%)	Settling Time (s)
10	1/4.53	2.96	1.875	30.0	4.5
20	1/4.53	2.93	3.97	52.0	4.0
30	1/4.53	2.93	5.94	69.0	4.5
20	2/9.1	2.93	2.52	50.0	8.8
30	2/9.1	2.93	3.92	62.0	8.0
40	2/9.1	2.93	5.27	71.0	6.2
20	3/24.42	2.96	2.01	50.0	9.2
40	3/24.43	2.93	4.3	69.5	9.6
60	3/24.43	2.90	6.5	80.0	8.5
20	3.5/53.9	2.90	1.98	52.0	9.6
40	3.5/53.9	2.85	4.2	70.0	10.6
60	3.5/53.9	2.85	6.3	80.0	15.6

Note: Limit cycle frequency predicted by Routh criterion is $\omega = \sqrt{9.06}$.

root locus since this is determined solely by the transfer function. In like manner, the nonlinear element and the describing function are the same in both cases and so cannot explain the differences in settling time. Knowing the specific nonlinearity (in this case saturation), we can evaluate the settling time of the *linear* system. Using physical analysis, i.e., observing what happens to signal amplitude, we can estimate the effect of the nonlinearity on the transient response, but for quantitative evaluation the best method seems to be simulation.

10.6 NYQUIST ANALYSIS FOR STABILITY AND LIMIT CYCLES

We have already defined the sinusoidal describing function and have developed the mathematical definition in Eq. (10.9). This process is a linearization in the frequency domain: We treat the describing function $G_D(A, \omega)$ as a transfer function, we form the loop transfer function (or the characteristic equation if one prefers to think in these terms), and we apply the Nyquist criterion. For a single-loop system such as that shown in Figure 10.6,

$$\text{loop transfer function} = G_D(A, \omega)G(j\omega)$$
$$\text{characteristic equation is } 0 = 1 + G_D(A, \omega)G(j\omega).$$

Rearranging in conventional form for the Nyquist criterion,

$$G_D(A, \omega)G(j\omega) = -1.0.$$

To apply the Nyquist criterion, we would normally make a polar plot of the loop transfer function. For nonlinear systems, a *single* polar plot does not result because the describing function is a function of both A and ω. One can plot a family of curves with ω as the running variable and A as a family parameter, but interpretation of stability and limit cycles is not always easy.

We note that the loop transfer function can be rearranged in the form:

$$G(j\omega) = -\frac{1}{G_D(A, \omega)}.$$

The left side of this equation plots as a single curve on the polar plane. If the nonlinearity is a function of A only, then the right side of the equation also plots as a single curve. If these curves intersect, the equation is exactly satisfied at the point of intersection and that point defines a limit cycle. If the curves do not intersect, then (with some exceptions) there is no limit cycle—the system is either stable or unstable. We will use examples to demonstrate the interpretation.

 EXAMPLE 10.4

For the system of Figure 10.3(a), the linear transfer function is

$$G(j\omega) = \frac{K}{(j\omega)(j\omega + 1)(0.5j\omega + 1)}.$$

We let $K_1 = 3.0$. The block diagram and polar plot of this transfer function are shown in Figures 10.8(a) and (b), respectively. If the system were linear, with amplifier gain of 1.0, stability would be interpreted with respect to the $-1.0 + j0$ point and the system would be at the limit of stability. If $K_1 = 1.0$ instead of 3.0, the system is stable and well damped. If the system were linear but with an amplifier gain of $K = 10.0$, the same plot may be used with the scale changed by a factor of 10, and the system would be unstable.

When the amplifier saturates, as in the case shown in Figure 10.3, $G_D(A, \omega)$ is a function of A only, and the curve obtained for $-1/G_D(A), 0 \le |A| \le +\infty$ is a straight

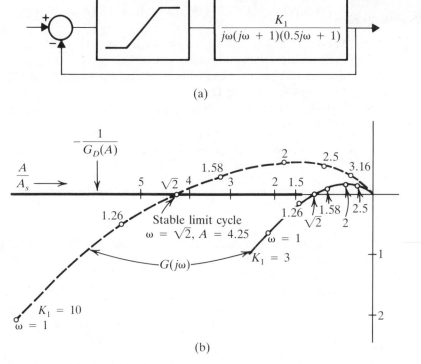

FIGURE 10.8 Nyquist Analysis of a Saturating Servo: (a) Block Diagram and (b) Polar Plot.

line lying on the negative real axis as shown in Figure 10.8(b). It is plotted for an unsaturated gain of 1.0 and a saturation limit at $A = 1.0$.

Since the two curves in Figure 10.8(b) intersect, there will be a limit cycle. The frequency of the limit cycle is the value of ω at the intersection, which is $\omega = \sqrt{2}$. The amplitude of the limit cycle (amplitude of the sinusoidal signal at the input to the nonlinearity) is the value of A at the intersection, which is $A = 4.25$.

Next we would like to determine whether the limit cycle is stable or unstable. To do this, we observe that the right side of our loop transfer function equation is $-1/G_D(A)$, rather than -1.0 as it is for linear systems. It is easy to see that $-1/G_D(A)$ may be considered a *moving* or *variable* -1.0 point. Then, if we choose any value of $A = A_1$ and determine the polar plane location of $-1/G_D(A_1)$, we may consider that point to be the critical point (i.e., the instantaneous -1 point) and use it to determine the stability of the system with respect to that value of amplitude A_1!

For the system of Figure 10.8, if we choose a large value for A_1, the critical point lies *outside* the $G(j\omega)$ curve. By normal Nyquist analysis, this means that the system is stable for that value of A and as time progresses the signal amplitude A will decrease. Inspection of the plot shows that the critical point will move *toward* the limit cycle point. In like manner, if we initially choose a small value for A_1, the critical point will lie *inside* the $G(j\omega)$ curve and the system will be unstable for that value of A; the amplitude A will increase and the critical point will move toward the limit cycle point again. The system is thus seen to *seek* the limit cycle as a steady-state operating condition and we say that the limit cycle is stable.

◼ EXAMPLE 10.5

To demonstrate an unstable limit cycle, we need only replace the saturating amplifier with one that has dead zone but no saturation. The block diagram is shown in Figure 10.9(a) and the polar plot in Figure 10.9(b). Note that the describing function curve again lies exactly on the negative real axis. Comparison of the dead zone curve of Figure 10.9(b) with the saturation curve of Figure 10.8(b) shows that the essential difference between them is the *direction* of the variation in A.

It should be noted that all single-valued, frequency-independent nonlinearities provide describing function plots on the negative real axis of the polar plot because the describing functions do not introduce any phase shift. Double- or multiple-valued, frequency-independent nonlinearities plot as single curves but not necessarily on the

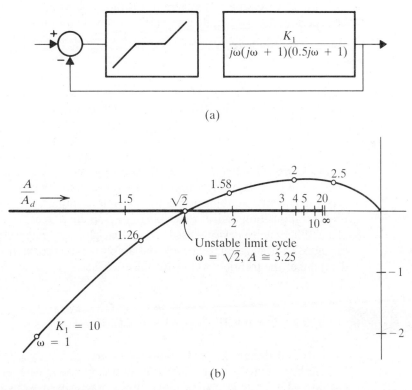

(a)

(b)

FIGURE 10.9 Nyquist Analysis of a Servo with Dead Zone: (a) Block Diagram and (b) Polar Plot.

axis because the describing functions do introduce phase shift. If the describing function is frequency dependent, a family curves is needed for $-1/G_D(A, \omega)$.

In Figure 10.9(b), analysis of the limit cycle condition proceeds similarly to that of Figure 10.8(b). The limit cycle frequency is the value of ω at the intersection, which is $\omega = \sqrt{2}$; the amplitude is also determined at the intersection and is $A = 4.25$. Selecting two test values of $A = A_1$—one smaller and one larger than the limit cycle value—we find that the operating conditions diverge from the limit cycle value and, therefore, the limit cycle is unstable.

A heuristic interpretation gives some insight into the limit cycle condition for this system with dead zone. We observe that the limit cycle cannot exist unless the linear (no dead zone) gain is high enough to make the linear system unstable. For small signals, i.e., with amplitude less than that of the dead zone, the actual amplifier gain is zero, and of course the system is not perturbed. For slightly larger signals, there is an output from the amplifier but its equivalent gain is small and the signals damp out. For very large signals, the loop gain is above the stability limit and the system oscillates with increasing amplitude. Between these extremes there exists a signal amplitude for which the equivalent loop gain is exactly at the stability limit, so oscillations remain at a fixed amplitude. This is the limit cycle and it is a very delicate equilibrium condition. One would not expect any real system to operate in such a limit cycle except possibly for brief periods, but a computer model can be made to do so!

For both the saturation nonlinearity and the dead zone nonlinearity, we observe that the describing function curve terminates at a finite point on the polar plane. Clearly, if the curve for $G(j\omega)$ can be reshaped so that it curves around this terminal point without intersecting the $-1/G_D(A)$ curve, then there will be no limit cycle and the system will be stable. Such reshaping can be accomplished with filters or with state feedback, etc. The design of linear stabilizing networks (compensators) is discussed later in this chapter. It must be noted, however, that all nonlinear devices do not have such convenient describing functions. A simple example is the two-position relay; it has no neutral position and its describing function curve reaches the origin of the polar plane. When such a device is used in a feedback loop, a limit cycle usually exists, but we can hope to design the frequency and amplitude of this limit cycle and perhaps the damping of the transient following a disturbance.

10.7 DESCRIBING FUNCTIONS ON THE NICHOLS PLOT

Analysis and design on polar coordinates are laborious and inconvenient. Use of Nichols coordinates ($|G|$ in decibels as ordinate, $\angle G$ in degrees as abscissa, origin of plot at $-180°$, 0 dB) proves much more practical so we will use this system for convenience. The negative real axis of the polar Nyquist becomes the vertical axis of the Nichols plot. The loop transfer function equation, which we used for the polar plot,

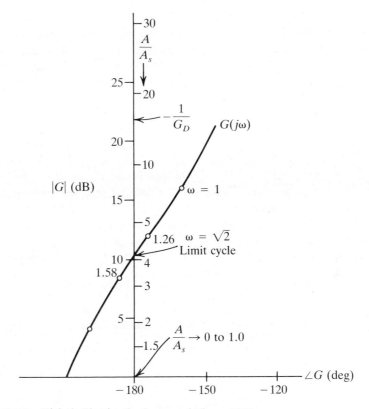

FIGURE 10.10 Nichols Plot for the System of Figure 10.8.

may be rewritten as

$$[20\log_{10}|G(j\omega)|]\angle G(j\omega) = 20\log_{10}\left|\frac{1}{G_D(A,\omega)}\right|\angle[-180° - G_D(A,\omega)].$$

The left side of this expression is the transfer function of the linear system; the right side is the negative reciprocal of the describing function. Each gives rise to a separate curve. Figure 10.10 shows the Nichols plot for the system of Figure 10.8 with linear gain of 10, so that the specific equation for this system is

$$20\log_{10}\left|\frac{10}{j\omega(j\omega+1)(0.5j\omega+1)}\right|\angle\frac{K}{j(j+1)(0.5j+1)} = 20\log_{10}\left|\frac{1}{g}\right|\angle -180° - \angle g,$$

$$g = \frac{2}{\pi}\left[\sin^{-1}\frac{A_s}{A} + \frac{A_s}{A}\left(1 - \frac{A_s^2}{A^2}\right)\right],$$

$$A_s = \text{signal level for saturation,}$$

$$A = \text{amplitude of input.}$$

We note that there is one unusual feature about this relationship: A transfer of a gain constant from one side of the equation to the other displaces *both* curves by the same amount in the ordinate direction; thus, their relative positions remain unchanged. For example, if the transfer function gain were 10 000 and calculated ordinate values were undesirably large, we divide both sides of the equation by (perhaps) 100. This lowers both curves by 40 dB and, if there is a limit cycle intersection, it is unchanged by this shift.

To eliminate the limit cycle, we must eliminate the intersection between the $G(j\omega)$ and the $-1/G_D(A)$ curves. An obvious solution is to reduce the gain. If we reduce the gain associated with $G(j\omega)$, this curve can be lowered until it does not intersect the $-1/G_D(A)$ curve. Most applications, however, will not permit gain reduction because of specifications on accuracy and speed of response. Alternative solutions required that we reshape either the $-1/G_D(A)$ curve or the $G(j\omega)$ curve. Both methods have been used, but as a general rule one chooses to reshape the $G(j\omega)$ curve because we know how to do it easily, whereas reshaping the $-1/G_D(A)$ curve requires a detailed design of a special nonlinear element and we have little understanding of the problem and no generally applicable techniques—development of such a nonlinear compensator would usually be classified as an invention.

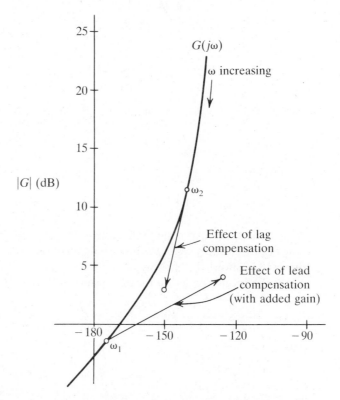

FIGURE 10.11 Sketch Showing Effect of Compensator in Reshaping $G(j\omega)$.

Any of the normal methods for design of linear compensators may be used to reshape $G(j\omega)$, and the compensator may be a cascaded filter or a feedback loop around the linear part of the system or around the nonlinear element—in fact, there is no restriction on the choice of structure providing one recognizes that the dynamic behavior of the system may be significantly different for various possible structures. One should also recognizes that many trials may be necessary and simulation studies would be essential.

It is of value to understand, qualitatively, the effects of phase lead (high-pass) and phase lag (low-pass) filters on the shape of the Nichols plot of $G(j\omega)$. Consider the sketch of Figure 10.11 and the effect of a phase lead compensator at frequency ω_1. Because the lead filter introduces positive phase shift, the filter will move ω_1 to the right. Assuming that the Bode gain is kept constant when the filter is inserted, the $|G(j\omega_1)|$ is increased so the point ω_1 is moved *up*. We may analyze the effect of a lag filter similarly. Consider the effect of such a filter on the point ω_2. Because the filter introduces negative phase shift, point ω_2 will be moved to the left, and because the filter attenuates,

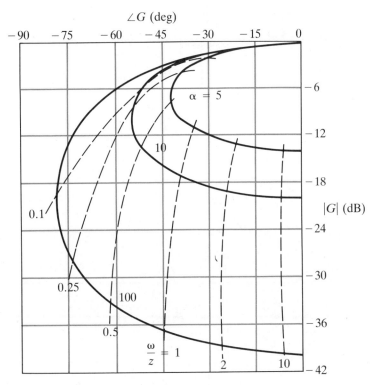

FIGURE 10.12 Nichols Plot for a Lag Compensator:

$$G_c = \frac{P}{Z}\left(\frac{s+Z}{s+P}\right) \qquad \alpha = \frac{Z}{P}(\alpha > 1.0).$$

ω_2 will be moved down. The design engineer may then look at the Nichols plot of the uncompensated system and decide whether he should try to move a set of high-frequency points to the right and up, or a set of low-frequency points to the left and down. In either case, the calculations are made easier by use of nomographs, which are available in textbooks[1] and elsewhere. Calculation of such a filter to eliminate a limit cycle is not difficult, providing the nature of the transient response is not closely specified.

EXAMPLE 10.6

Consider the system of Figure 10.3 with a gain of 10. There is a limit cycle at $\omega = \sqrt{2}$, as shown on the Nichols plot of Figure 10.8. We choose to stabilize this system with a lag filter having the frequency response characteristics shown in Figure 10.12. We accomplish the stabilization by *lowering* (reducing gain at) the point $\omega = \sqrt{2}$ until it is below the describing function curve. This is shown in Figure 10.13. In the range of

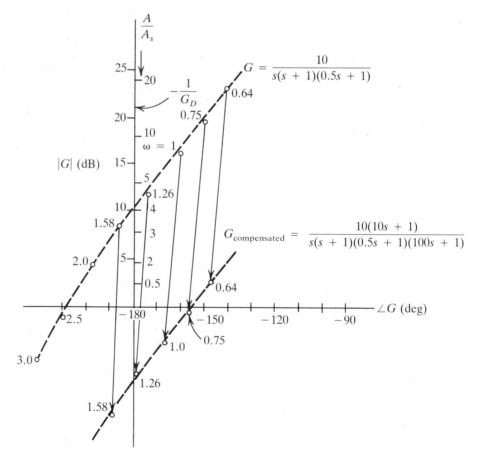

FIGURE 10.13   Stabilization Using a Phase Lag Compensator.

frequencies about the design frequency ($\omega = \sqrt{2}$), the $G(j\omega)$ curve is simply lowered by the chosen number of decibels, and only a slight change in shape is caused by the small phase lag in that range of frequencies. At lower frequencies, however, considerable phase lag is introduced (not shown in Figure 10.13). Note from Figure 10.13 that the phase margin and gain margin achieved with respect to the lower end of the $-1/G_D(A)$ curve are quite small. This means that the system, in linear operation, will not be well damped although it will be stable. For saturated operation, the system may be adequately damped. Evaluation of such features requires simulation results that can be checked against specifications.

We can also stabilize the system with a phase lead (high-pass) filter. The frequency response characteristics are shown in Figure 10.14. In using this filter for stabilization,

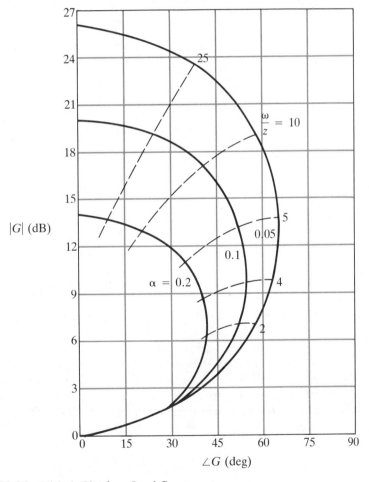

FIGURE 10.14 Nichols Plot for a Lead Compensator:

$$G_c = \frac{P}{Z}\left(\frac{s+Z}{s+P}\right) \qquad \alpha = \frac{Z}{P}(\alpha < 1.0).$$

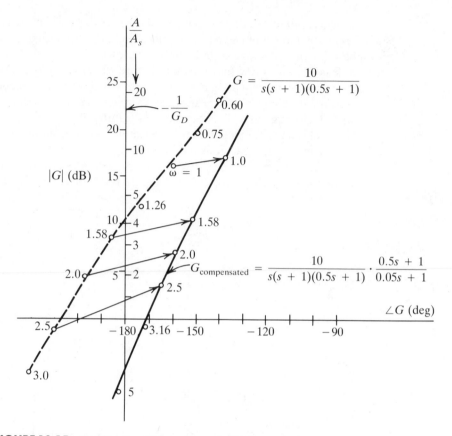

$$G = \frac{10}{s(s + 1)(0.5s + 1)}$$

$$G_{\text{compensated}} = \frac{10}{s(s + 1)(0.5s + 1)} \cdot \frac{0.5s + 1}{0.05s + 1}$$

FIGURE 10.15 Stabilization Using a Phase Lead Compensator.

the objective is to use the phase shift characteristics to move points to the *right* on the Nichols plot. As can be seen from Figure 10.14, this will also move the point upward because of the gain characteristic. Choice of the values for the Z and P of the compensator is not quite as easy as for the phase lag filter; more trial and error may be required. A reasonable starting point in most designs is to select a frequency at approximately $\omega = 2\omega_{lc}$ and choose a phase shift that will make the net phase at that frequency less negative than $-180°$. Since many zero-pole pairs can provide the desired phase, trial and error is used and the final design is usually chosen on the basis of transient characteristics rather than stabilization only. A possible design is shown in Figure 10.15.

In considering Figures 10.13 and 10.15, the student should recognize that neither is a finished design, but merely the result of a first trial. In both cases, the system has been stabilized, but certainly not as well stabilized as would be required for a linear (nonsaturating) system. To obtain more phase margin and gain margin, the pole-zero spacing would be increased and the design procedure repeated, or a second section of filter would be cascaded with the first.

For this particular problem, because the nonlinearity is saturation, design with respect to the finite end of the describing function curve is in fact design of the linear operation of the system. Therefore, compensation design could be done (and more easily so) on the Bode diagram. However, for nonlinear characteristics other than saturation, the Nichol plot seems preferable.

10.8   FALSE INDICATIONS

The describing function is a linear approximation of a nonlinear function, so it is not surprising that answers obtained with the describing function are not precisely accurate. The following inaccuracies are possible:

1. An existing limit cycle may not be predicted.
2. A predicted limit cycle may not exist.
3. The amplitude and/or frequency of a limit cycle may not be evaluated accurately.

I do not have a specific example for item 1 but the classic example for 2 is the case of a second-order servo with backlash.[2] Describing function analysis predicts two limit cycles—a stable one at high frequency and an unstable one at low frequency. Nichols proves that the low-frequency unstable limit cycle does not exist.

Since the describing function itself is an approximation, it is probable that predicted amplitude and frequency are *never* precisely correct. If we need precise values, simulation studies should be a satisfactory way of obtaining them. It is important, however, to understand the reasons for false and inaccurate results, and it is important that we be able to recognize on the plots those graphical conditions that point out the possibility of errors.

The fundamental constraint on describing function theory is that the signal input to the nonlinear element must be a pure sine wave. When this constraint is satisfied, the results of describing function analysis are sufficiently accurate for engineering purposes, but this constraint is *never* completely satisfied. We note that the signal output from a nonlinear element is nonsinusoidal. Since this signal is transmitted around the feedback loop, it contributes to the signal which is the *input* to the nonlinear element. Thus, in a feedback system with a nonlinear element inside the loop, the signal going into the nonlinear element is never a pure sine wave since it contains harmonic frequencies generated by the nonlinear element itself. When the linear elements in the loop constitute a low-pass filter, the harmonics are attenuated and may not contribute significantly to the signal feeding the nonlinear element. When this is the case, i.e., the linear part of the feedback system is a low-pass filter, describing function predictions are trustworthy and usually pretty accurate. We must then define what we mean by a "low-pass filter," but this is not really possible in the context of describing function theory. Usually the describing function analysis will correctly predict the *existence* and *nature* of any limit cycles, even for second-order (all pole) systems, but the accuracy of the predicted amplitude and frequency is subject to other conditions.

The danger of inaccurate predictions can usually be recognized by inspection of the Nichols plot (or polar plot, if one prefers). Note—on Figures 10.8 or 10.9 or 10.10—that the intersection between the describing function curve and the transfer function curve is a well defined intersection; while the two curves are not perpendicular, the angle of intersection is substantial and the curves separate (diverge) on both sides of the intersection. When this kind of geometric configuration occurs, the analysis usually gives accurate predictions. The reasoning is as follows: The transfer function curve is correct for all frequencies and therefore is not affected by harmonics in the signals, but the describing function curve is known to be correct only for a pure sine wave into the nonlinearity, and therefore is not correct when the harmonics are transmitted. If we knew how to alter the describing function curve to include the effects of the harmonics, we should expect some small (because harmonic content is small) displacement of the describing function curve. However, small distortions of the describing function curve would cause little change in the predicted limit cycle given by Figures 10.8, 10.9, and

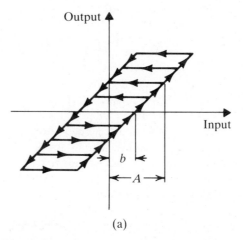

(a)

FIGURE 10.16   Describing Function Analysis for a Servo with Backlash: (a) Backlash Gain $= 1.0$, $R = b/A$

$G(j\omega) = K/j\omega(j\omega + 1)$
$K = 2.0$; no limit cycle
$K = 4.0$; stable limit cycle at $\omega \simeq 1.22$
$K = 8.0$; stable limit cycle at $\omega \simeq 2.3$
Unstable limit cycle; predictions are false.

(b) Nichol's Plot.

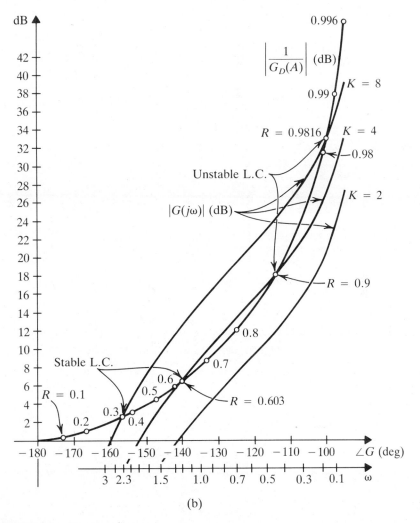

(b)

FIGURE 10.16 (*Continued*)

10.10, but could cause substantial changes if the intersection was "flat," i.e., if the two curves were nearly parallel in the vicinity of the intersection. Note that, in the case of the second-order system with backlash as given in Figure 10.16, *both* intersections are *flat* and the prediction of an unstable limit cycle is incorrect.

Figures 10.16(c) through (g) show the simulation results. For $K = 2$, the oscillations die out and the error goes to zero. For $K = 4$ and $K = 8$, there is a limit cycle. The amplitudes and frequencies of the limit cycles are marked on the curves. Table 10.4 compares predicted values with those provided by the simulation.

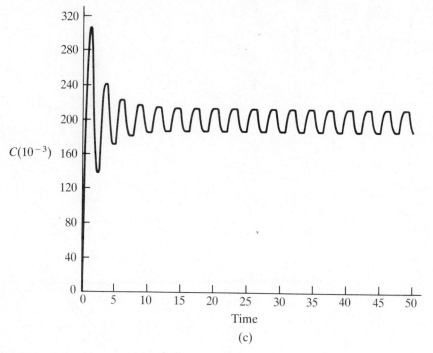

FIGURE 10.16(c) Step Response at Output Showing Limit Cycle where $K = 8$.

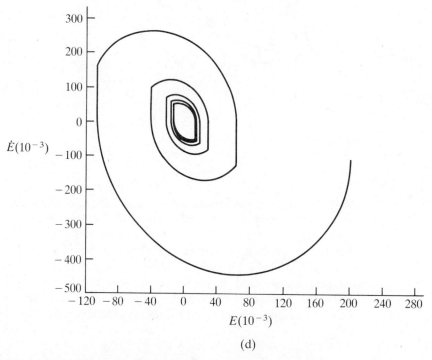

FIGURE 10.16(d) Limit Cycle on the Phase Plane where $K = 8$.

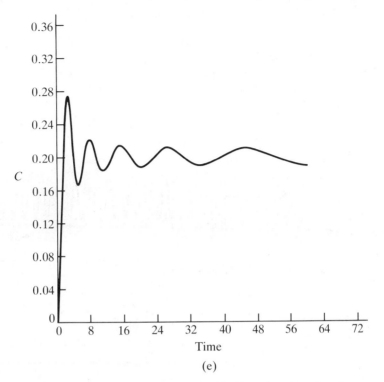

(e)

FIGURE 10.16(e) Step Response at Input to the Nonlinearity where $K = 2$.

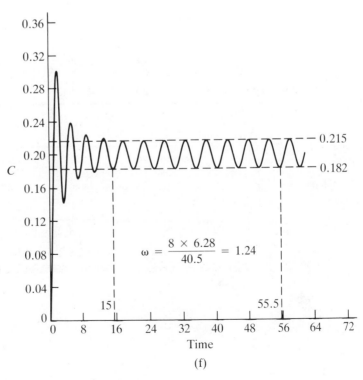

$$\omega = \frac{8 \times 6.28}{40.5} = 1.24$$

(f)

FIGURE 10.16(f) Step Response at Input to the Nonlinearity where $K = 4$.

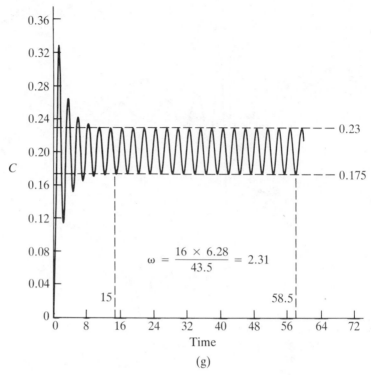

FIGURE 10.16(g)   Step Response at Input to the Nonlinearity where $K = 8$.

TABLE 10.4   Limit Cycles Due to Hysteresis

Gain	Describing Function Predictions		Simulation Results	
	ω	R	ω	R
$K = 2$	No limit cycle		No limit cycle	
$K = 4$	1.22	0.61	1.24	0.61
$K = 8$	2.3	0.364	2.31	0.364

$R = b/2I$,   for $K = 4$, $2I = (0.215 - 0.182) = 0.033$, $b = 0.02$,

$R = 0.02/0.033 = 0.61$ for $K = 8$, $2I = (0.23 - 0.175) = 0.055$,

$b = 0.02$, $R = 0.02/0.055 = 0.364$

10.9 MORE ON DESCRIBING FUNCTIONS

In the preceding discussions, the nonlinearities we have considered have been sensitive to the *amplitude* of an input signal, but completely independent of frequency. In most cases, we have been able to describe the nonlinear characteristics with a curve relating the output of the nonlinear device to its input. There are at least two situations in which the input-output relationship for the nonlinearity is a function of *both* signal amplitude and signal frequency. In both cases, of course, the nonlinearity is a dynamic device, i.e., it contains energy storage elements. It is these elements that generate the functional dependency on frequency. Examples of components that are frequency dependent are a two-phase motor[3] and various hydraulic devices.[4] In addition, when two amplitude-dependent nonlinearities are separated by a linear dynamic component, the combination leads to an equivalent describing function that is both amplitude and frequency dependent.

In practice, many devices (such as the two-phase motor) are difficult to model mathematically, but an input-output relationship can be obtained by *test*, i.e., we can measure the sinusoidal input and the periodic output over a finite range of frequencies and amplitudes. Typically, we use an oscillator for input with the signal amplitude fixed and measure the output over a selected frequency range, and repeat for selected input signal amplitudes. If the output signal appears to be nearly sinusoidal, we determine gain and phase as usual; if the output is badly distorted, special techniques may be required to get acceptable values for magnitude ratio and phase. The final result is a family of curves for the describing function $G_D(A, \omega)$, which then must be manipulated to get the desired $-1/G_D(A, \omega)$ curves.

The second situation usually occurs when there are two or more nonlinearities in the system. A fundamental requirement for the validity of describing function analysis is that the signal at the input of the nonlinearity must be a pure sine wave. Describing functions work well for control system analysis largely because such systems are normally low-pass filters and signals distorted by nonlinearities are filtered as they go around the feedback loop, thus providing nearly sinusoidal signals at the nonlinear element. When there are several nonlinear elements in the system, the low-pass filtering between them may not be sufficient to remove harmonics from the wave shape, in which case the describing function of the individual nonlinearities cannot be used individually and a composite describing function must be obtained.

Consider the block diagram of Figure 10.17(a), and assume that the input signal E is a pure sine wave. If the linear block G_1 is a good low-pass filter, then the input b to the second nonlinear block will be approximately a sine wave, and the input-output transfer function may be written

$$\frac{C}{E} = \frac{C}{b} \times \frac{b}{a} \times \frac{a}{E} = G_{D2} \cdot G_1 \cdot G_{D1},$$

where G_{D1} and G_{D2} are the conventional describing functions of the two nonlinear elements. Theoretically then, one can use these describing functions directly; indeed,

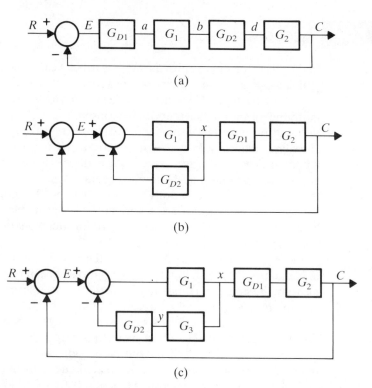

FIGURE 10.17   Systems with Two Nonlinear Components: (a) Two nonlinear Blocks Separated by Linear Blocks, (b) Two Nonlinear Blocks Fed by the Same Signal, and (c) a More Complex Case.

one can obtain a useful $G_D(A, \omega)$ for the combination using the normal definitions for G_{D1} and G_{D2}. However, if G_1 is *not* a good low-pass filter, then the describing functions G_{D1} and G_{D2} are not useful at all. One must obtain a composite describing function for the three blocks, $C/E = G_D(A, \omega)$, and calculation of this function is not a trivial problem; in fact, it may be easier and better to obtain it experimentally, as by driving the combination with a pure sine wave and feeding the output into a Fourier analyzer.

The net result for any of these cases is a family of curves for the describing function. We shall consider one such case in order to demonstrate the interpretation of the graphical relationships on the Nichols plot.

Figure 10.17(b) illustrates another case in which two nonlinearities in the system are in different paths but are fed by the same signal. As long as this signal is sinusoidal, the individual describing functions may be used, and their interpretation is relatively simple because the signal amplitude into the blocks is the same [which is not the case for Figure 10.17(a)]. In like manner, the system of Figure 10.17(c) has two non-linearities in different paths, that are *not* fed by the same signal. In this case, it is not at all clear whether or not the basic graphical analysis techniques can be used.

In all cases, the algebraic manipulations can be helpful in defining the problem, and a convenient starting point is the formulation of the characteristic equation. For the systems of Figure 10.17, there are:

1. $$1 + G_{D1}(E)G_1(j\omega)G_{D2}(b)G_2(j\omega) = 0$$

2. $$1 + G_1(j\omega)G_{D1}(x)G_2(j\omega) + G_1(j\omega)G_{D2}(x) = 0$$

3. $$1 + G_1(j\omega)G_{D1}(x)G_2(j\omega) + G_1(j\omega)G_3(j\omega)G_{D2}(y) = 0.$$

Since the *describing function* is a *linear* approximation of the nonlinear function, it may be treated as any other linear function. Thus, the characteristic equation can be analyzed by root locus methods, as was done for systems with only one nonlinear element, or by frequency response methods, as was also demonstrated for the case of one nonlinear element. When multiple nonlinearities are encountered, however, it is difficult to apply the basic graphical methods because the characteristic equation, though linearized, is a function of several variables so methods that required only one curve for the simpler cases may require curve families. If we choose suitable families, we can interpret the results. If our choice is poor, interpretation may be impossible. In either case, the computational labor is greatly increased.

EXAMPLE 10.7

Consider the system of Figure 10.18(a), where the two nonlinearities are both saturating amplifiers with unit gain. The linear system is unstable and, using the Routh criterion, the frequency at the stability limit is $\omega = 2.54$. Since both of the nonlinearities are saturation, the *small signal* system is *linear*, and we would therefore expect the limit cycle frequency to be this value. Figure 10.18(b) shows the describing function plot on the Nichols chart and, indeed, the indicated limit cycle is at $\omega = 2.54$. However, we cannot put an amplitude scale on our $-1/G_{D1} G_{D2}$ plot.

We observe that $G_1(j\omega)$ is a low-pass filter with poles at 10 and 20. As a consequence, signals in the frequency range $0 \le \omega < 5$ will not be attenuated appreciably, but rather will be amplified by the gain of 20. Thus, the *second amplifier* G_{D2} will probably be unsaturated in the limit cycle. Thus, we could—for this problem—evaluate the amplitude of the limit cycle by treating the system as if there

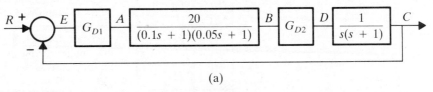

(a)

FIGURE 10.18 (a) A Specific System with Two Nonlinearities, (b) the Nichols Plot for this System, and (c) the Nichols Plot of $G_2(j\omega)$ and of

$$-\frac{1}{G_{D1}(E)G_1(j\omega)G_{D2}(B)}.$$

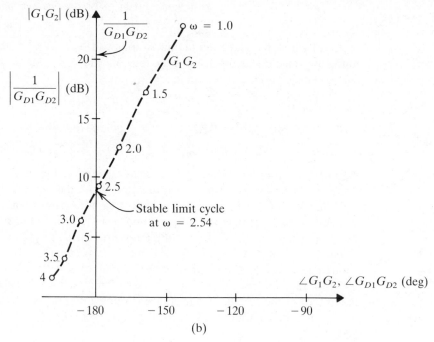

(b)

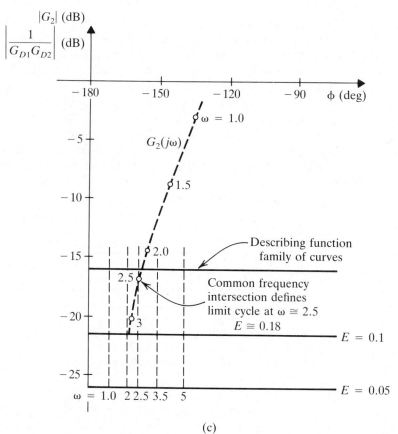

(c)

FIGURE 10.18 (*Continued*)

were only one nonlinearity. However, to illustrate the methods we will proceed with the solution.

Figure 10.18(c) is a Nichols plot showing the limit cycle prediction when we plot $G_2(j\omega)$ and $-1/G_{D1}(E)G_1(j\omega)G_{D2}(B)$. This arrangement gives a family of describing function curves thus clarifying the effect of signal amplitude and permitting prediction of both the amplitude and the frequency of the limit cycle.

As previously indicated, when the linear element separating two nonlinearities is not low pass, the signal input to the second nonlinearity will not be sinusoidal and, thus, the conventional describing function cannot be used for that nonlinearity and a new describing function must be defined for the cascade combination of both nonlinear elements with intermediate linear element. While various techniques may be applied, all must satisfy the basic definition, i.e., the Fourier fundamental for the output wave is determined and the describing function for the cascade combination is the ratio of this Fourier fundamental to the input sine wave.

Consider the block diagram of Figure 10.16a with somewhat different linear functions and nonlinearities as shown in Figure 10.19a-c. Note particularly that the compensator is a *high-pass filter*. To derive a suitable describing function, we consider a sinusoidal input to the open loop as shown in Figure 10.19a-c.

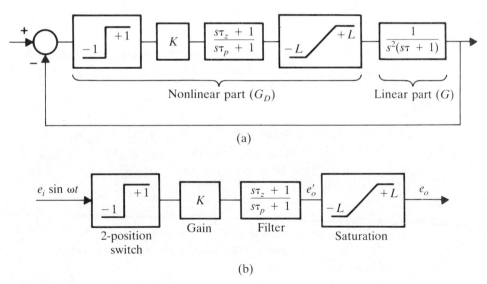

(a)

(b)

FIGURE 10.19 (a) Block Diagram Defining the Linear and Nonlinear Parts of the System, and (b) the Diagram for Derivation of Describing Function, and (c) the Output Response to a Sine Wave Input.

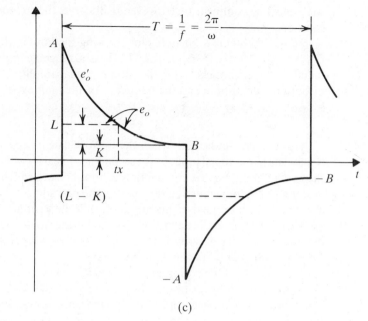

(c)

FIGURE 10.19   (*Continued*)

We can derive the equation of this repetitive transient by inspection (at least for the steady-state condition, which is what we need to derive a describing function):

1. Every time the switch reverses, the applied signal changes by a factor of 2.0.[a] If, at the time of reversal, the output level is some value $e'_0 = -B$, then

$$A = -B + 2.0 \times K \times \frac{\tau_z}{\tau_p}.$$

   The jump is due to the fact that the filter is a high-pass (lead) filter.

2. The transient decay from value A is exponential with time constant τ_p. Decay is toward the steady-state value (which is actually K for this system.). The equation of the decay is

$$e'_0 = K + (A - K)e^{-t/\tau_p}.$$

3. The time of the next switch reversal is at $t = T/2$, where $T = 1/f = 2\pi/\omega =$ period of the sinusoidal forcing function, and t is measured from the preceding

---

[a] This number is based on the assumption of ∓ 1.0 for the switch levels. If the physical system value is different, we need only adjust the value of K.

reversal instant. Then, at $t = T/2$,

$$e'_0 = K + (A - K)e^{-T/2\tau_p} = B.$$

4. Since the output amplifier saturates at a level $e_o = L$, the initial part of the e_o wave shape is clipped. The duration of this clipped portion is a time t_x. The magnitude can be calculated as follows:

$$L = (A - K)e^{-t_x/\tau_p} + K,$$

$$e^{-t_x/\tau_p} = (L - K)/(A - K),$$

$$-t_x/\tau_p = \ln\frac{L - K}{A - K},$$

$$t_x = -\tau_p \ln\frac{L - K}{A - K}.$$

We have thus described the output wave shape completely, and shall use this description of the wave shape to define our describing function:

$$\frac{e_o}{e_i}(j\omega) \triangleq G_D(j\omega).$$

The output wave shape can also be described over a half-period time interval as follows:

$$e_o = Lu(t) - Lu(t - t_x) + Ku(t - t_x) + (L - K)u(t - t_x)\exp[-(t - t_x)/\tau_p].$$

To obtain a describing function, e_o can be expanded in a Fourier series and all terms discarded except the fundamental. The coefficients of the fundamental component are evaluated from:

$$A_1 = \frac{2}{\pi}\int_0^\pi e_o \cos\omega t\, d\omega t,$$

$$B_1 = \frac{2}{\pi}\int_0^\pi e_o \sin\omega t\, d\omega t.$$

Then, by substitution,

$$A_1 = \frac{2}{\pi}\int_0^{\omega t_x} L(\cos\omega t)\, d\omega t + \frac{2}{\pi}\int_{\omega t_x}^\pi K(\cos\omega t)\, d\omega t$$

$$+ \frac{2}{\pi}\int_{\omega t_x}^\pi (L - K)\exp[-(t - t_x)/\tau_p]\cos\omega t\, d\omega t,$$

$$B_1 = \frac{2}{\pi}\int_0^{\omega t_x} L(\sin\omega t)\, d\omega t + \frac{2}{\pi}\int_{\omega t_x}^\pi K(\sin\omega t)\, d\omega t$$

$$+ \frac{2}{\pi}\int_{\omega t_x}^\pi (L - K)\exp[-(t - t_x)/\tau_p]\sin\omega t\, d\omega t.$$

Evaluating the integrals yields the results,

$$A_1 \frac{2(L-K)}{\pi} \sin \omega t_x + \frac{2(L-K)}{\pi} \frac{\omega^2 \tau_p^2}{1 + \omega^2 \tau_p^2}$$

$$\left(\frac{\exp[(-\frac{\pi}{\omega} - t_x)/\tau_p]}{\omega \tau_p} \sin \omega t_x + \frac{\cos \omega t_x}{\omega \tau_p} \right),$$

$$B_1 = \frac{2(K-L)}{\pi} \cos \omega t_x + \frac{2(K+L)}{\pi} + \frac{2(L-K)}{\pi}$$

$$\left(\exp[(-\frac{\pi}{\omega} - t_x)/\tau_p] + \frac{\sin \omega t x}{\omega \tau_p} + \cos \omega t_x \right),$$

where

$$e_o = A_1 \cos \omega t + B_1 \sin \omega t.$$

The computation of A_1 and B_1 for given values of K, τ_p and τ_z, and ω are obtained from the above equations using the previously derived value of t_x, where

$$t_x = -\tau_p \ln \frac{L-K}{A-K},$$

$$A = \frac{Ke^{-T/2\tau_p} - 1 + 2K\tau_z/\tau_p}{1 + e^{-T/2\tau_p}}.$$

The describing function then can be computed as follows:

$$\frac{e_o}{e_i}(j\omega) = G_D(j\omega) = \frac{\sqrt{A_1^2 + B_1^2} \tan^{-1}\left(\frac{A_1}{B_1}\right)}{|e_i| \angle 0°}$$

10.10 SUMMARY

Nonlinearities in a feedback control system affect its accuracy, stability, and dynamic response. An important phenomenon encountered only in nonlinear systems is the limit cycle—a constant amplitude, constant frequency oscillation, which may impair the usefulness of a system or, in special cases, may be used to improve control. A mathematical description of nonlinear elements that is useful in analysis and design of limit cycle characteristics is the describing function. The describing function is a transfer function for the nonlinear element, obtained by linearizing in the frequency domain.

The describing function may be used with root locus methods and Bode-Nichols methods (and others) to analyze the stability of a nonlinear system, i.e., to test for the existence of limit cycles, for their nature (stable or unstable), and for their amplitude and frequency. If the nonlinearity is single-valued, the root locus method is a convenient tool, and examples have been given. For more complex nonlinearities, the Nyquist-Bode-Nichols methods are used.

Design procedures are available using the same tools. Compensators may be designed that eliminate the limit cycle or change its frequency and amplitude. When the nonlinear system does not have a limit cycle, designing to adjust the dynamic response may be possible.

REFERENCES

1. Thaler, G. J.; and Pastel, M. P. *Analysis and Design of Nonlinear Feedback Control Systems.* Chapter 4, New York: McGraw-Hill Book Co. (1960).

2. Nichols, N. B. "Backlash in a Velocity Lag Servomechanism." *AIEE Trans Appl. Ind.* (Part II) 73: 462–67 (1954).

3. Stein, W. A.; and Thaler, G. J. "Obtaining the Frequency Response Characteristics of a Nonlinear Servomechanism from an Amplitude and Frequency-Sensitive Describing Function." *AIEE Trans.* (Part II) 77: 91–96 (1958).

4. Zaborsky, J.; and Harrington, H. J. "Describing Function for Hydraulic Valves," *AIEE Trans.* (Part I) 76: 183–98 (1957).

BIBLIOGRAPHY

Dutilh, J. "Theorie des Servomechanisms a Relais." *Onde Elec.* 438–45 (1950).

Goldfarb, L. C. "On Some Nonlinear Phenomena in Regulatory System." *Automatika i Telemekhanika.* 349–83 (1947); translation: R. Oldenburger, Ed. *Frequency Response.* New York: MacMillan Co. (1956).

Grief, H. D. "Describing Function Method of Servomechanism Analysis Applied to Most Commonly Encountered Nonlinearities." *AIEE Trans. Appl. Ind.* (Part II) 72: 243–48 (1953).

Kochenburger, R. J. "A Frequency Response Method for Analyzing and Synthesizing Contactor Servomechanisms." *AIEE Trans.* 69: 270–83 (1950).

Oppelt, W. "Locus Curve Method for Regulators with Friction." *Z. Duet Ingr. Berlin* 90: 179–83 (1948); translation: Report 1691. Washington, D.C.: National Bureau of Standards (1952).

Thaler, G. J.; and Brown R. G. *Analysis and Design of Feedback Control Systems.* 2nd ed. Chapter 13. New York: McGraw-Hill Book Co. (1960).

Tustin, A. "The Effects of Backlash and of Speed Dependent Friction on the Stability of Closed Control Systems." *JIEE* (Part II) 94: 143–51 (1947).

PROBLEMS

10.1   A nonlinear electronic device produces an output that is the cube of its input (i.e., $y = x^3$). Derive the describing function of the device.

10.2   An amplifier has a gain $K = 1$, but saturation limits are $+2$ V and -1.5 V. How does this affect the describing function of the amplifier?

10.3   Tachometer feedback is used to stabilize a servo, but the amplifier in the tachometer channel has a small dead zone. Would you expect a limit cycle? Analyze the system below.

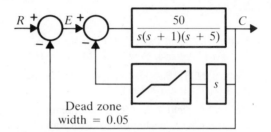

10.4   An instrument servo has a small amount of backlash in its gear train that affects the position feedback signal. The servo has second-order dynamics, i.e., $G(s) = K/s(s + P)$. Analyze the situation with describing functions and simulation. Theory shows that a limit cycle will exist if $\zeta \leq 0.29$. Can you verify this?

10.5   The following system is a high-gain regulator. A tachometer is inserted to provide damping, but an error is made and the amplifier in the tachometer channel has extremely high gain and also saturates. Under what circumstances will there be a limit cycle?

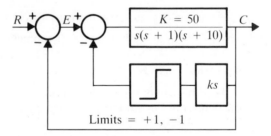

10.6   In the system of Problem 10.3, an R-C differentiation replaces the tachometer, as shown below. How do the values of k and p affect the performance?

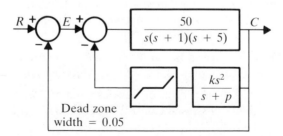

10.7   Given:
Analyze the system for possible limit cycles. How do they depend on the value of k?

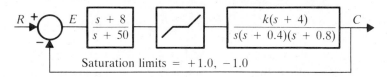

10.8   Given:
Will a limit cycle exist? If so, what are the frequency and the amplitude?

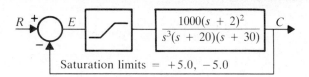

10.9   A servo using an ideal relay has only second-order dynamics. Describe and explain the expected step response.

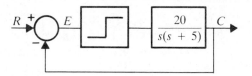

10.10   Given:

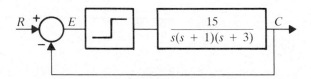

Determine the limit cycle frequency and amplitude.

10.11   Given:

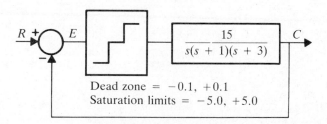

Will a limit cycle exist if the system is subjected to a unit step input? If so, determine its frequency and amplitude.

10.12   A small instrument servo is shown:

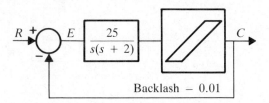

Will there be a limit cycle? If so, predict the frequency and amplitude. The gain is reduced from 25 to 16; what is the result?

10.13   The motor in the servo of Problem 10.12 has a third pole, so that

$$G(s) = \frac{250}{s(s + 2)(s + 10)}.$$

Will there be a limit cycle? What happens if the gain is reduced? Increased?

10.14   An unstable system is to be stabilized with velocity feedback as shown:
The amplifier in the feedback path has a gain of 1.0 when unsaturated. Can a limit cycle exist?

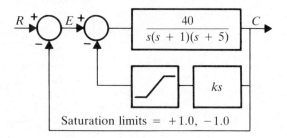

Under what circumstances?

11 Nonlinear Systems: The Phase Plane and Discontinuous Systems

INTRODUCTION

The purpose of this chapter is to develop the concepts of the phase plane and phase (state) space, and to develop associated techniques that are useful in the analysis and design of nonlinear control systems. Because the digital computer is commonly used to solve the nonlinear equations and provide the phase trajectories, the various methods of graphical integration formerly used to obtain phase portraits will not be developed here. However, we will show how auxiliary lines may be added to the phase portrait to improve interpretation, facilitate analysis, and guide design efforts. Such lines are switching lines, dividing lines, limit lines, eigenvectors, etc. We shall also need to know something about singular points, how to locate them, and how to interpret them. To discuss singular points and eigenvectors, it is convenient to introduce the method of isoclines. We do not use this method to perform the usual graphical integration, but only to aid in the discussion of analysis and design; we will rely on computer solutions to provide the phase trajectories as needed.

The basic concept may be developed from the general differential equation:

$$A_N \frac{d^N x}{dt^N} + A_{N-1} \frac{d^{N-1} x}{dt^{N-1}} + \cdots + A_2 \ddot{x} + A_1 \dot{x} + A_0 x = 0. \tag{11.1}$$

For a given set of initial conditions, the solution to Eq. (11.1) may be represented by a single curve in the N-dimensional phase space for which the coordinates are x, $\dot{x}$, $\ddot{x}, \ldots d^{N-1} x/dt^{N-1}$. The curve traced out by the state point is called the *phase* or *state* trajectory, and the family of all possible such trajectories is called the *phase portrait*. Normally, a finite number of trajectories, defined in a finite region, is considered a portrait.

399

The coefficients of Eq. (11.1), designated by the A's, may be constants or functions of time, in which case this differential equation is considered to be linear. The A's may also be functions of other state variables, in which case the equation is nonlinear. If the differential equation is of the n'th order, then an n-dimensional space is required for the trajectory. Thus, our computer solutions, while providing all n coordinates of the state point in numerical form, provide only two-dimensional projections of the trajectory on a selected phase plane. Computer graphics is usually capable of three dimensional plots, which can be useful, but are rarely worthwhile. When the differential equation is of second (or first) order, then the phase trajectory is two-dimensional and the correct solution curve can be plotted rather than just a projection.

For second-order systems, the two-dimensional nature of the problem permits a variety of graphical integration methods. They are not useful for higher order systems (which comprise the majority of practical nonlinear problems) but at least one of these methods provides interesting insights that permit better understanding of higher order systems. We include some development and application of this one method.

11.2   SLOPES, ISOCLINES, SINGULAR POINTS, AND LIMIT CYCLES

Let us consider the differential equation of a linear second-order servo:

$$\ddot{C} + 2\zeta\omega_n\dot{C} + \omega_n^2 C = \omega_n^2 R. \tag{11.2a}$$

Rearranging,

$$\ddot{C} = \omega_n^2 R - \omega_n^2 C - 2\zeta\omega_n\dot{C}. \tag{11.2b}$$

Letting $\ddot{C} \triangleq d\dot{C}/dt$ and dividing by $\dot{C} = dC/dt$,

$$\frac{\ddot{C}}{\dot{C}} = \frac{d\dot{C}/dt}{dC/dt} = \frac{d\dot{C}}{dC} = \frac{\omega_n^2 R - \omega_n^2 C - 2\zeta\omega_n\dot{C}}{\dot{C}}. \tag{11.3}$$

The quantity $d\dot{C}/dC$ is the tangent to the solution curve on the C versus $\dot{C}$ plane. Thus, if we choose a particular initial condition point on the C versus $\dot{C}$ plane and substitute the numerical values of the coordinates into the right side of Eq. (11.3), the number obtained is the direction of the trajectory at that point. We can repeat this procedure as often as we wish, defining the slope of the trajectory at any point, as shown in Figure 11.1. It is clear that this can be done whether the equation is linear or nonlinear. Inspection of such a family of slope markers clearly shows the general shape of the trajectory (and other features such as singular points) and we can use them for graphical integration if we wish. For the purposes of this text, however, we will proceed to the method of isoclines.

The slope of the trajectory, $d\dot{C}/dC$, is defined by the right side of Eq. (11.3). It seems reasonable that there will be many points on the C versus $\dot{C}$ plane at which the slope of the trajectory is the same value. If we work the problem backwards, i.e., choose a number for the slope of

$$\frac{d\dot{C}}{dC} \triangleq N,$$

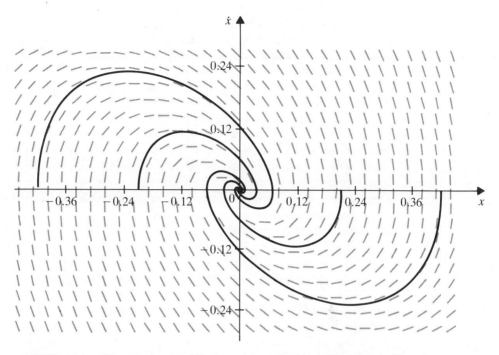

FIGURE 11.1 Slope Markers and Trajectories on the Phase Plane for the Equation:
$$\ddot{x} + 2AB\dot{x} + B^2x = 0 \qquad (A = 0.5,\ B = 1.0).$$

then Eq. (11.3) becomes

$$N = \frac{\omega_n^2 R - \omega_n^2 C - 2\zeta\omega_n C}{\dot{C}}. \tag{11.4a}$$

Rearranging $\dot{C}(N + 2\zeta\omega_n) = -\omega_n^2 C + \omega_n^2 R,$

$$\dot{C} = \frac{-\omega_n^2}{N + 2\zeta\omega_n}C + \frac{\omega_n^2}{N + 2\zeta\omega_n}R, \tag{11.4b}$$

which is the equation of a line on the C versus $\dot{C}$ plane. For the linear system of
Eq. (11.4b), the line defined is a straight line with slope $-\omega_n^2/(N + 2\zeta\omega_n)$ and inter-
cept on the $\dot{C}$ axis at $\dot{C} = \omega_n^2 R/(N + 2\zeta\omega_n)$. If the system were nonlinear, the equation would be that of a curve. In either case, the line is called an *isocline* because it is
the locus of all points at which the slope of the trajectory is a single fixed number N.
Obviously, we can generate a family of isoclines by choosing many different numerical values for N. We shall illustrate this with examples after a brief discussion of
singular points.

By inspection of Eq. (11.3), it is easily seen that for selected values of C and $\dot{C}$,
the right side of Eq. (11.3) may become indeterminate. Specifically, if $\dot{C} = 0$ and $C = R$,
then Eq. (11.3) becomes:

$$\frac{d\dot{C}}{dC} = \frac{0}{0}.$$

Thus, the slope of the trajectory *is not defined* at this point, which is called a *singular point*. We need to know how the trajectory of a given system behaves in the vicinity of its singular points, and we are particularly interested in the singular points of nonlinear systems. We find, however, that it is easiest to study the singular points of linear systems and then extend the results to the analysis of nonlinear systems.

The first question of interest is: Where will the singular points be, and how many will there be? This question is really answered by our definition! There will be singular points at all points for which the slope of the trajectory is undefined. For our linear system, there is only one point, but for nonlinear systems there may be many singular points. The interesting feature is that all such points seem to be points of potential equilibrium for the system. For the specific case of our second-order linear servo, the singular point is at $\dot{C} = 0$ and $C = R$, i.e., output position at commanded value. Thus, the singular point for this system is the desired steady-state equilibrium point. As we shall see, the multiple singular points of some nonlinear systems are also equilibrium points.

The second question of interest is: What is the behavior of the trajectories (i.e., the system) in the vicinity of the singular point? We can discuss this behavior—the *NATURE* of the singular point—in terms of our linear equation, Eq. (11.2). We note that the coefficients of this equation have all been defined algebraically, but not numerically, so we do not know where the roots are! The *location* of the singular point is not affected by this, and so we conclude that the characteristics of the trajectories near the singular point are functions of the locations of the roots.

For a second-order system, the characteristics for all of the possible combinations of two roots are given in Figure 11.2. Considering each in some detail:

1. Stable system with complex roots. The transient is an exponentially damped sinusoid. Phase trajectory is a logarithmic spiral *into* the singular point. This type of trajectory is called a *stable focus*.

2. Unstable system with complex roots. The transient is an exponentially increasing sinusoid. The phase trajectory is a logarithmic spiral expanding out of the singular point and is called an *unstable focus*.

3. Stable system with real, finite, and unequal roots. The transient response is the sum of two negative exponentials. It is called an *overdamped* response, and the phase portrait at the singular point is called a *stable node*. The phase portrait contains two trajectories that are exactly straight lines. They are the *eigenvectors* of the system, and the slopes of these lines are numerically equal to the root values, which will be shown later in the text. The eigenvector due to the smaller root is called the *slow* eigenvector, and that due to the larger root is the *fast* eigenvector. All trajectories except this fast eigenvector approach the singular point asymptotic to the slow eigenvector.

4. Unstable system with both roots, real, finite, unequal, and positive. This, too, is called a node, but an *unstable node*. There are two eigenvectors, as for the stable node, and all trajectories emerge from the singular point and go to infinity.
    Note: For case 3, the eigenvectors lie in the second and fourth quadrants, but for case 4 they are in the first and third quadrants.

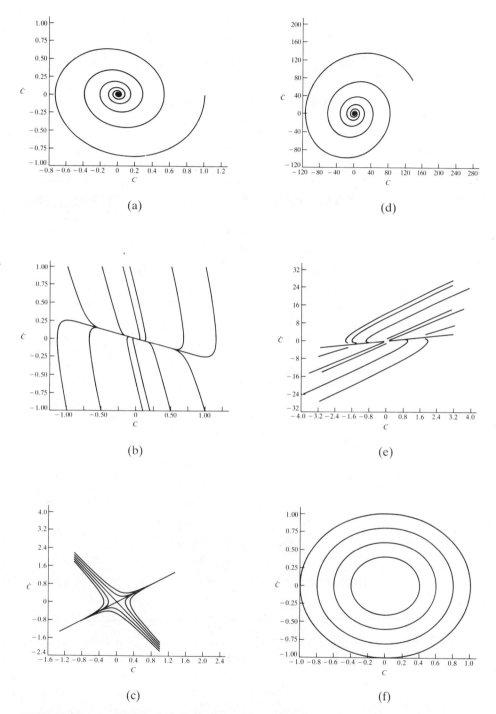

FIGURE 11.2 Definition of Singular Point Characteristics on the Phase Plane of a Linear Second-Order System: (a) Stable Focus, (b) Stable Node, (c) Saddle, (d) Unstable Focus, (e) Unstable Node, and (f) Center.

5. Systems with repeated real roots. In such cases, the two eigenvectors coalesce into a single eigenvector, again with slope determined by the root value. The phase portrait is still called a *node* (stable or unstable!).

6. Systems with one negative real root and one positive real root. Such systems are unstable because there is a positive root. Again, there are two eigenvectors with slopes defined by the root values. The eigenvectors due to the negative root provide a trajectory that *enters* the singular point, while the trajectory due to the positive root *leaves* the singular point. All other trajectories approach the singular point adjacent to the incoming eigenvector, then curve away and leave the vicinity of the singular point, eventually approaching the second eigenvector asymptotically. This configuration of the trajectories is called a *saddle*.

7. If the complex roots happen to have a real part of zero, i.e., if they are exactly on the imaginary axis, then the phase trajectories are closed curves *concentric* with the singular point, and this portrait is called a *center*.

It is also possible (mathematically speaking) to have one root at the origin with the second root on either the positive or negative real axis. Two roots at the origin are also possible. Such conditions are not expected in close-loop control so the portraits are not included here.

To relate these characteristics to nonlinear systems, we note that it is not difficult to locate the singular points from the differential equation. Once the singular point is located, we can linearize the equation at the singular point and determine its nature, noting that for a sufficiently small region about the singular point the nonlinear system behaves as predicted by the linearized analysis. Extension of this information to predict the behavior of the nonlinear system over a somewhat larger region about the singular point is not particularly difficult.

 EXAMPLE 11.1

A simple example is the equation of the Van der Pol oscillator:

$$\ddot{x} - \mu(1 - x^2)\dot{x} + x = 0, \tag{11.5}$$

where μ is a "small" number. From this equation,

$$\frac{\ddot{x}}{\dot{x}} = \frac{d\dot{x}}{dx} = \frac{\mu(1 - x^2)\dot{x} - x}{\dot{x}}, \tag{11.6}$$

for which there is a singular point at $x = 0$ and $\dot{x} = 0$. The equation of the isoclines is

$$\dot{x} = \frac{-x}{N - \mu(1 - x^2)}. \tag{11.7}$$

The family of slope markers showing the singular point and a trajectory with stable limit cycle is given in Figure 11.3.

To demonstrate that the singular point is an unstable focus, we linearize the differential equation by noting that for small x, $1 - x^2 \simeq 1$; thus, the equation becomes

$$\ddot{x} - \mu\dot{x} + x = 0. \tag{11.8}$$

Because there are two sign changes, it is clear that both roots are in the right half plane, and for μ (a small number) both roots are complex, so the singular point is indeed an

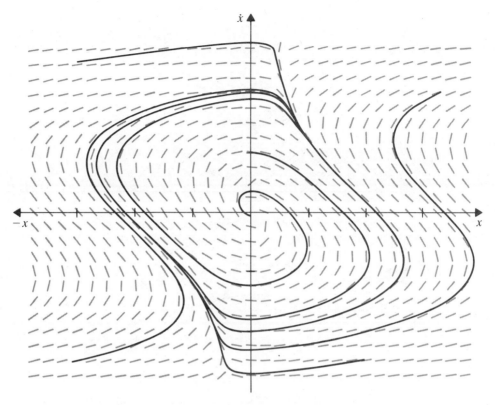

FIGURE 11.3 Slope Markers and Trajectories for Van der Pol's Equation:

$$\ddot{x} - A(1 - x^2)\dot{x} + Bx = 0 \qquad (A = 1, B = 1).$$

unstable focus. Therefore, the trajectory will spiral out of the singular point in a clockwise fashion. In this case, we can also predict the existence of the limit cycle by noting that if x is large, then $1 - x^2 \simeq -x^2$. The equation then becomes

$$\ddot{x} + \mu x^2 \dot{x} + x = 0 \tag{11.9}$$

Considering μx^2 to be the coefficient of the $\dot{x}$ term, this equation has both roots in the left half plane. For very large x the roots will be real, for smaller (but still "large") x they may be complex. In either case the trajectory proceeds inward in a clockwise direction toward the singular point. Thus at some intermediate values of x there must be an "equilibrium" condition such that the trajectory neither diverges nor converges, but forms a closed trajectory (limit cycle) around the singular point.

 EXAMPLE 11.2

Figure 11.4 gives the phase portrait of an undamped pendulum, for which the differential equation is

$$\ddot{\theta} + \frac{g}{l}\sin\theta = 0. \tag{11.10}$$

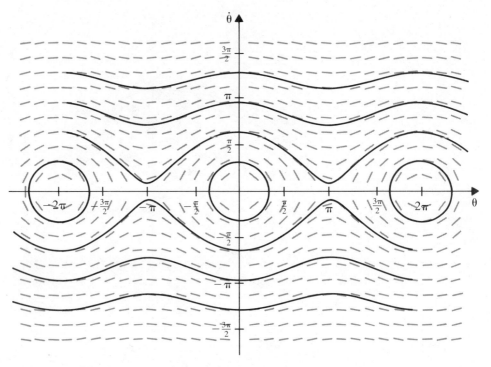

FIGURE 11.4 Phase Portrait of an Undamped Pendulum:

$$\ddot{\theta} + \frac{g}{l}\sin\theta = 0 \qquad (g/l = 1.0).$$

This system has two kinds of singular points, a center and a saddle point. It also permits definition of a new term: *separatrix,* which we define very simply to be the limiting condition that subdivides the phase plane into two regions in which the trajectories have different characteristics. Thus, the separatrix is a kind of dividing line (more on dividing lines later). Physical interpretation of this case may be instructive: inside this separatrix, the trajectories are characteristic of the "center"; they are closed curves about the singular point, and this corresponds to "small" amplitude oscillations of the pendulum about its rest point. Above and below the separatrix, the trajectories look almost like sine waves as they progress across the phase plane in the θ (or $-\theta$) direction; this corresponds to the pendulum swinging "over the top." Its rotation is greater than $\pm 180°$; in fact, it rotates unidirectionally. The separatrix is the line separating these motions. Note that the separatrix goes through the saddle point, and for the case of the pendulum the physical meaning of the saddle point is now clear. It corresponds to the pendulum being stationary (zero velocity) but oriented vertically upward from its pivot point.

It is interesting to note at this point that the equation of a *damped* pendulum,

$$\ddot{\theta} + k\dot{\theta} + \frac{g}{l}\sin\theta = 0, \tag{11.11}$$

is essentially the equation of a stepper motor. The damping term converts the centers into stable foci, and the *sinusoidal-type* trajectories eventually reach a focus, but the general appearance of the phase portrait is quite similar to that of Figure 11.4 because the stepper motor usually has little damping.

11.3   ISOCLINES AND EIGENVECTORS

Figure 11.5(a) shows a second-order linear system with isoclines and phase trajectory on the E versus $\dot{E}$ plane. We chose to use the E versus $\dot{E}$ plane for feedback control studies for two reasons:

1. We are vitally concerned with accuracy, and the error is clearly displayed for all instants in time.

2. The singular point is at or near the origin of the coordinate system, which aids us in the choice of scales, etc. For initial condition problems and for step inputs, a type-one or type-two system has its singular point exactly at the origin, and a type-zero system should have its singular point near the origin. For a ramp input, the type-one system has a finite steady-state error so its singular point will be near the origin; while for a ramp input, the type-two system has its singular point exactly at the origin. If nonlinearities are present, the singular point remains *near* the origin, since we require small steady-state errors.

The transformation of coordinates from the C versus $\dot{C}$ plane to the E versus $\dot{E}$ plane is simple; note that

$$\ddot{C} + P\dot{C} + KC = KR,$$
$$E = R - C,$$
$$\dot{E} = \dot{R} - \dot{C}, \tag{11.12}$$
$$\ddot{E} = \ddot{R} - \ddot{C}.$$

For initial condition problems or a step input, $R = 0$ or $R = $ constant. Thus, $\dot{R} = 0$, $\ddot{R} = 0$, and

$$E = R - C,$$
$$\dot{E} = -\dot{C}, \tag{11.13}$$
$$\ddot{E} = -\ddot{C}.$$

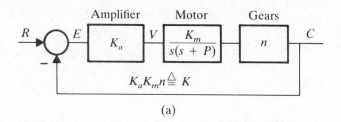

Amplifier, Motor, Gears

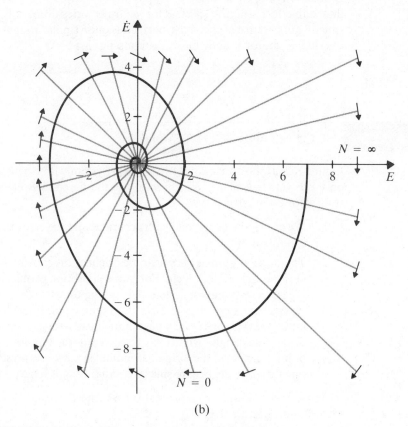

(b)

FIGURE 11.5 (a) Block Diagram for a Second-Order Servo:

$$\frac{C}{R}(s) = \frac{K}{s^2 + Ps + K}.$$

For a unit step input,

$$\ddot{C} + P\dot{C} + KC = KR; \qquad \ddot{E} + P\dot{E} + KE = 0,$$

$$N \triangleq \frac{\ddot{E}}{\dot{E}} = \frac{-P\dot{E} - KE}{\dot{E}} \rightarrow \begin{array}{l} \text{singular point} \\ \text{at } E = 0, \dot{E} = 0. \end{array}$$

The isocline equation is

$$\dot{E} = -\frac{KE}{N + P}.$$

(b) Isoclines and a Trajectory for the Second-Order Servo.

Substituting,

$$-\ddot{E} - P\dot{E} - K(E - R) = KR, \tag{11.14}$$

from which

$$\ddot{E} + P\dot{E} + KE = 0. \tag{11.15}$$

For a ramp input, $R = at$, $\dot{R} = a$, $\ddot{R} = 0$, and

$$
\begin{aligned}
E &= R - C, \\
\dot{E} &= a - \dot{C}, \\
\ddot{E} &= -\ddot{C},
\end{aligned}
\tag{11.16}
$$

from which

$$\ddot{E} + P\dot{E} + KE = -Pa. \tag{11.17}$$

Returning to Figure 11.5(a), we observe that one of the isoclines has been drawn for $N = 0$. By inspection, this is the line along which the system behavior changes from acceleration to deceleration. If this line were rotated clockwise to the vertical axis, deceleration would not start until the error was reduced to zero and, clearly, the amount of overshoot and oscillation would increase so that clockwise rotation of the $N = 0$ isocline corresponds to decreased damping. In fact, when this isocline lies on the vertical axis, the damping is zero and the singular point becomes a *center*. Conversely, counterclockwise rotation of the $N = 0$ isocline corresponds to increased damping. Figure 11.6(a) considers the simple servo with velocity feedback introduced. It is clear that the introduction of the feedback has rotated the $N = 0$ isocline counterclockwise, and we observe the reduced overshoot and oscillation. Velocity feedback (and also the introduction of a derivative of error signal) is said to *anticipate* the overshoot and reduce it. From the phase portrait, we see that this is true—the rotation of the $N = 0$ isocline causes an earlier transition from accelerate to decelerate, thus in a sense anticipating the arrival of the desired steady-state position.

Increased amounts of velocity feedback can overdamp the system. The roots then become real (either repeated or different) and the singular point becomes a node. As previously indicated, the phase portrait then contains two straight line trajectories, one for each real root, and these are the system eigenvectors by relating them to the isoclines. Since the isoclines for the linear system are straight lines, and so too are the eigenvectors, it follows that the eigenvectors must lie on special isoclines. By inspection it also follows that the necessary and sufficient condition is that the slope of the isocline and the slope of the trajectory must be the same. Now the equation of the isocline is, from Eq. (11.4b),

$$\dot{C} = \frac{-\omega_n^2}{N + 2\zeta\omega_n} C + \frac{\omega_n^2}{N + 2\zeta\omega_n} R, \tag{11.4b}$$

and the slope of this line is $-\omega_n^2/(N + 2\zeta\omega_n)$. For the isocline to be an eigenvector, it is

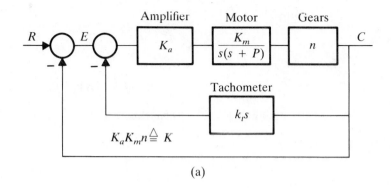

(a)

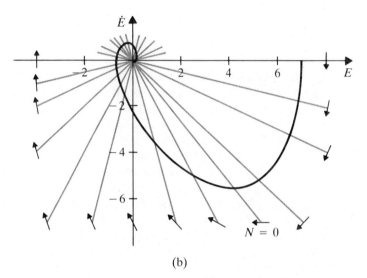

(b)

FIGURE 11.6   (a) Block Diagram for a Second-Order Servo with Velocity Feedback:

$$\frac{C}{R}(s) = \frac{K}{s^2 + (P + Kk_t)s + K}.$$

For a unit step input,

$$\ddot{C} + (P + Kk_t)\dot{C} + KC = KR; \qquad \ddot{E} + (P + Kkt)\dot{E} + KE - 0$$

$$N \triangleq \frac{\ddot{E}}{\dot{E}} = \frac{-(P + Kk_t)\dot{E} - KE}{\dot{E}} \rightarrow \begin{array}{l}\text{singular point}\\ \text{at } E = 0, \dot{E} = 0.\end{array}$$

The isocline equation is

$$\dot{E} = \frac{-KE}{N + P + Kk_t}.$$

(b) Effect of Velocity Feedback on Slopes, Isoclines, and Trajectory.

necessary that

$$N = \frac{-\omega_n^2}{N + 2\zeta\omega_n} \tag{11.18}$$

or

$$N^2 + 2\zeta\omega_n N + \omega_n^2 = 0. \tag{11.19}$$

Using the quadratic formula,

$$N = \frac{-2\zeta\omega_n \pm \sqrt{(2\zeta\omega_n)^2 - 4\omega_n^2}}{2}. \tag{11.20}$$

For the overdamped case, $\zeta > 1.0$; thus, N_1 and N_2 are real and are numerically equal to the real roots of the characteristic equation. Thus, we can determine and construct the eigenvectors without any information other than the root values. A phase portrait of a system with two real roots is given in Figure 11.7. The eigenvectors are marked; the one that is due to the smaller magnitude root is closest to the E axis and is called the *slow* eigenvector. It is seen that all trajectories except the *fast* eigenvector tend to reach the singular point asymptotic to this slow eigenvector. No phase trajectory can cross either eigenvector.

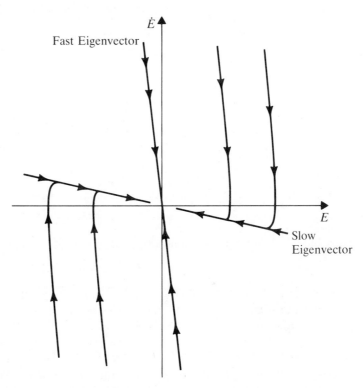

FIGURE 11.7 Stable Node, Showing Eigenvectors ($\zeta = 3.0$).

11.4   SWITCHING LINES AND DIVIDING LINES

A common type of "nonlinear" system is the discontinuous system, in which a relay or a switch is thrown to change the operating characteristics. We may use a relay to replace a power amplifier, and the switch then changes the power level (or signal level) from *on* to *off* or from *forward* to *reverse*. Figure 11.8(a) is such a system. The relay is operated by the error signal and reverses the signal applied to the plant as the error goes through zero. The vertical axis of the phase plane is then a switching line because the switch (relay) is operated whenever the state point crosses this line. The switch line is also a *dividing line* in this case, in the sense that it divides the phase plane in half such that one differential equation applies when the state point is to the right of the line; the other equation applies when the state point is to the left of the line. The system does not have a singular point since the slope of the trajectory is defined everywhere but on the switch line itself, where there are two possible slopes depending on the switch position. Thus, the slope of the trajectory changes discontinuously as the trajectory crosses the switch line.

Figure 11.9(a) shows the rotation of the switch lines when velocity feedback is used and is fed around the switch. For negative feedback, the switch line is rotated counterclockwise, and the feedback produces *equivalent* damping in the sense that overshoot and oscillation are reduced. Actually it *anticipates* arrival at the desired location and switches to reverse (decelerate), reducing the velocity as the origin is approached. For the system of Figure 11.9, we would say that overdamping has been accomplished; if the relay remained in the reverse position, the system could reach zero velocity before it reduced the error to zero, so there would be *no overshoot*. As actually shown, the trajectory intersects the switch line near the origin. Mathematically speaking, the motion beyond that point is *undefined* because at every point on the switch line two slopes are defined (for the two relay positions) and both force the state point back to the switch line. The result is that the state point *slides* or *chatters* down the switch line to the origin.

The equation of the switch line is easily determined from the block diagram. Note that the relay reverses when its control signal goes through zero. This control signal is

$$E^1 = E - k_t \dot{C}, \tag{11.21}$$

but $\dot{C} = -\dot{E}$ for a step input (or an initial condition). Thus,

$$E^1 = E + k_t \dot{E} = 0, \tag{11.22}$$

which is the equation of the switching line on the phase plane. Note that for this physical system the relay reverses every time the state point crosses the switching line. For other systems this may not be true.

Consider the possibility of a linear system with switched velocity feedback as shown in Figure 11.10. When the switch is open the system is lightly damped, perhaps undamped (or even unstable if the system is higher order). We choose k_t such that when the switch is closed the system is critically damped, or overdamped. Then we choose a

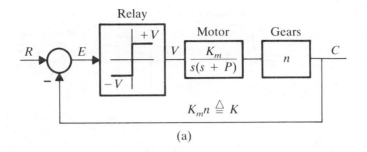

(a)

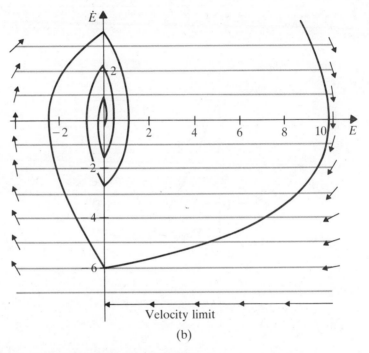

(b)

FIGURE 11.8 (a) Block Diagram of a Relay Servo:

$$\frac{C}{V}(s) = \frac{K}{s(s+P)} \qquad \ddot{C} + P\dot{C} + \mp KV \qquad \ddot{E} + P\dot{E} = \mp KV$$

$$N \triangleq \frac{\ddot{E}}{\dot{E}} = \frac{-P\ddot{E} \mp KV}{\dot{E}}.$$

The Equation of the switch line is $E = 0$ and the isocline equation is

$$\dot{E} = \frac{\mp KV}{N+P}.$$

(b) Phase Trajectory of the Relay Servo.

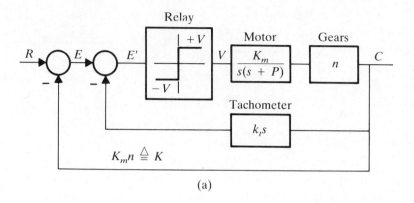

(a)

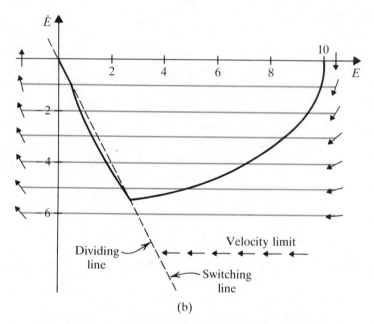

(b)

FIGURE 11.9 (a) Block Diagram of a Relay Servo with Velocity Feedback:

$$E = R - C, \qquad E' = E - k_t\dot{C} = E + k_t\dot{E} \text{ (step)}.$$

The equation of the switchline is

$$E' = 0 = E + k_t\dot{E}.$$

(b) Rotation of Switch Line by Velocity Feedback.

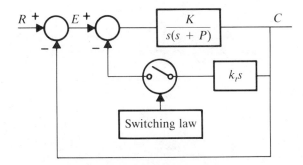

FIGURE 11.10 A Servo with Switched Velocity Feedback.

switching law (i.e., when do we close the switch?). Note that this is a different kind of switching operation:

1. We are switching in a component and thus changing the differential equation of the system, we are *not* reversing the signal to the plant (though reversed sign is probable) and there is no limit on signal amplitude.
2. The switching law we choose is arbitrary, i.e., it can be anything we want, and does not depend on the value of k_t.
3. The system has two different differential equations, which apply before and after switching, but each in its turn applies to the entire phase plane. This switching line is *not* a dividing line.
4. The switch need not operate every time the switch line is crossed; that is a design option.

This type of switching can be quite interesting when combined with the eigenvector concept, and we return to it at a later point.

Many types of nonlinearities give rise to dividing lines. We will consider a few. Figure 11.11 shows the effect of a saturating amplifier. We represent the saturating element with three straight line segments, which is common practice because this is sufficiently accurate for most problems and simplifies the model. The intersections of the straight lines provide dividing lines on the phase plane. When the signal into the amplifier is smaller in magnitude than the saturation level E_s, the system is linear, has a singular point at the origin of the E versus $\dot{E}$ plane for step inputs, and the isoclines and trajectory are as shown. For $|E| \geq |E_s|$ the amplifier is saturated. Thus, we have dividing lines at $E = \pm E_s$, and the typical isoclines for saturation when $|E| > |E_s|$. It is interesting to note that for this case the isoclines are continuous across the dividing line, and the $N = 0$ isocline essentially designates the velocity limit of the plant (initial conditions at higher velocity are theoretically possible but unlikely in practice). The dividing lines can be rotated by velocity feedback, which also increases the damping in the linear zone as shown in Figure 11.12. Note that the isoclines are rotated, but are still continuous across the dividing lines.

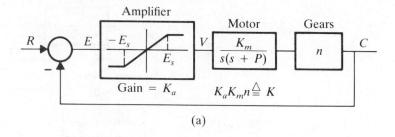

(a)

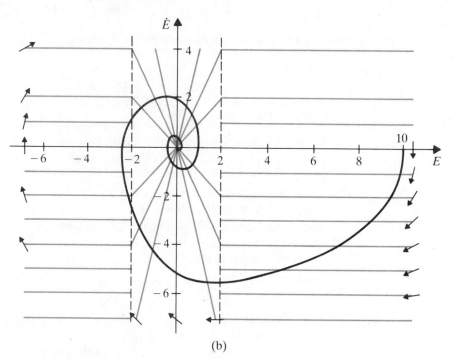

(b)

FIGURE 11.11 (a) A Servo with Saturating Amplifier:

$$\text{If} \quad -E_s < E < E_s, \quad N = (-P\dot{E} = KE)/\dot{E} \quad \text{and} \quad \dot{E} = \frac{-KE}{N+P}.$$

$$\text{If} \quad E < -E_s, \quad V = -K_a E_s N = (-P\dot{E} = KE_s)/\dot{E} \quad \text{and} \quad \dot{E} = \frac{KE_s}{N+P}.$$

$$\text{If} \quad E > +E_s, \quad V = +K_a E_s N = (-P\dot{E} = KE_s)/\dot{E} \quad \text{and} \quad \dot{E} = \frac{-KE_s}{N+P}.$$

(b) Isoclines, Dividing Lines, and Trajectory for the Saturating Servo.

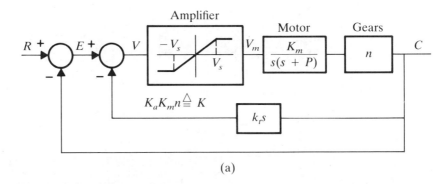

(a)

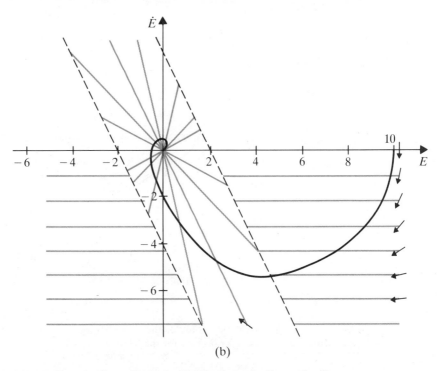

(b)

FIGURE 11.12   (a) Effect of Velocity Feedback on the Saturating Servo:

If $-V_s < V < V_s$, $N = [-(P + Kk_t)\dot{E} - KE]/\dot{E}$ and $\dot{E} = \dfrac{-KE}{N + (P + Kk_t)}$.

If $V < -V_s$, $N = (-PE + KV_s)/\dot{E}$ and $\dot{E} = \dfrac{KV_s}{N + P}$.

If $V > +V_s$, $N = (-PE - KV_s)/\dot{E}$ and $\dot{E} = \dfrac{-KV_s}{N + P}$.

Dividing lines: $E + k_t\dot{E} = \mp V_s$.

(b) Isoclines, Dividing Lines, and Trajectory for the Saturating Servo with Velocity Feedback.

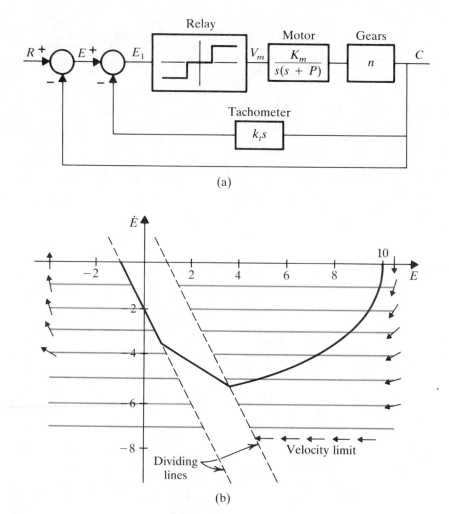

FIGURE 11.13 (a) Block Diagram for the Servo with Velocity Feedback and Relay with Dead Zone and (b) Phase Trajectory for the Servo.

Figure 11.13 shows the system with a dead zone nonlinearity (also modeled with straight line segments). The dividing lines are defined in the usual fashion. Note that the trajectory in the dead zone has constant slope everywhere between the dividing lines (verify this theoretically).

The presence of Coulomb friction in the system also gives rise to dividing lines on the phase plane, as does certain types of backlash. Coulomb friction is defined as that component of the total friction that opposes motion with a constant force (or torque). We may then write the Newton's law force summation at any free body as

$$M\ddot{x} + f\dot{x} \pm D + Kx = 0, \tag{11.23}$$

where the friction force is $f\dot{x} \pm D$, the constant (Coulomb) component being D. Since this force is always directed opposite to the velocity, one may account for the $\pm$ sign by writing $M\ddot{x} + f\dot{x} + D$ sign $\dot{x} + Kx = 0$. It is clear that the Coulomb friction is a reaction force and if $\dot{x} \equiv 0$, D sign $\dot{x}$ is not defined. However, in programming a computer to study this phenomenon, one must remember that the computer is not likely to recognize $\dot{x} \equiv 0$, and may try to *drive* the system with D. Some care is needed in the simulation.

We note also that, when two surfaces are in rubbing (or sliding) contact, a finite force is required to initiate motion and the surfaces are said to *stick*. The magnitude of the sticking force or *stiction* is larger than the Coulomb friction force (sometimes several times larger). Its value depends on the specific materials in contact, their surface finish, the normal force, etc., and usually can only be determined by testing. Thus, the force used to accelerate the mass must exceed the stiction force to initiate motion. Once motion starts, the stiction force decreases (as a function of *velocity*) and normally approaches the Coulomb level at relatively low velocities. Again, the functional relationship of drag to velocity (over this low-velocity range) can only be determined by test.

For a second-order servo with stiction, Coulomb friction, and viscous friction the equations are:

$$J\ddot{C} + f\dot{C} + S \text{ sign } \dot{C} + KC = KR, \qquad (11.24)$$

where

C = output position
R = input (commanded) position

and it is assumed that:

1. The nonlinear sliding friction occurs between the output member C and some stationary member.

2. The symbol S represents the sticking level, which changes to a Coulomb level D at very low velocity. For convenience, we usually consider the transition to be discontinuous and to occur at $\dot{C} = \pm\epsilon$ where ε is *almost* zero.

In formulating the phase plane equations, it is important to note that the Coulomb force depends on $\dot{C}$, especially if we wish to use error coordinates, and the input R is a ramp of position. The slope of the phase trajectory is

$$\frac{d\dot{C}}{dC} = \frac{KR - KC - f\dot{C} - S \text{ sign } \dot{C}}{J\dot{C}}. \qquad (11.25)$$

In error coordinates, this becomes:

$$\frac{d\dot{E}}{dE} = \frac{KE - f(\dot{R} - \dot{E}) - S \text{ sign } (\dot{R} - \dot{E})}{J(\dot{R} - \dot{E})}. \qquad (11.26)$$

If R is a constant or a step, then $\dot{R} = 0$ and

$$\frac{d\dot{E}}{dE} = \frac{KE + f\dot{E} + S \text{ sign } \dot{E}}{-J\dot{E}}. \qquad (11.27)$$

The singular points are found by equating numerator and denominator to zero:

$$\frac{K}{J} E + \frac{f}{J} \dot{E} + \frac{S}{J} \text{ sign } \dot{E} = 0$$

$$\dot{E} = 0. \tag{11.28}$$

Substituting $\dot{E} = 0$ into the numerator equation,

$$KE + S \text{ sign } \dot{E} = 0, \tag{11.29}$$

from which

$$E = \pm \frac{S}{K}. \tag{11.30}$$

Thus, there are *two* singular points. Their location can be interpreted physically—they are at a value of E such that $|KE| = |S|$, i.e., the drive torque from the motor is exactly equal to the stiction torque. We note that both singular points are on the E axis ($\dot{E} \equiv 0$) and that one of them is related to the upper half plane ($\dot{E} = +$), the other to the lower half plane ($\dot{E} = -$). Thus, the horizontal axis is a dividing line. The phase portrait is shown in Figure 11.14. This portrait actually shows *four* singular points, $\pm S/K$ and $\pm D/K$. Initially ($t < 0$), the state point may be anywhere, but if the system is at rest it is

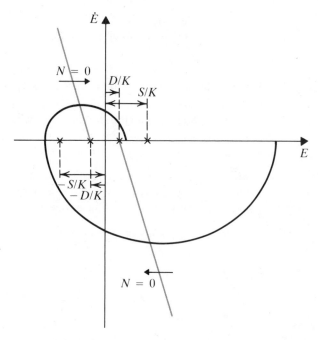

FIGURE 11.14   Phase Plane for a Servo with Coulomb Friction and Stiction, Showing Singular Points.

between the two stiction singular points. If a step input is applied, the state point jumps to the value $E = R - C$ and $\dot{E} = 0$. If this value of E is outside the sticking range, motion is initiated, and the first small increment of trajectory behaves as if the singular point is at S/K. Once motion starts ($\dot{E}$ very small), the sticking friction disappears, and the nonlinear drag drops to the Coulomb level D. On the phase plane, the singular points move to $\pm D/K$ to define the trajectory in the plane. We must note, however, that whenever the state point reaches the horizontal axis, $\dot{E} = 0$. Since, in this case, $\dot{C}$ also becomes zero, the system *sticks* momentarily, the singular points return to S/K and, if the state point is between the two S/K singular points, the drive for that value of E will not be sufficient to break free from the stiction so the motion *stops*.

If the command R is a ramp of position, then the dividing line moves and all of the singular points move on the E versus $\dot{E}$ phase plane. Note that the basic differential equation is unchanged, but the forcing function is now $R = \omega t$. Therefore

$$E = \omega t - C,$$
$$\dot{E} = \omega - \dot{C},$$
$$\ddot{E} = -\ddot{C}. \tag{11.31}$$

Substituting for C, $\dot{C}$, and $\ddot{C}$,

$$-J\ddot{E} + f(\omega - \dot{E}) + K(\omega t - E) + S \text{ sign } (\omega - \dot{E}) = K(\omega t), \tag{11.32}$$

from which

$$J\ddot{E} + f\dot{E} + KE - S \text{ sign } (\omega - \dot{E}) = f\omega, \tag{11.33}$$

$$\frac{d\dot{E}}{dE} = \frac{1}{J} \frac{f\omega + S \text{ sign } (\omega - \dot{E}) - KE - f\dot{E}}{\dot{E}}. \tag{11.34}$$

The singular points are obtained by setting numerator and denominator equal to zero:

$$\dot{E} = 0,$$
$$f\omega + S \text{ sign } (\omega - \dot{E}) - KE - f\dot{E} = 0, \tag{11.35}$$

and from the second equation,

$$E = \frac{f\omega}{K} + \frac{S}{K} \text{ sign } (\omega - \dot{E}). \tag{11.36}$$

Clearly, the singular points for the system with ramp input are at different locations from the singular points with step input. They are still on the E axis and since the term sign $(\omega - \dot{E})$ just determines whether a plus or minus sign applies, the location of these singular points is at $\dot{E} = 0$:

$$E = \frac{f\omega}{K} \pm \frac{S}{K} \quad \text{(for the sticking case)}$$

$$= \frac{f\omega}{K} \pm \frac{D}{K} \quad \text{(for the Coulomb case).} \tag{11.37}$$

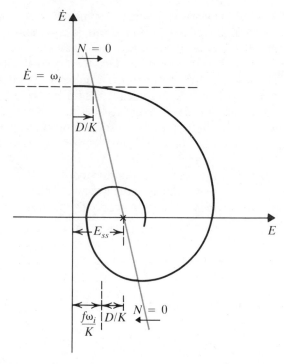

FIGURE 11.15   Phase Plane for a Servo with Coulomb Friction and Stiction: Ramp Input.

However, the term sign $(\omega - \dot{E})$ warns us that the *dividing line* has also moved, i.e., the Coulomb friction reverses direction when $\dot{C}$ reverses, but we are on the E versus $\dot{E}$ plane, and $\dot{C} = \omega - \dot{E}$; therefore, $\dot{C}$ goes to zero when $\dot{E} = \omega$, which defines the dividing line as a horizontal straight line at $\dot{E} = \omega$. The phase plane is shown in Figure 11.15.

The meaning of the dividing line is unchanged. It divides the phase plane into regions where different differential equations apply. In this case, the half plane *above* the dividing line refers to the equation which provides the left-hand singular point. This singular point is then a *virtual* singular point. Its location is defined mathematically but the state point can never approach it because the singular point exists only when the state point is above the dividing line.

In most cases, the effect of the friction is to provide some damping in the transient period and increased error in steady state. However, if the static friction S is much larger than the Coulomb friction D, then a relaxation limit cycle can be excited if ω is a small number. This is demonstrated in Figure 11.16. Such limit cycles were observed in the 16-in. guns on battleships when operated by the fire control servos and tracking a slowly moving distant target. The gun would rotate through a small angle—stick— and remain stuck until target motion generated enough error to pull it free from the stiction. Then it would overshoot—stick again—and repeat!!

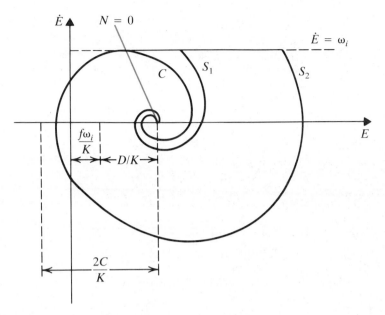

FIGURE 11.16 Limit Cycle for the Servo of Figure 11.15.

The mechanism of the limit cycle can be explained in terms of Figure 11.16. Assume that the system is initially at rest and ramp command is given, with rate ω. The surfaces are held by the stiction, so the state point starts at $\dot{E} = \omega$; then $E = 0$ and E increases along the dividing line (because the output does not move) until enough torque (KE) is developed to overcome the stiction. This occurs at $E = S/K$. The state point then leaves the dividing line and follows a trajectory about the singular point at $E = (f\omega + D)/K$ (the sticking friction reduces to Coulomb friction once motion starts). When $S \gg D$, as shown in Figure 11.16, the state point is far from the singular point and overshoots in both position and velocity. If the positive velocity overshoot is large so that $\dot{E}_{max} \geq \omega$, then the output velocity $\dot{C}$ goes to zero (when $\dot{E} = \omega$) and the system *sticks*. This occurs at the dividing line, and the state point must again follow the dividing line until $E = S/K$, at which point the drive torque breaks the system free from the stiction. The trajectory is seen to be a closed loop, designating a limit cycle. The time response is given with the simulation results of Figure 11.17.

Limit cycles often appear when a multivalued nonlinearity is in the system. Hysteresis and backlash are typical, common, two-valued nonlinearities. If such a nonlinearity is present in a second-order control system, the dynamic behavior can be analyzed on the phase plane, though prediction of the limit cycle may require a great deal of labor. We will discuss the procedures here without completing the details.

Let us consider the case of backlash (free play) in a gear train or in a mechanical linkage. In such cases, we do not have a rigid connection between the inertia (mass) of the driving member and that of the load. The behavior of the closed-loop system depends not so much on the amount of backlash as on the characteristics of the drive

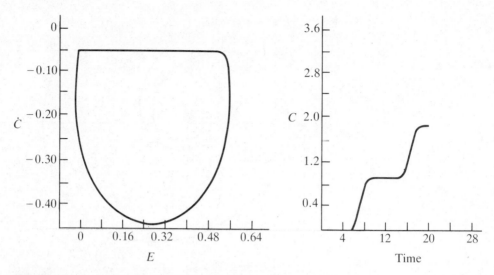

FIGURE 11.17 Simulation Results for the Servo of Figure 11.15.

member and of the load, and on the location of the gear train in the system. The drive member usually has appreciable inertia and some friction. The load, however, may have much inertia or little inertia compared with that of the driver, and in like manner the load friction may be large or small. When the backlash has been "taken-up" so that the gears are all in contact and there is a solid connection from driver to load, then the system is effectively *linear*. However, when driver and load lose this solid connection, each obeys its own differential equation! In particular, if the velocity differential is significant and the displacement of the load exceeds that of the driver by an amount equal to the backlash, then the gear teeth collide, and at impact we must consider both conservation of energy and conservation of momentum.

EXAMPLE 11.3

A simple case of backlash that is encountered in some instrument servos is illustrated in Figure 11.18. Because the load inertia is negligible but there is enough Coulomb friction to prevent the indicator from drifting, two differential equations are enough to describe the system:

$$\ddot{C} + 2\zeta\omega_n\dot{C} + KC = KR, \tag{11.38}$$

which is the equation of the linear system, and C is the displacement of the output shaft. Also,

$$\ddot{\theta}_m + 2\zeta\omega_n\dot{\theta}_m = \omega_n^2 E_1, \tag{11.39}$$

where θ_m is the displacement of the motor shaft (The gear ratio is taken as unity in this

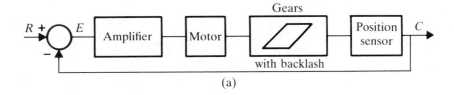

(a)

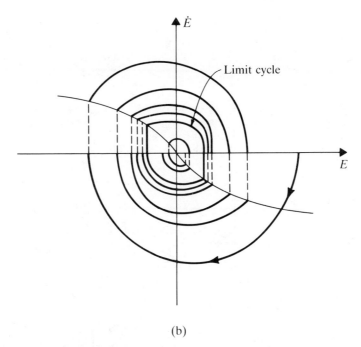

(b)

FIGURE 11.18 (a) Block Diagram for an Instrument Servo with Backlash and (b) Phase Portrait Showing Limit Cycle.

case for simplicity) and E_1 is the value of the error at which the output stops when the motor reverses, and, of course:

$$E = R - C.$$

Figure 11.18(b) shows the phase plane with dividing lines and the limit cycle trajectory. The E axis is the dividing line at which the drive reverses to take up the backlash. The curved dividing line is the locus of points at which the backlash is "taken-up." It is calculated by integrating the equation

$$\ddot{\theta}_m + 2\zeta\omega_n\dot{\theta}_m = \omega_n^2 E_1 \tag{11.40}$$

(we must know Δ = amount of backlash expressed in terms of θ_m). Figure 11.19 shows

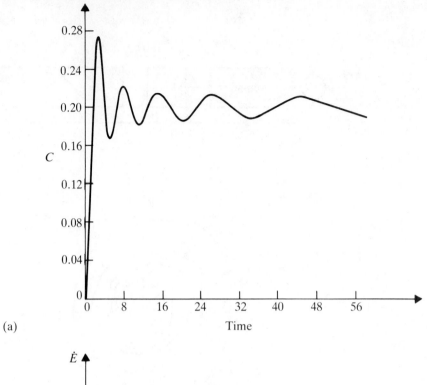

(a)

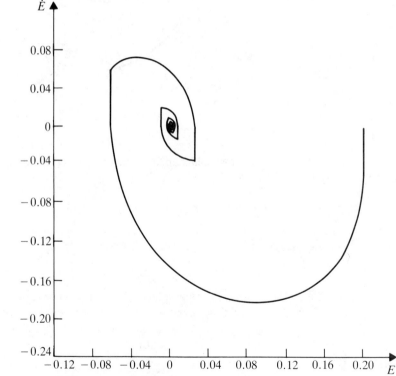

(b)

FIGURE 11.19 (a) Step Response for a Servo with Backlash:

$$G = \frac{2}{s(s + 1)}, \qquad \zeta = 0.327, \qquad \Delta = 0.01,$$

(b) Phase Trajectory $\zeta = 0.327$, (c) Step Response $\zeta = 0.25$, (d) Phase Trajectory $\zeta = 0.25$, (e) Step Response $\zeta \cong 0.17$, and (f) Phase Trajectory $\zeta = 0.17$.

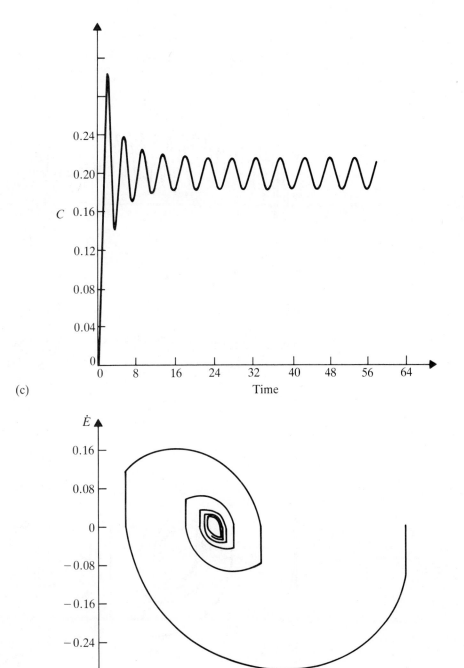

(c)

(d)

FIGURE 11.19 (*Continued*)

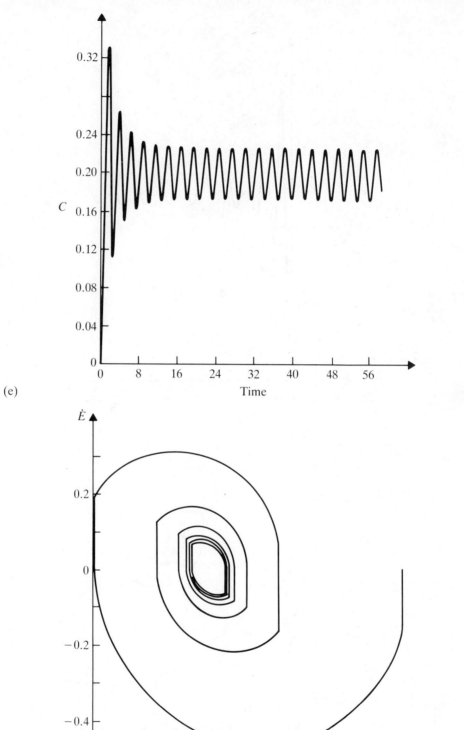

(e)

(f)

FIGURE 11.19 (*Continued*)

simulation results for various amounts of backlash and various values of damping for the linear system. It can be shown that the limit cycle will not exist for this system unless $\zeta < .029$.

For the case of example 11.3, we did not need to consider conservation of energy or momentum. If the second-order system shown in Figure 11.18(a) has a load inertia J_L and a load friction f_L, then three equations are needed to describe the motion:

$$J\ddot{C} + f\dot{C} + KC = KR \tag{11.41}$$

for the linear mode (gears in contact) and

$$J_L\ddot{C} + f_L\dot{C} = 0, \tag{11.42}$$

$$J_m\ddot{\theta} + f_m\dot{\theta}_m = K(R - C), \tag{11.43}$$

which describe the motion of the load and motor when in the backlash region. These equations can be used to define dividing lines and trajectories on the phase plane. Analysis of this system is beyond the scope of this text.

11.5 INTRODUCTION TO DISCONTINUOUS SYSTEMS

Any system that includes a switching operation in its normal behavior may be called a *discontinuous system*. In a general way, we may classify such switching as

1. power switching
2. mode switching.

In power switching systems, the primary function of the switch (or relay) is to apply, remove, or reverse the power source. The switching devices can be mechanical, such as cam-operated contacts, or friction clutches. They can be electromagnetic relays, or transistors or silicon-controlled rectifiers, or they can be hydraulic or pneumatic valves, etc. The basic characteristic is that they are low-powered devices (usually relatively inexpensive) that control the flow of large amounts of power. Control systems designed for such operation are usually *low-performance systems*, and the power switch is used in place of a power amplifier, which would provide much better control but at considerable increase in cost.

Power switching controls are common in the home—thermostatic control of heating and/or air conditioning, refrigerators, and freezer. They have been used in guided missiles, target drones, satellites, and similar applications. Early autopilots for aircraft used clutches, and perhaps the earliest relay servos were the positioning servos designed by Hazen (see bibliography) for Vannevar Bush's differential analyzer.

Figure 11.20 shows a simple electric motor servo using a relay (with dead zone) as a power amplifier. The behavior is shown on the phase plane. Figure 11.21 shows the

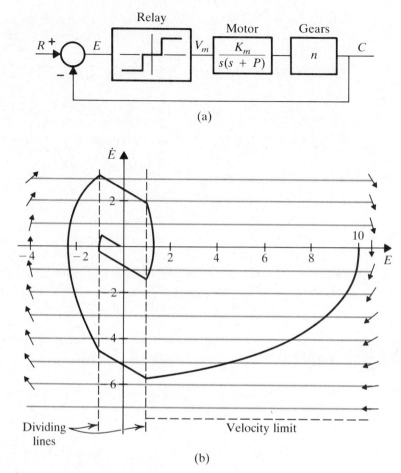

FIGURE 11.20   Servo with Three-Position Relay: (a) Block Diagram and (b) Phase Trajectory.

effect of adding velocity feedback. Figure 11.22 shows the same motor but with relay having both dead zone and hysteresis. The resulting limit cycle is shown on the phase plane. The phase trajectory of a servo with a two-position relay was shown on Figure 11.8(a) and (b). In general, the performance specifications of such systems are essentially the same as for a continuous servo with similar purpose. Thus, steady-state accuracy and settling time are important, and some overshoot and oscillation may be permissible. At the same time, these specifications are generous enough so that they may be satisfied with the power switching scheme.

Mode switching, on the other hand, cannot be as clearly defined. In a broad sense, it is any switching that changes the equations governing the operation of the system—yet even this definition does not necessarily include all of the types of

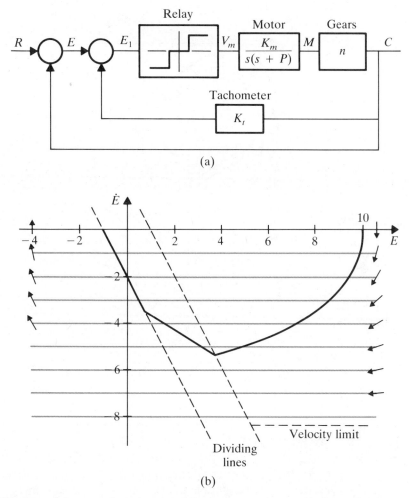

FIGURE 11.21 Servo with Three-Position Relay and Velocity Feedback: (a) Block Diagram and (b) Phase Trajectory.

switching that can be used in control systems. For example, it does not include input switching, since this changes only the forcing function but not the equations of motion. Most of the mode switching operations are designed to accomplish one or more of the following objectives:

1. fast response (often minimum time) with no overshoot

2. fast initial response (rise time) with accurate, well-damped steady-state operation

3. adjustment of system parameters to match system characteristics to operating conditions.

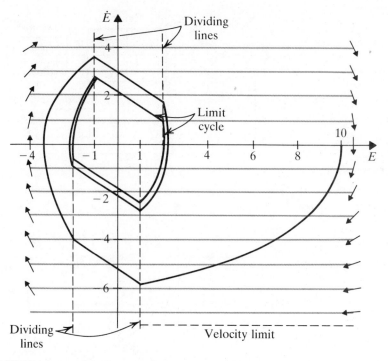

FIGURE 11.22 Servo with Three-Position Relay and Hysteresis.

BANG-BANG (MINIMUM TIME) CONTROL

For a *linear* control system, the response to a step input takes the same amount of time regardless of the magnitude of the step. This settling time can be reduced only by redesigning the system or by discontinuous operation (such as posicast control). For a system with signal limitations (saturating amplifiers) or power or torque limits, the time required to carry out a step command increases with the size of the step. In many applications of such systems, the response time must be made as short as possible. The question then arises: How shall such systems be operated in order to obtain minimum time response? Early analysis showed that the ideal operating procedure is to use maximum (saturated) drive at all times. This result has been confirmed and extended by optimal control theory, using the calculus of variations. In this text, we study the problem using physical reasoning, since this promotes understanding of the conditions that exist in hardware. We note that the equations of optimal control are needed if a high-order plant is to be controlled in optimal fashion using an on-line computer.

In its simplest form, the concept of minimum time response was derived from purely physical reasoning, and in early applications the concept was used with varying success. The basic concept is this:

> Given a system for which the drive is limited (has a maximum or saturation value), fastest response is obtained if maximum forward drive is applied at $t = 0$, and is reversed at a proper instant $t = t_R$ so that deceleration under maximum reverse drive reduces the velocity to zero at precisely the commanded value of the output. The drive is then set to zero.

Obviously, the concept was developed primarily for mechanical positioning systems subject to initial conditions or a step command. It works reasonably well if the system is of second order and if the range of commands is small. Difficulties are encountered due to parameter variations resulting from wear, temperature, humidity, and unknown causes! In addition, the terminal status (drive set to zero) is often unsatisfactory and a linear control mode is added.

Consider the idealized system of Figure 11.23. This represents a pure mass (or inertia) with pure force (torque) drive. The ideal relay permits only two conditions:

1. full accelerate
2. full decelerate.

We must choose a switching law to operate the relay, and for this particular system the law may be a function of error only. If we choose the law:

1. Full accelerate upon receiving the step command.
2. Switch to full decelerate when the error is reduced to one half of the initial value.
3. Disconnect drive when error reaches zero.

The system will reach exactly the commanded position with zero velocity and zero stored energy, and so will not have further motion after the drive is removed. Note that during full accelerate and/or decelerate,

$$J\ddot{C} = K \text{ sign } E,$$

converting to E coordinates, separating the variables, and integrating,

$$\dot{E}^2 + \frac{2KE}{J} + D1 = 0,$$

$$\dot{E}^2 - \frac{2KE}{J} + D2 = 0,$$

where $D1$ and $D2$ are constants of integration. These equations describe parabolas on the E versus $\dot{E}$ plane as shown in Figure 11.23(b). Also shown is a dotted parabola with vertex displaced from B to B' (at the origin). With the aid of this dotted trajectory, we determine the requirement for minimum time control: if the state point starts at D, full forward acceleration moves the state along the parabolic trajectory toward point X. If

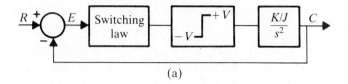

(a)

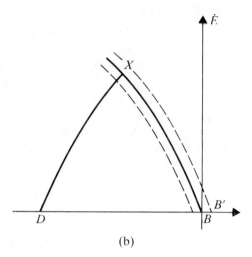

(b)

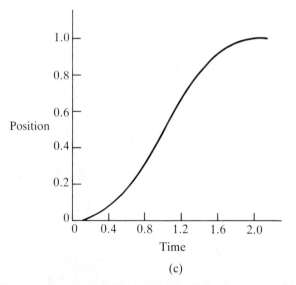

(c)

FIGURE 11.23 (a) Block Diagram for Idealized Second-Order Relay Servo, (b) Time Optimal Phase Trajectory, (c) Simulation: Position Versus Time, (d) Simulation: Velocity Versus Time, and (e) Simulation: Position Versus Velocity.

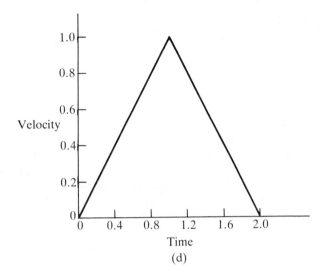

(d)

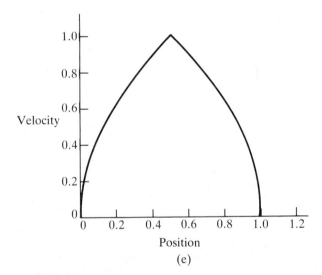

(e)

FIGURE 11.23   *(Continued)*

the relay is reversed at X, the state point transfers to the deceleration curve and follows it to the origin where the drive must be disconnected. There has been one period of full acceleration, followed by one period of full deceleration, and the state point has been driven to the desired final position. It is intuitively clear that no other sequence of events can drive the state to the origin in shorter time. If the relay is reversed at either of the dotted trajectories, the state point does not reach the origin.

For the conditions of Figure 11.23(b), the transit time is readily calculated:

$$J\ddot{E} = K(\text{at } t = 0).$$

Integrating,

$$\dot{E} = \frac{K}{J}t + k_1,$$

but at

$$t = 0 \qquad \dot{E} = 0, \qquad k_1 = 0, \quad \text{then} \quad \dot{E} = \frac{K}{J}t.$$

Integrating again,

$$E = \frac{Kt^2}{2J} + k_2,$$

but at

$$t = 0 \quad \text{and} \quad E = -E_1, \quad \text{then} \quad E = \frac{K}{2J}t^2 - E_1.$$

Solving for t,

$$t = \left[\frac{2J(E + E_1)}{K}\right]^{1/2}.$$

If we reverse the relay when $E = -E_1/2$, and call the time t_1,

$$t_1 = \left[\frac{2J}{K}\left(-\frac{E_1}{2} + E_1\right)\right]^{1/2} = \sqrt{\frac{JE_1}{K}}$$

and the total response time is

$$t_f = 2\sqrt{\frac{JE_1}{K}}.$$

For the case of Figure 11.23, the computation of the response time is simple because the system is assumed to have no damping. Before proceeding to more complex cases, let us note that for the frictionless system, switching is done at the halfway point—halfway in distance as well as halfway in time—but this is a particular solution, and in general one does not switch at the halfway point. The basic switching rule can be defined from Figure 11.23(b) by observing that, although we can reverse the relay whenever we wish, unless it is reversed at point X, the deceleration trajectory followed does not reach the origin. The general switching law for second-order systems may then be stated as follows: the state point follows the saturated acceleration trajectory until it reaches the one saturated deceleration trajectory that passes through the desired final point (origin). The relay must be reversed at this intersection to transfer the state point to the proper deceleration trajectory.

For the undamped system of Figure 11.23(a), simulation results are given in Figure 11.23(c), and the system is seen to be *dead beat* i.e., no overshoot, Figure 11.23(d) gives the velocity versus time trace. Note that this consists of two straight lines. The constant slope of the lines clearly indicates constant acceleration followed by constant deceleration and, in this case, the symmetry shows that acceleration and deceleration are equal. This pattern is a convenient one to use in adjusting prototypes and is popular with both technicians and engineers. Figure 11.23(e) shows the phase plane trajectory.

For linear systems of third and higher order, time optimal operation is theoretically possible but is usually considered impractical. Theoretical treatment is developed in books on optimal control theory and is outside the scope and intent of this text.

11.7 SWITCHED DAMPING

The term *switched damping* can (in a sense) be applied to both saturated (bang-bang) operation and to linear operation. We will discuss both.

Most of the mathematical theory of feedback controls assumes that the basic parameter values are fixed, i.e., the inertia, friction, gains, and time constants do not change during operation and voltage, current, and power limits are fixed, etc. Such assumptions are not mandatory, they are only convenient because the equations become quite nonlinear and virtually unsolvable (except by simulation) if the parameters change significantly. Of course, parameters can be changed, and if the engineer thinks there is a net advantage in some particular scheme that involves parameter changes he will usually try it.

When fast response to a step command is desired, there are a variety of ways to increase response speed many of which have been tried in practice—successfully— but discarded as the technology developed, either because they could not be used in applications demanding great accuracy or because better performing or less costly alternatives were developed. Some of them are still useful; others have not been fully developed and hold promise for the future.

In a general sense, if we use state feedback loops and switch them into the system or, alternatively, keep them in the system at all times but change the corresponding gains in a discontinuous manner, we introduce damping into the system in the sense that we relocate the roots of the characteristic equation. By proper choice of the gains, we can adjust the roots to provide better damping, but care must be taken to provide a suitable switching law to make the transient characteristics acceptable.

To introduce the details of discontinuous system design, consider the second-order system of Figure 11.24(a). Note that the amplifier saturates; we will consider both saturated and unsaturated operation. First we assume that the switch is open so there is no velocity feedback and the closed position loop has characteristic roots on the imaginary axis at $\omega = \pm 1$. For a large initial condition or a large step input in this particular system, the system overshoot is 100% and a fixed amplitude oscillation results.

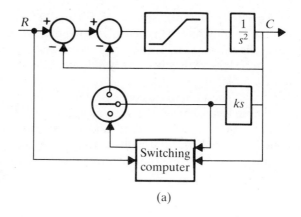

(a)

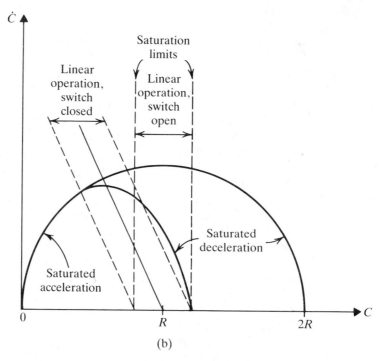

(b)

FIGURE 11.24 One Type of Discontinuous Damping: (a) Second-Order Discontinuous System and (b) Trajectory Change Due to Switched Damping.

If we close the switch (permanently) and adjust k, we introduce anticipation and substantially reduce the overshoot as also shown in Figure 11.24. When this is done, there is also damping due to the velocity signal, but not necessarily a great deal.

Note that we can operate the switch independently of the amount of velocity signal fed back. The switching computer is fed all available states, plus the command signal R. We can make the switch operate according to a linear law such as

$$E + \alpha\dot{E} = 0,$$

while the velocity feedback signal provides the law

$$E - k\dot{C} = E_1$$

and the numerical values of α and k may be considerably different because of the design objectives.

11.8 A SECOND-ORDER LINEAR SYSTEM WITH DISCONTINUOUS VELOCITY FEEDBACK

A second-order system is defined in Figure 11.25 and it is assumed that the feedback switch can be closed according to any desired law, but once closed it remains closed until reset. The differential equation of the system without the feedback is

$$\ddot{E} + 2.6\dot{E} + 64E = 0,$$

for which

$$\zeta = 0.162 \quad \text{and} \quad \omega_n = 8.$$

The choice of a value for the feedback gain H determines the ζ for the compensated system, and we may choose to make the system underdamped, critically damped, or overdamped depending on our objectives. Common specifications that lead to the switched damping design might be the requirement for very fast response with no overshoot. Such specifications eliminate an underdamped condition. Let us choose overdamping (in the limit this becomes critical damping). Figures 11.25(b), (c), and (d) show eigenvectors and phase trajectories for the discontinuous system with different amounts of overdamping. For Figure 11.25(b), the gain $H = 0.3$, which gives

$$\ddot{E} + 21.8\dot{E} + 64E = 0$$

and the roots are real at -3.5 and -18.3. For Figure 11.25(d), $H = 0.96$ and $\ddot{E} + 64\dot{E} + 64E = 0$ with roots at -1.016 and -62.984.

For a linear system, each real root provides an eigenvector that is a straight line. From Figure 11.25, it is seen that, in terms of isoclines, the eigenvectors are those isoclines for which the slope of the isocline and the slope of the trajectory (slope marker) are identical. To illustrate for this system [Figure 11.25(d)]:

$$\ddot{E} = -64E - 64\dot{E}$$

$$\frac{\ddot{E}}{\dot{E}} = \frac{d\dot{E}}{dE} = N = \frac{-64E - 64\ddot{E}}{\dot{E}},$$

where N is the slope of the phase trajectory. However,

$$N\dot{E} + 64\dot{E} = -64E$$

$$\dot{E} = \frac{-64}{N + 64}E,$$

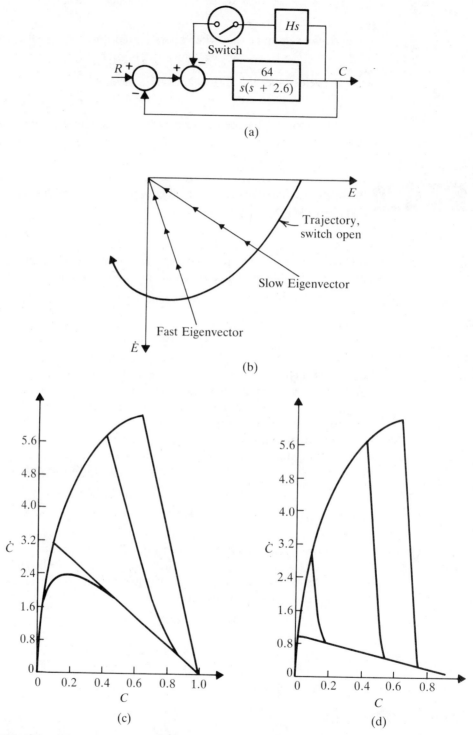

FIGURE 11.25 Linear Second-Order System with Switched Damping: (a) Block Diagram, (b) Phase Plane Showing Location of Eigenvectors when Switch is Closed, (c) Phase Portrait, $H = 0.3$, and (d) Phase Portrait, $H = 0.96$.

which is the equation of the isocline and the slope of the isocline is $-64/(N + 64)$. Then the eigenvector is the line for which

$$N = \frac{-64}{N + 64}$$

or

$$N^2 + 64N + 64 = 0$$

and we see that the values of N are precisely the real roots of the characteristic equation.

The portrait also permits us to interpret some salient features of system response, and helps us choose a suitable value for H and an effective switching law—by inspection we see that when the velocity feedback path is closed so that we have eigenvectors in the portrait, it is not possible for a trajectory to cross an eigenvector—the trajectory must approach the singular point asymptotic to the eigenvector. We also see that if we can provide an initial condition which is *on* the eigenvector, then the state point will go to the singular point along the eigenvector—it cannot leave it. We observe that if we leave the velocity feedback path permanently closed and apply a step command, then the phase trajectory starts on the E axis and becomes asymptotic to the closest eigenvector. The response to the step does not over-shoot, but is very slow (high velocity is never achieved). The second eigenvector cannot be reached in this fashion. However, if we open the switch initially and apply the step command, then the system accelerates and crosses the *location* of the first eigenvector since neither eigenvector exists with the switch open. Thus, the state point can be driven to the second eigenvector and the switch closed when the state point reaches it. Deceleration along this eigenvector reaches the singular point. Note that use of the second eigenvector provides a no-overshoot response, which is much faster than use of the first eigenvector. We therefore designate as the *slow* eigenvector that one which is defined by the smaller real root, and the one associated with the larger real root is the *fast* eigenvector. Figures 11.25(b), (c), and (d) also show that the precision of the switching law is not critical; if the switch is closed a little early or a little late, the responses are essentially unchanged. Note that the equation of the switching law is

$$\dot{E} = \frac{64}{N + 64} E = 0 = \dot{E} + 62.99E = 0$$

for $N = -62.99$. The signal applied to the plant is

$$E_p = E - 0.96\dot{C},$$

but for a step input $C = -E$; therefore,

$$E_p = E + 0.96\dot{E}$$

or

$$\dot{E} + 1.042E = 1.042E_p.$$

11.9   SOME EXTENSION OF BANG-BANG CONTROL

Many modern applications of control require very fast response to commands and, as a result, amplifiers are driven into saturation for all except the smallest excursions. Bang-bang control concepts thus become very attractive, not only because we are interested in minimum response time, but because the system is being driven into saturation anyway. The basic laws of bang-bang control must be modified to be practical. A commonly used modification is called *curve following*, and usually involves three modes of operation in following a step position command:

1. an initial full accelerate mode
2. a curve following mode
3. a terminal linear servo mode.

Several very successful applications of this modification are its use to position the read/write heads in high-performance magnetic disk memories and its use in automatic typewriters to position the carriage and daisy wheel. The head servo has all three of the indicated modes, but commonly used terminology refers to a *track-seek* mode and a *track-follow* mode. Figure 11.26 shows the basic structure of the seek mode. The internal loop is called a *velocity loop*; it is designed with very high gain, the concept being that its input is a commanded velocity and the loop must force the output velocity to duplicate the command with very little error. The purpose of the curve, then, is to generate the velocity command, which is made a function of the error (or "distance to go"). One chooses a curve that approximates the natural deceleration curve[a] of the motor to be used. In general, the chosen curve must be below the natural curve or the system will not curve follow. For the servos used in disk memories, the ordinate of

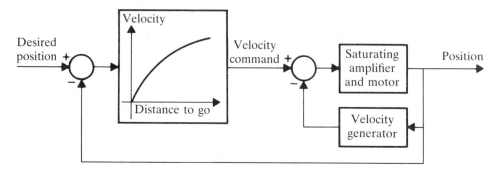

FIGURE 11.26   A Curve Following Minimal Time Servo Loop.

---

[a]   The deceleration curve for a dc armature-controlled motor with constant applied voltage is a parabola on the position versus velocity plane. When the motor transfer function is K/s^2 the equation of the parabola may be written as $\dot{E} = K_1 \sqrt{E}$, where E(the error) is the "distance to go," and K_1 depends on the motor-load parameters.

the natural curve is reduced about 20% to arrive at a practical solution that includes consideration of motor tolerances.

To relate the curve choice to phase plane theory, note that the curve is both a switching line and a dividing line. At each point on the line, two trajectory slopes are defined, one for $+V$ and the other for $-V$. Thus, if the state point is displaced from the line, full drive is applied to drive it back. Move time is not the absolute minimum, of course, because the chosen curve is below (at lower velocity than) the ideal curve. However, the difference in time buys reliability; the positioning system (which includes the linear terminal mode) is able to meet specifications of no more than one seek error in 10^7 seeks.

<h2>11.10   SUMMARY</h2>

The *state* of an N-dimensional system at any instant in time may be represented by a single point in the N-dimensional state space. When the system dimension is $N = 2$, the space becomes a plane, and if the states are defined as a variable and its first derivative (e.g., position and velocity) the plane is called the *phase plane*.

For a second-order system, as its state changes, a curve (trajectory) is traced out. The *slope* of this trajectory at any point is found easily, a line of constant slope (isocline) may be drawn. If the system is stable, the trajectory terminates at a *singular* point, and the behavior of the trajectory near the singular point is defined in terms of the root location for the equivalent linear system. If the system is unstable, the trajectory goes to infinity. For nonlinear systems, the trajectory may go to a limit cycle, which is a closed curve on the phase plane.

When the second-order system has real roots, the phase portrait contains two trajectories defined by the eigenvectors. For a linear system, these are straight lines.

For nonlinear and discontinuous systems, lines can be drawn on the phase plane that:

1.   divide the plane into areas where different equations apply
2.   define the points at which a switching operation may take place.

Switching is often used in practical systems to reverse the power or to introduce (or remove) damping or to change operating modes.

When fast operation is needed, maximum available drive is used at all times, with power reversed by switching at the appropriate point.

A more practical application of this idea is to design the system to reverse power repeatedly so as to follow a curve to the desired terminal point.

<h2>BIBLIOGRAPHY</h2>

Bogner, I.; and Kazda, L. F. "An Investigation of the Switching Criteria for Higher Order Contactor Servomechanisms." *AIEE, Trans.* Part II, 72: 118–27 (1954).

Chang, S. S. L. "Optimum Switching Criteria for Higher Order Contactor Servo with Interrupted Circuits." *AIEE, Trans.* Part II, 73: 273–76 (1955).

Harris, W. L., Jr. "Discontinuous Damping of Relay Servomechanisms." M.S. Thesis, U.S. Naval Postgraduate School, Monterey, Calif., 1956.

Harris, W. L., Jr.; McDonald, C.; and Thaler, G. J. "Quasi-optimization of Relay Servos by Use of Discontinuous Damping." *AIEE, Trans.* Part II, 75: 292–296 (1957).

Hazen, H. L. "Theory of Servomechanisms." J. Franklin Institute. 218: 279–330 (1934).

Hopkins, A. M. "A Phase-plane Approach to the Compensation of Saturating Servomechanisms." *AIEE, Trans.* Vol. 69 (1951).

Kalman, R. E., "Analysis and Design Principles of Second and Higher Order Saturating Servomechanisms." *AIEE, Trans.* Part II, 73: 294–310 (1955).

Lutkenhouse, W. J., "Dividing Lines for Backlash in the Phase Plane." M.S. Thesis, U.S. Naval Postgraduate School, Monterey, Calif., (1959).

McDonald, Carlton A. K., "Quasi-optimization of Relay Servos." M.S. thesis, U.S. Naval Postgraduate School, Monterey, Calif. (1957).

———; and Thaler, G. J. "Quasi-optimization of Relay Servos by Use of Stored Energy for Braking." *AIEE, Trans.* Part II, 77: 628–634 (1959).

McDonald, D. "Multiple Mode Operation of Servomechanisms." Rev. Sci. Instrum. (1952).

———. "Nonlinear Techniques for Improving Servo Performance." Proc. Natl. Electronics Conf., (1950).

Rauch, L. I.; and Howe, R. M. "A Servo with Linear Operation in a Region about the Optimum Discontinuous Switching Curve." Presented at Symposium on Nonlinear Circuit Analysis, Brooklyn Polytechnic Institute, New York, 1956.

PROBLEMS

11.1 Given:

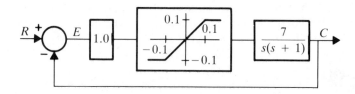

a. From the block diagram, derive the applicable differential equations.
b. On the E versus $\dot{E}$ plane, where are any singular points?

c. What are the equations of the dividing lines?
d. What are the equations of the isoclines?
e. Obtain several trajectories on the E versus $\dot{E}$ plane. Show singular points, dividing lines, and some isoclines.

11.2 Tachometer feedback is added to the system of Problem 11.1. Repeat Problem 11.1 for several values of k_t.

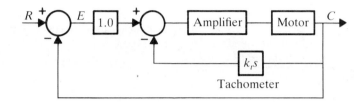

11.3 Repeat Problem 11.1 with dead zone replacing saturation.

11.4 Repeat Problem 11.2 with dead zone replacing saturation.

11.5 Given a servo with Coulomb friction. Repeat Problem 11.1.

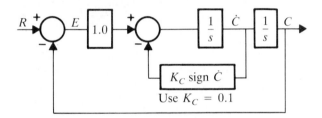

11.6 The system of Problem 11.5 is driven with a ramp input of 0.1 rad/s. How does this affect system behavior? Show on the phase plane.

11.7 In the system of Problem 11.1, the saturating amplifier is replaced by an ideal (two-position) relay. Repeat Problems 11.1 and 11.2.

11.8 An instrument servo with backlash in its gear train may be represented as: Analyze on the phase plane.

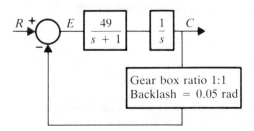

11.9 Given a system with two nonlinearities:

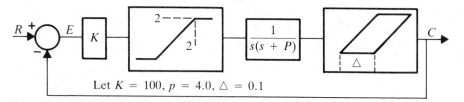

Let $K = 100$, $p = 4.0$, $\triangle = 0.1$

a. Determine singular points, dividing lines, and isoclines on the E versus $\dot{E}$ plane.
b. Construct (or simulate) several trajectories.
c. Discuss results.

11.10 A tachometer is added to the system of Problem 11.9 (note change in plant). Repeat Problem 11.9.

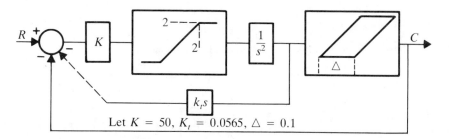

Let $K = 50$, $K_t = 0.0565$, $\triangle = 0.1$

11.11 A positioning servo consists of a torque motor driving an inertial load. The basic system has essentially no friction so a viscous damper is connected to the load with a centrifugal clutch:

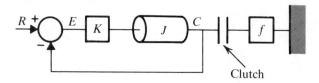

The clutch operates whenever the velocity $\dot{C}$ exceeds 1.0 rad/s. Let $J = 1$, $K = 9$, $f = 0.6$, and $R =$ step of 5 rad.
a. Write the differential equations.
b. Show singular points, dividing (or switching) lines, and isoclines on the E versus $\dot{E}$ plane.
c. Sketch the trajectory.

11.12 A linear undamped servo is defined by:

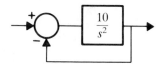

It is at rest and is subjected to a step input of 1 rad. When it reaches maximum velocity, a clutch mechanism connects to this system a load consisting of one unit of inertia (assume that the inertia of the original system was also one unit) and one unit of Coulomb friction($C = 1.0$). What is the maximum overshoot?

11.13 A single-loop, linear system has unity feedback and a forward transfer function $G(s) = 10/s(s + 10)$. The forcing function is:

$$R = 1.0(0 < t < 1)$$
$$R = 0.5(1 < t < 2)$$
$$R = -t(2 < t < 3).$$

What are the error and error rate at $t = 3$.

11.14 Given:

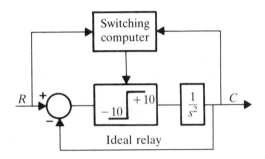

a. A switching law is used reversing the relay when $C = 0.9R$. If $R = 1$ rad, will there be an overshoot? How much?
b. If it is desirable to have no overshoot at all, at what point should the relay be reversed?
c. What is the response time if the relay is switched as in (b).

11.15 Repeat Problem 11.14 if the plant transfer function is

$$G(s) = \frac{10}{s(s + 10)}.$$

11.16 Given:

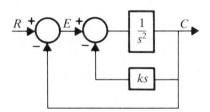

a. Find the value of k required for repeated real roots.
b. Show the resulting eigenvector on the E versus $\dot{E}$ plane.
c. If $k = 3$, show the eigenvectors on the E versus $\dot{E}$ plane.

11.17 If the forward transfer function in Problem 11.16 is $10/s(s + 1)$, repeat (a) and (b). Choose some large value of k and repeat (c).

11.18   The velocity feedback of Problem 11.16 is added with a switch:

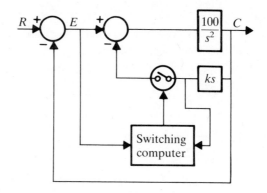

a.   Choose k to give distinct eigenvectors (when switch is closed).
b.   Devise a *linear* switching law such that the trajectory passes *over* the slow eigenvector and switches *on* (or near) the fast eigenvector.
c.   Show eigenvectors, switchline, and trajectories on the plane.

11.19   Given:

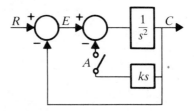

If $R = 0.05t$ and A is initially open:
a.   Compute the response on the E versus $\dot{E}$ plane.
b.   The switch A is closed at $t = 2$ s. Show the effect on the E versus $\dot{E}$ plane.

11.20   In the system of Problem 11.19, the switch is open and the forcing function is

$$R = 0.2t \qquad (0 < t < 1.0)$$

$$R = 0.2 \qquad (t > 1.0).$$

a.   Compute the response the E versus $\dot{E}$ plane.
b.   The switch A is closed at $t = 2.0$ s. Show the effect on the E versus $\dot{E}$ plane.

11.21   A simple linear servo has unity feedback and

$$G(s) = \frac{10}{s(s + 1)}.$$

At some time t_0, the following conditions exist in the system:

$$R = 1.0, \qquad E = 0.5, \qquad \dot{E} = -1.0.$$

At time t_0, the input is changed to $R = 2.0 + (t - t_0)$. Show the path of the state point on the E versus $\dot{E}$ plane and give the coordinates of the state point at $t = t_0 + 1$.

A Matrix Theory

A.1   DEFINITIONS AND BASIC OPERATIONS

A matrix is a collection of elements (numbers) arranged in a rectangular array. When used in mathematics, the matrix is most commonly a collection of the coefficients of linear simultaneous equations. For example the simultaneous equations

$$a_{11}x_1 + a_{12}x_2 = y_1,$$
$$a_{21}x_2 + a_{22}x_2 = y_2 \tag{1}$$

may be rewritten in matrix form

$$\begin{bmatrix} a_{11} & a_{12} \\ a_{21} & a_{22} \end{bmatrix} \begin{bmatrix} x_1 \\ x_2 \end{bmatrix} = \begin{bmatrix} y_1 \\ y_2 \end{bmatrix}, \tag{2}$$

where each of the arrays within brackets is called a *matrix*. Matrices with a single column or a single row are also called *vectors*. The matrix equation (2) may be abbreviated to

$$\mathbf{A}X = Y \tag{3}$$

where the definitions of **A**, X, and Y are obvious.

In general, a matrix is a rectangular array with m rows and n columns. For many physical problems, the coefficient matrix is square, that is $m = n$. When subscript notation is used, as in Eq. (2), the first subscript designates the row, the second subscript designates the column. The basic mathematical operations of addition (subtraction) and multiplication can be carried out with matrices, using suitable precautions. Division of matrices is not defined.

■ **EXAMPLE 1   ADDITION OF MATRICES**

$$\text{Let } \mathbf{A}(ij) = \begin{bmatrix} A_{11} & A_{12} \\ A_{21} & A_{22} \end{bmatrix}$$

$$\mathbf{B}(ij) = \begin{bmatrix} B_{11} & B_{12} \\ B_{21} & B_{22} \end{bmatrix}$$

$$\mathbf{C} = \mathbf{A} + \mathbf{B} = \mathbf{B} + \mathbf{A} = \begin{bmatrix} A_{11} + B_{11} & A_{12} + B_{12} \\ A_{21} + B_{21} & A_{22} + B_{22} \end{bmatrix}$$

In general, $\mathbf{C}(ij) = a_{ij} + b_{ij}$.

■ **EXAMPLE 2   PRODUCT OF MATRICES**

$$\mathbf{D} = \mathbf{AB} \neq \mathbf{BA} = \begin{bmatrix} A_{11}B_{11} + A_{12}B_{21} & A_{11}B_{12} + B_{12}A_{22} \\ A_{21}B_{11} + A_{12}B_{21} & A_{21}B_{12} + A_{22}B_{22} \end{bmatrix}.$$

If the matrices are not square it is necessary that the number of columns of the first must be equal to the number of rows of the second, i.e., if $\mathbf{A}$ is an $i \times j$ matrix and $\mathbf{B}$ is a $k \times l$ matrix, then it is required that $j = k$. The resulting matrix may be designated $\mathbf{D}(i, j)$ where any element is

$$d_{ij} = \sum_{k=1}^{n} a_{ik}b_{kj}.$$

A *square* matrix $\mathbf{A}(m, n)$ has equal numbers of rows and columns.

A *diagonal* matrix is a square matrix for which all elements are zero except those on the main diagonal, i.e., $A_{ij}(i = j) \neq 0$.

The *identity* matrix is a diagonal matrix for which all of the nonzero elements have a value of 1.0, i.e., $A_{ij}(i = j) = 1.0$.

The *transpose* of a matrix is obtained by interchanging the rows and columns.

■ **EXAMPLE 3**

If

$$\mathbf{A} = \begin{bmatrix} a_{11} & a_{12} \\ a_{21} & a_{22} \end{bmatrix} \quad \text{then} \quad \mathbf{A}^T = \begin{bmatrix} a_{11} & a_{21} \\ a_{12} & a_{22} \end{bmatrix}.$$

The transpose of the product of two matrices is the product of their transposes in reverse order: $[\mathbf{AB}]^T = \mathbf{B}^T\mathbf{A}^T$.

The *determinant* of a square matrix is a scalar number evaluated by expanding the matrix by *minors*. Common ways of designating a determinant are

$$\Delta = \det \mathbf{A} = \begin{vmatrix} a_{11} & a_{12} & a_{13} \\ a_{21} & a_{22} & a_{23} \\ a_{31} & a_{32} & a_{33} \end{vmatrix}. \tag{4}$$

Note the use of vertical bars rather than brackets. The *minor* of an element a_{ij} in an $n \times n$ determinant is a determinant of dimensions $(n-1) \times (n-1)$ defined by deleting the row and column containing element a_{ij}.

EXAMPLE 4

For the determinant of Eq. (4), the minor of the element a_{12} is:

$$\begin{vmatrix} a_{11} & a_{12} & a_{13} \\ a_{21} & a_{22} & a_{23} \\ a_{31} & a_{32} & a_{33} \end{vmatrix} \Rightarrow \begin{vmatrix} a_{21} & a_{23} \\ a_{31} & a_{33} \end{vmatrix}.$$

The *cofactor* of the element a_{ij} is the minor of a_{ij} with proper sign introduced. The proper sign is determined by the location of the element in the matrix, and is $(-1)^{i+j}$. Thus,

$$\text{cofactor of } a_{12} = (-1)^{1+2} \begin{vmatrix} a_{21} & a_{23} \\ a_{31} & a_{33} \end{vmatrix}.$$

To expand the determinant and obtain its scalar value, one finds the cofactor of each element in a chosen row or column, multiplies each element by its respective cofactor, and sums all of these products.

EXAMPLE 5

For the determinant of Eq. (4), expanding along the first row,

$$\det \mathbf{A} = (-1)^{1+1} a_{11} \begin{vmatrix} a_{22} & a_{23} \\ a_{32} & a_{33} \end{vmatrix} (-1)^{1+2} a_{12} \begin{vmatrix} a_{21} & a_{23} \\ a_{31} & a_{33} \end{vmatrix} (-1)^{1+3} a_{13} \begin{vmatrix} a_{21} & a_{22} \\ a_{31} & a_{32} \end{vmatrix}$$

The value of a 2×2 determinant is found by taking the product of the main diagonal terms and subtracting the product of the crossdiagonal terms;

$$\begin{vmatrix} a_{22} & a_{23} \\ a_{32} & a_{33} \end{vmatrix} = a_{22}a_{33} - a_{23}a_{32}$$

Note that if the order of the cofactor is greater than 2×2, it must be expanded to permit evaluation.

A *singular* matrix is one for which the value of the determinant is zero.

The *adjoint* of a square matrix is found by replacing each element of the matrix by its cofactor, then taking the transpose of this cofactor matrix. If we define cofactor $a_{ij} \triangleq \alpha_{ij}$, then

$$\text{adjoint } \mathbf{A} = \begin{bmatrix} \alpha_{11} & \alpha_{12} & \alpha_{13} \\ \alpha_{21} & \alpha_{22} & \alpha_{23} \\ \alpha_{31} & \alpha_{32} & \alpha_{33} \end{bmatrix}^T = \begin{bmatrix} \alpha_{11} & \alpha_{21} & \alpha_{31} \\ \alpha_{12} & \alpha_{23} & \alpha_{32} \\ \alpha_{13} & \alpha_{23} & \alpha_{33} \end{bmatrix}.$$

Division is not defined for matrix algebra. The matrix operation corresponding to division is defined in terms of the *inverse* matrix. The *inverse* of matrix $\mathbf{A}$ is designated $\mathbf{A}^{-1}$ and is defined such that

$$\mathbf{A}\mathbf{A}^{-1} = 1 = \mathbf{A}^{-1}\mathbf{A}.$$

To find the inverse matrix, one applies Cramer's rule.

$$\mathbf{A}^{-1} = \frac{\text{adjoint } \mathbf{A}}{\det \mathbf{A}}.$$

The *trace* of a square matrix is the sum of its diagonal elements.

The *rank* of a matrix is the number of its rows (or columns) that are linearly independent.

The eigenvectors e and eigenvalues λ of a matrix $\mathbf{A}$ must satisfy the vector matrix equation

$$\mathbf{A}e = \lambda e$$

or, in alternate form,

$$(\mathbf{A} - \lambda \mathbf{I})e = 0. \tag{5}$$

In order that a nontrivial solution exist, the matrix $(\mathbf{A} - \lambda \mathbf{I})$ must be singular i.e.,

$$|\mathbf{A} - \lambda \mathbf{I}| = 0. \tag{6}$$

Expansion of this determinant results in an n'th order polynomial in λ, which is called the *characteristic equation*. The roots of this polynomial are the eigenvalues of $\mathbf{A}$, and for each eigenvalue there is a corresponding eigenvector. The collection of eigenvalues and eigenvectors is called the *eigenstructure*.

Index

A

Acceleration feedback, 266, 272, 273, 274, 275, 276, 277, 288
Accelerometer, 276
Accuracy, 7, 26, 30, 31, 32, 34, 42, 235, 244, 249, 260, 300, 337, 407
A/D, 321, 322, 340
AGC, 352
Aircraft, 3
Alias frequencies, 325, 332
Antennas, 3, 37, 295
    satellite, 3
    tracking, 3, 37
Asymptote, 131, 135, 143, 146, 149, 150, 155, 156, 240, 263, 268, 279, 280
Asymptote centroid, 178, 183, 268
Attenuation, 238, 251
Automatic typewriters, 353, 442
Autopilots, 3

B

Backlash, 351, 382, 383, 423, 424, 425, 426, 427, 428
Bandlimited, 325, 326
Bandwidth, 111, 129, 143, 235, 255, 260, 273, 341
Bang-bang, 432
Bilinear transformation, 342, 344, 345
Black H. S., 74
Block diagrams, 7, 13, 14, 15, 16, 19, 20, 21, 25, 26, 30, 41, 66, 278, 290, 294, 296, 297, 339

Bode diagram, 129, 130, 132, 137, 140, 141, 143, 146, 149, 150, 151, 152, 155, 156, 236, 237, 238, 239, 240, 247, 249, 255, 261, 279, 280, 282, 284, 328, 341
Bode gain, 29
Branch, 19
Break frequency, 131

C

$C, \dot{C}, \ddot{C}$, 11
Calibrated, 3
Cauchy, 74
Characteristic equation, 65, 71, 73, 170, 183, 192, 229, 244, 246, 254, 266, 268, 270, 271, 272, 273, 274, 277, 293, 299, 327
Closed loop, 1, 2, 3, 304
Closed loop frequency response, 93, 94, 151, 154, 255
Compensation, 235, 236, 240
    Feedback, 264
Compensator, 238, 242, 246, 248, 254
Compensator, digital, see Digital compensator
Complementary frequencies, 325
Control, Digital, see Digital control
Controllability, 223, 225, 230
Controlled variables, 1, 2, 3, 4
Controller, 2, 237, 261, 321
Control system, 12, 105, 106, 143
Corner frequency, 131, 135, 143, 145, 149, 150, 156, 240
Coulomb friction, 30, 35, 351, 418, 419, 420, 421, 422, 423
Curve following, 442

Damping, 106, 114, 131, 143, 238, 242, 260, 264, 272, 273, 278, 341
   Switched, 437, 438, 439, 440
D. C. gain, 29
Deadzone, 30, 35, 351, 418
Decibel, 130, 131, 140
Describing functions, 351, 354, 355, 359, 374, 387
   Sinusoidal, 357, 358
Design, 235, 236
Determinant, 450
   Compensator, 340
Digital control, 6, 340
   Filter, 340
Discontinuous systems, 351, 353, 399, 429
Disturbance, 3, 17, 29, 34, 36, 295
   Load, 17, 34, 36, 278, 295
Dividing lines, 353, 399, 412, 415, 416, 417, 418, 420, 425
Dominance, 106, 110, 143, 146
Dominant, 106, 117, 122, 145, 236, 276, 300
Dynamic error coefficients, 39
Dynamic response, 63

e, 11
Eigenvalues, 63
Eigenvector, 353, 399, 407, 409, 411, 439
   Fast, 411, 440
   Slow, 411, 440, 441
Error, 2, 17, 26, 28, 29, 407
   Steady state, 26, 28, 29, 30, 33, 34, 35, 36, 407
Error bounds, 39
Error coefficients, 29, 117, 156, 238, 242, 244, 245, 278, 280, 295, 300, 338
   Dynamic, 39
Error constants, 27, 28
   Acceleration, 29
   Position, 28, 29
   Velocity, 28, 29
Evans, Walter, 170, 198

f, 11, 209
Fast eigenvector, *see* Eigenvector
Feedback, 1, 2, 3, 4, 12, 16, 26, 29
Feedback controls, 1, 2, 26, 29, 74
Feedback loop, 12, 15, 228

Feedback path, 16
   Tachometer, 12
Filter, digital, *see* Digital filter
Final value theorem, 9, 26, 27, 39, 42, 337
Forward path, 16, 20, 23
Fourier transform, 5
Frequency response, 5, 74, 93, 94, 95, 106, 110, 112, 113, 118, 119, 126, 129, 144, 145, 151, 154, 156, 236, 237
Friction, 11
   Torque, 18

Gain crossover, 114, 117, 119, 122, 125, 140, 141, 145, 146, 149, 150, 238, 240, 249, 251, 255, 259, 281, 284, 285, 288
Gain margin, 87, 88, 89, 90, 91, 92, 93, 97, 105, 111, 129, 140, 141, 143, 145, 146, 149, 152, 156, 238, 261, 381
Gain distribution, 368

Hold, 322
Hydraulic, 264
Hysteresis, 351

I, 11, 209
Inertia, 11, 209
Isoclines, 353, 399, 400, 401, 407, 408, 409, 410, 413

J_2, 11, 209
Jump resonance, *see* Resonance

K_a, 29, 150
k_B, 11, 209
K-curves, 179, 180, 181
K_o, 28
K_x, 27
k_T, 11, 209
K_v, 28, 137, 149
Kirchhoff's law, 10, 209, 210

L, 11, 209
Laplace transformation, 5, 7, 26
  Final value theorem, 9, 26, 27, 39, 42
  Tables, 8, 9
L'Hospital, 302
Limit cycles, 351, 359, 361, 362, 363, 372, 400,
    423, 424, 425, 426, 427, 428, 432
  Amplitude, 365, 367, 368, 372
  Frequency, 365, 368, 373
  Stable, 361, 362, 372
  Unstable, 352, 362, 372, 373
Limit lines, 399
Liquid level system, 215
Load, 17
  Disturbance, 17, 29, 34, 36, 42, 264, 278
  Torque, 17, 37
  Wind, 295
Loop, 20, 23, 25
  Feedback, 12, 13, 228
  Gain, 20
  Major, 278, 279, 281
  Minor, 12, 15, 278, 281, 282, 291, 293
  Self, 19, 20
  Single, 26
Lyaponuv, 353

M, 95, 109, 155
M-circles, 95, 96
M_{pt}, 109, 110, 116, 118, 119, 121, 122, 125,
    143
$M_{p\omega}$, 111, 113, 115, 116, 118, 119, 120, 121,
    125, 259
Magnetic disk memory, 3, 340, 353
Major loop, *see* Loop
Mason's gain rule, 19, 23, 25, 41
Matrix, 449, 450, 451, 452
  Adjoint, 451
  Cofactor, 451
  Diagonal, 450
  Identity, 450
  Rank, 452
  Singular, 451
  Square, 450
  Trace, 452
  Transpose, 450
Microprocessor, 6, 340
Minor, 450
Minor loop, 12, 15, 278, 281, 282
Model, 7, 10
Modeling, 4
Monic polynomial, 72
Motor, 10, 18, 30, 209, 264
  DC, 16, 17, 18, 20, 21, 25, 210

Servo, 264
Speed Control, 30
Torque constant, 11, 209

N, 27, 28, 29, 95, 155
N-circles, 95, 96
Newton's law, 10, 209
Nichols, N. B., 261, 306
Nichols chart, 95, 151, 152, 153, 154, 155, 156,
    255, 258, 259, 375, 376, 377, 378, 379, 380,
    383
Node, 19, 22
  Input, 20
  Output, 20
  Sink, 20
  Source, 20
Noise, 278, 295
Nonlinear, 6, 29, 34, 278, 294, 351, 353, 355,
    399
Non-minimum phase, 85, 86, 87, 282
Nyquist, H. S., 74, 97
  Rate, 325
Nyquist stability criterion, 5, 64, 74, 85, 87, 92,
    93, 105, 129, 130, 140, 156, 327, 333, 359,
    371
Nyquist plot, 75, 76, 78, 79, 80, 81, 85, 88, 89,
    90, 92, 93, 115, 117, 119, 120, 121, 122,
    151, 372, 373

Observability, 223, 226
Observers, 226, 227, 228, 229, 230
ω_b, 111, 114, 119, 120, 121, 123, 125, 259, 260
ω_ϕ, 111, 123, 140
ω_n, 107, 109, 114, 117, 145, 255, 260, 266, 273
ω_r, 111, 119, 120, 121, 123, 125, 259, 260
ω_t, 109
ω_x, 115, 117, 119, 120, 121, 123, 125, 140, 145,
    242
Open loop, 1, 145, 151, 152, 156
Output equation, 218
Overshoot, 105
  Maximum, 109, 122, 143
  Peak, 122

Parabola, 27
Partial fraction expansion, 331, 343
Partitioning, 193, 194, 195, 244, 254, 266, 270,
    271, 273, 275

Path, 20, 23
  Feedback, 16
  Forward, 16, 20
  Gain, 20
PD, 260, 261
Peak overshoot, *see* Overshoot
Phase crossover, 140, 141, 238
Phase margin, 87, 88, 89, 90, 91, 92, 93, 97, 105
    106, 111, 113, 115, 117, 120, 122, 123, 125,
    129, 140, 141, 149, 150, 152, 156, 238, 240,
    242, 249, 251, 254, 255, 259, 261, 282, 285,
    288, 381
Phase plane, 353, 354, 362, 363, 399
  Portrait, 399, 405, 409, 411
  Trajectory, 399, 404, 405
Phase variables, 207
PI, 260, 261
PID, 260, 261
Pneumatic, 264
Pole, 27, 132, 135, 146, 149, 150, 171, 172, 174,
    183, 240
  Complex, 125, 143
Pole placement, 226, 237
Positioning systems, 106
Primary strip, 332, 333, 334
Principle of argument, 74
Process, 264
Pulse width modulation, 322

R, 11, 209
Radar tracking, 36
  Range, 321
Ramp, 27, 33
Rate, 260
Regulator, 106
Reset, 260
Relative stability, 87, 90, 92
Resonance, 111, 150, 151, 285
  Jump, 352
  Subharmonic, 352
Resonant frequency, 145
Robot, 2
Robust, 39
Root, 63, 105, 106, 110, 114, 116, 119, 121,
    122, 125, 126, 170, 188, 266, 270
REAL, 172
Root locus, 5, 64, 97, 169–172, 198, 236–238,
    244, 245, 251, 255, 261, 266, 268, 333,
    359, 360, 364, 367, 368, 370
  Angles, 175–177
  Asymptote centroid, 183, 184
  Asymptotes, 175, 183, 184
  Branches, 174
  Construction rules, 173–178
  Emergence from real axis, 178–180
  Into real axis, 179, 180

$j\omega$-axis crossing, 181
  Of a parameter, 193, 196
  On the real axis, 172, 178, 183
  On the Z-plane, 343
  Terminal points, 173
  Two parameters on the Z plane, 343
Root relocation zone, 246
Routh, 64, 67, 69, 70, 71, 87, 97, 182, 293,
    333
Routh array, 67, 70, 71, 182, 293

Sampled, 321
  Data, 321, 337
Sampling, 321, 322, 324, 332
  Frequency, 324
  Method, 322
  Period, 322
  Rate, 322, 341
Satellite tracking, 3, 17, 31
Saturation, 351, 354, 355, 356, 360
Second order system, 105, 106, 107, 108, 110,
    112, 113, 114, 116, 117, 122, 126, 143, 146,
    255, 268
Self loop, *see* loop
Sensitivity, 37, 39, 300
Sensor, 264, 321
Settling time, 105, 109, 143, 149, 150, 228, 236,
    242, 246, 254, 255, 273, 341
Shannon's theorem, 325
Ship, 219
Signal flow graph, 7, 19, 20, 21, 22, 23, 24
Simulation, 353, 354
Single loop, *see* Loop
Singular points, 349, 400, 402, 403, 404, 405,
    406, 407, 420, 421
Slopes, 400
Slope marker, 353, 401
Slow Eigenvector, *see* Eigenvector
Spectrum analyzer, 151
Speed control, 31
Sprinkler, 3
Spirule, 183
Stable, 4, 63, 68, 80, 87, 89
Stability, 4, 63, 64, 67, 68, 70, 72, 74, 79, 87,
    89, 90, 92, 97, 223, 235, 236, 341
State, 7, 22, 207, 230
  Equations, 7, 22, 208, 210, 211, 215, 225,
    235
  Feedback, 237, 266, 278, 296, 437
  Variables, 7, 64, 207, 208, 213, 216, 271,
    272, 306
  Vector, 209
Static error coefficients, 29
Steady state, 7
  Accuracy, 26, 235, 278, 337
  Errors, 29, 30, 31, 32, 34, 295, 338

Step, 27, 143
　Response, 106, 115, 117, 118, 119, 120, 123, 124, 126, 143
Stepper motor, 2
Subharmonic resonance, *see* Resonance
Summer, 17
Summing junction, 17, 18
Superposition, 29, 37, 351, 352
Submarine, 3
Switching lines, 353, 399, 412, 413, 414, 415

Tachometer, 12, 264, 266, 282, 284
τ, 109
Telemetering, 321
Temperature, 1, 2, 3, 294
Template, 137, 138, 139
Thermostatic control, 2
Threshold, 351
Time constant, 12
Time delay, 323, 340
Time optimal, 353, 432, 434
Tolerances, 294
Torque, 18, 36, 37
　Coulomb, 37
　Friction, 18
　Summing junction, 18
　Wind, 36, 37
Tracking antennas, *see* Antennas
Transference, 19
Transfer function, 5, 7, 9, 12, 15, 17, 26, 27, 29, 30, 64, 77, 81, 85, 94, 129, 171, 211, 213, 214, 217, 221, 228, 235, 237, 242, 261, 278, 282
Transient oscillation, 105
Transient response, *also see* Step response, 5, 89, 106, 129, 143, 236, 254, 295
T_s, 109, 116, 119, 121, 125, 143, 145, 149, 150, 255

Type number, 27, 28, 135, 295
Type One, 28, 29, 31, 32, 33, 34, 36, 77, 79, 81, 106, 126, 135, 137, 149, 301, 338
Type Two, 29, 32, 33, 36, 80, 81, 135, 137, 149, 338, 340
Type Zero, 28, 29, 30, 34, 77, 78, 79, 135, 146, 301, 338

V, 11, 209
Valve, 3, 264
Van der Pol, 404, 405
Velocity feedback, 194, 264, 266, 268, 270, 271, 287, 409, 410, 417, 418, 430, 431, 439
Voltage controlled oscillator, 340

Word length, 326

Zero, 119, 122, 132, 146, 149, 150, 171, 172, 183, 240
ζ, 109, 114, 126, 143, 145, 149, 150, 151, 236, 266, 273
Ziegler, J. G., 261, 306
ZOH, 322
　Transfer function of, 322
Z-plane, 335, 336
Z-transform, 327, 328, 329, 341
　Matched, 344
　Transfer function, 328, 331, 336